LIPOSOMES as TOOLS

in

BASIC RESEARCH

and

INDUSTRY

Edited by

Jean R. Philippot, Ph.D.

CNRS Research Director
Membrane Interactions Laboratory
University of Montpellier II
Montpellier, France

Francis Schuber, Ph.D.

CNRS Research Director
Bioorganic Chemistry Laboratory
Loius Pasteur University
Strasbourg-Illkirch, France

CRC Press
Taylor & Francis Group
Boca Raton London New York

CRC Press is an imprint of the
Taylor & Francis Group, an **informa** business

First published 1995 by CRC Press
Taylor & Francis Group
6000 Broken Sound Parkway NW, Suite 300
Boca Raton, FL 33487-2742

Reissued 2018 by CRC Press

© 1995 by Taylor & Francis
CRC Press is an imprint of Taylor & Francis Group, an Informa business

No claim to original U.S. Government works

A Library of Congress record exists under LC control number: 94022192

Publisher's Note
The publisher has gone to great lengths to ensure the quality of this reprint but points out that some imperfections in the original copies may be apparent.

Disclaimer
The publisher has made every effort to trace copyright holders and welcomes correspondence from those they have been unable to contact.

ISBN 13: 978-1-138-10605-5 (hbk)
ISBN 13: 978-1-138-56059-8 (pbk)
ISBN 13: 978-0-203-71325-9 (ebk)

Visit the Taylor & Francis Web site at http://www.taylorandfrancis.com and the
CRC Press Web site at http://www.crcpress.com

PREFACE

This book, devoted to a broader understanding of liposomes, is not aimed at people engaged in current uses of these vesicles in fields such as drug delivery, targeting, and other pharmacological fields. The liposome is a versatile tool which can be used in many domains, including basic research and applied technology. We aim to present these different domains, and to describe some of the main results already obtained. Most of the liposome applications which are already well documented in other books will be excluded from this presentation, with the exception of some very recent developments.

Liposomes have been widely used to reconstitute membrane functions due to the similarity between the liposome bilayer and biological membranes. However, recent progress (e.g., mild solubilization of membranes) has completely changed our perspectives in this field. Liposomes are also increasingly used in studies performed to understand such important phenomena as endocytosis, membrane fusion, antigen processing, and T cell presentation.

Over the past few years new types of liposomes have appeared, e.g. made of nonphospholipid molecules, which have opened new perspectives of applications because of their high resistance in body fluids, their possibility of carrying oil, and their cheapness. These lipid vesicles already used in cosmetology have begun to be manufactured for many other industrial and agricultural uses. On the other hand, pH-sensitive liposomes, Stealth® liposomes, and cationic liposomes have enlarged and improved the application field of liposomes in clinical research.

This book should cover these different uses of liposomes with particular attention to new formulations and new applications. It will be of interest to people involved in the field of cellular biology and other related biological disciplines, as well as those having to improve their knowledge on the different possiblilities of the liposome concept.

THE EDITORS

Jean R. Philippot, Ph.D., is a Research Director at the Centre National de la Recherche Scientifique (CNRS), France and Director of the Membrane Interactions Laboratory (URA-CNRS 530) at the University of Montpellier II. He received his graduate degree from the University of Sciences of Montpellier in 1958 and obtained his Ph.D. degree at Paris-Sorbonne University in 1962. He is an engineer in Organic Chemistry from the Ecole Nationale Superieure de Chimie de Montpellier. In 1963 he was a postdoctoral fellow at the Physiological Department of Cambridge University, England.

In 1958, Dr. Philippot was appointed to the CNRS. He served as Visiting Scientist at the Pharmalogical Laboratory of the Bern Veterinary School, Switzerland, from 1967 to 1968; as Professor of biophysics at Montpellier Medical School, from 1968 to 1972; and as Visiting Professor in the Radiobiology Department of Tufts Medical School in Boston, from 1974 to 1976.

Dr. Philippot is a founding member of the Groupe Thématique de Recherche sur les Vecteurs (GTRV), the French Society for Drug Delivery. He also serves as a consultant for several French and American pharmaceutical companies.

He has carried out research in such diverse fields as the active transport of ions in erythrocytes and feedback regulation of cholesterol biosynthesis. Since 1980, he has worked, published, and lectured on the targeting of drugs with liposomes and has written over 100 papers and book chapters, as well as co-editing one book on liposomes.

Francis Schuber, Ph.D., is a Research Director at the Centre National de la Recherche Scientifique Department (CNRS) and is Head of the Bioorganic Chemistry Laboratory (associated with CNRS) at the Pharmacy of the University Louis Pasteur in Strasbourg-Ilkirch, France.

Dr. Schuber graduated from the University of Strasbourg in 1963 with a B.S. in Chemistry. In 1967, he obtained his Ph.D. in Chemistry from the same institution. He was a postdoctoral fellow in the Chemistry Department of the University of Ottawa, Canada, from 1968 to 1970 and in 1973 was a Visiting Professor at McGill University in Montreal, Canada. From 1981 to 1982, he stayed at the Cancer Research Institute at the University of California, San Francisco.

Dr. Schuber was appointed to the CNRS in 1966 where he has spent most of his career. He is a member of the American Chemical Society, the Société Française de Chimie, and the Société Française de Biochimie et Biologie Moléculaire. In 1976, Dr. Schuber, along with Professor P. Couvreur, founded the Groupe Thématique de Recherche sur les Vecteurs (GTRV), the French Society for Drug Delivery. He also serves as a consultant for several pharmaceutical companies.

Dr. Schuber is the author or co-author of about 100 papers and book chapters and has co-edited one book on liposomes. His current research interests relate to bioorganic chemistry: chemical enzymology, bioconjugation chemistry, synthesis of ligands for targeting (liposomes, genes), and synthetic vaccines.

CONTRIBUTORS

Jean-Paul Behr, Ph.D.
Lab de Chimie Génétique
CNRS URA 1386
Faculté de Pharmacie de Strasbourg
Illkirch, France

David Collins, Ph.D.
Department of Pharmaceutics and
  Drug Delivery
AMGEN Center
Thousand Oaks, California

Robert J. Debs, Ph.D.
Cancer Research Institute
University of California
San Francisco, California

Nejat Düzgüneş, Ph.D.
Department of Microbiology
University of the Pacific
School of Dentistry
San Francisco, California

Martin Friede, Ph.D.
Université Louis Pasteur
CNRS URA 1386
Faculté de Pharmacie de Strasbourg
Strasbourg-Illkirch, France

Alberto A. Gabizon, M.D., Ph.D.
Sharett Institute of Oncology
Hadassah Medical Center
Jerusalem, Israel

Dick Hoekstra, Ph.D.
Department of Physiological Chemistry
University of Gröningen
Gröningen, The Netherlands

S. Kumar, Ph.D.
Vineland Laboratories
Vineland, New Jersey

Lee Leserman, M.D., Ph.D.
Centre d'Immunologie
INSERM-CNRS de Marseille-Luminy
Marseille, France

Maria C.P. de Lima, Ph.D.
Department of Biochemistry
Center for Neurosciences
  and Cell Biology
University of Coïmbra
Coïmbra, Portugal

Christan Marechal, Ph.D.
LVMH Service Propriété Industrielle
Paris, France

Rajiv Mathur
Micro Vesicular System, Inc.
Nashua, New Hampshire

Pierre G. Milhaud, Ph.D.
Institut de Génétique Moléculaire
CNRS
Montpellier, France

Roger R.C. New, Ph.D.
Cortecs Research Laboratory
London School of Pharmacy
London, England

Shlomo Nir, Ph.D.
The Seagram Centre
Faculty of Agriculture
Hebrew University of Jerusalem
Rehovot, Israel

Demetrios Papahadjopoulos, Ph.D.
University of California
Cancer Research Institute and
  Department of Pharmacology
San Francisco, California

Pierre Perrier, Ph.D.
Parfums Christian Dior
Saint Jean de Braye, France

Jean R. Philippot, Ph.D.
Université Monpellier II
Department Biologie-Santé
CNRS URA 530
Montpellier, France

Bruno Pitard, Ph.D.
Section de Bioénergétique
Centre d'Etudes Nucléaire de Saclay
Gif-sur-Yvette, France

Christophe O. Puyal
Université Montpellier II
Department Biologie-Santé
CNRS-URA 530
Montpellier, France

Gérard Redziniak, Ph.D.
Parfums Christian Dior
Saint Jean de Braye, France

Jean-Serge Remy, Ph.D.
Lab de Chimie Génétique
CNRS URA 1386
Faculté de Pharmacie de Strasbourg
Illkirch, France

Jean-Louis Rigaud, Ph.D.
Section de Bioénergéteque
Centre d'Etudes Nucléaire de Saclay
Gif-sur-Yvette, France

Francis Schuber, Ph.D.
Université Louis Pasteur
CNRS URA 1386
Faculté de Pharmacie de Strasbourg
Strasbourg-Illkirch, France

Colin Tilcock, Ph.D.
Department of Radiology
University of British Columbia
Vancouver, Canada

Carole Varanelli
Micro Vesicular System, Inc.
Nashua, New Hampshire

Peter Walden, Ph.D.
Abteilung Immungenetik
Max-Plank-Institut für Biologie
Tübingen, Germany

Donald F.H. Wallach, M.D., Ph.D.
Micro Vesicular System, Inc.
Nashua, New Hampshire

Claude Sirlin, Ph.D.
Lab de Chimie Génétique
CNRS URA 1386
Faculté de Pharmacie de Strasbourg
Illkirch, France

CONTENTS

PART I. THE LIPOSOMES

PART II. LIPOSOMES AS CELL OR MEMBRANE MODELS

PART III. NEW DEVELOPMENTS OF LIPOSOMES

PART I
The Liposomes

Chapter 1

Influence of Liposome Characteristics on Their Properties and Fate

Roger R.C. New

CONTENTS

0-8493-4569-3/95/$0.00+$.50

I. INTRODUCTION

Three key elements serve to define liposomes as unique: (1) they are closed vesicles, (2) they enclose an internal aqueous space, and (3) this internal compartment is separated from the external medium by a bilayer membrane composed of discrete lipid molecules. Aside from this, liposomes can be any size, shape, and composition; it is the variety of different qualities that liposomes can possess, while still conforming to the basic definition, that makes them such a powerful tool in a wide range of disciplines.

From the above, it is clear that liposomes are essentially particulate systems, which can be clearly distinguished from other classes of microparticle often used for similar purposes. They are thus distinct from oil-in-water emulsions in that the latter do not contain an aqueous phase, and the layer of surfactant at the interface between the two phases is a monolayer rather than a bilayer. Even a double emulsion consisting of water-in-oil-in-water phases fails to meet the third criterion, although, as the quantity of oil phase is progressively reduced, the boundary between emulsion and liposomes becomes decidedly blurred. Other structures, which viewed from the outside might appear identical to liposomes but possess a significantly different internal make-up, arise as a result of depositing a coat of phospholipids onto the hydrophobic surface of solid, polymeric spheres, in which the bulk of the internal phase is aqueous. Although lipid-coated nylon microcapsules and SupraMolecular BioVectors (SMBVs) will not be included here as classical liposomes, much of what is said in this chapter may apply to them, and the observation of differences between these entities and liposomes can throw light on how liposome structure influences the characteristic properties of liposomes.

Traditionally, the membrane components of liposomes have been phospholipids, particularly phosphatidyl cholines, partly because they are the building blocks that nature itself uses to form membranes and partly because the common phospholipids are lamella-forming lipids under all conditions and can do so easily without the aid of additional components. Such single component systems lend themselves to study without needing consideration of complicating factors such as interactions between different chemical species. Recently, however, bilayer membrane vesicles have been constructed using single-chain amphiphiles (e.g., UFAsomes) or non-ionic surfactants in which the principles of formation and physical properties are so similar to conventional liposomes that there is no reason why they should not be taken into the fold and treated as subsets of the general case.

Other subsets of liposomes, where the original concept is modified slightly, include those where the "bilayer" membrane is composed of membrane-spanning lipids in which the membrane components are long lipidic chains with polar moieties at either end. In theory, each molecule has one polar group at the internal surface of the membrane and the other at the external face, although in practice a number of molecules will have bent back upon themselves to be associated with one or other of the faces exclusively. A second subtype stretches the concept of discrete molecules making up the membrane, when the membrane compounds are chemically linked to each other by polymerization, to form an extensive network of macromolecules inextricably interlinked. To form these structures, however, a conventional liposome is first constructed from individual monomer units making up a fluid membrane, and the chemical cross-linking may be thought of as a natural extension of the non-covalent interactions between membrane components (van der Waals, hydrogen bonding, electrostatic), which are essential for maintaining stability of the membrane in the first place.

Finally, one should pay homage to the ultimate liposome, the single cell, which fits the definition given above and, indeed, has been the inspiration and raison d'être for much of the work that has been conducted with liposomes, as models for cells in a very much simpler form.

II. LIPOSOME PROPERTIES

Breaking things down to their simplest form, the two properties of liposomes that one wants to control are membrane stability and vesicle size, because these are the overriding factors that govern longevity *in vitro* and interactions in biological or other systems. Membrane stability, which includes permeability, fluidity, and fusogenicity, is determined by choice of lipid composition, while the size of liposomes is governed by the method of preparation.

A. CHOICE OF LIPID

The decisions as to what lipids to use can be taken sequentially, and an attempt to do so is given below, with the justification and implications explained for each possible decision. The determination of the

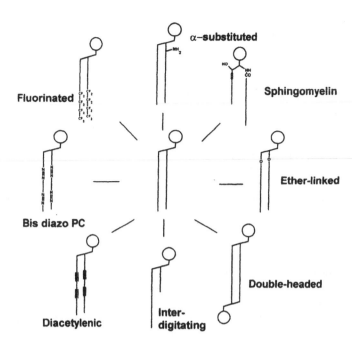

Figure 1 Variations in structure of lipid portions of synthetic phospholipids.

composition is based on the principle that the bulk of the liposome membrane can be composed of a single neutral phospholipid acting as the structural "backbone" of the bilayer, against the background of which other "minor" lipids can be overlaid, to confer specific properties that may modulate the behavior in any desired fashion. The first choice, therefore, is the nature of this basic lipid.

1. Phospholipid (PL) or Non-Ionic Surfactant (NIOS)?

As far as the robustness of the membrane is concerned, there appears to be little difference between liposomes made with either of these components. This is not too unexpected since the interactions between the lipid chains should be similar, and attachment of PEG and polysaccharides (typical headgroups used in NIOSs) to the surface of phospholipid vesicles is now becoming a common practice. It would be surprising, however, if the differential permeability characteristics displayed by PL membranes to ions such as Na^+ and Cl^- were to be mirrored by NIOS membranes, since such properties must be related to the net dipole across PL membranes, which is the result of orientation of the ester carbonyls and the polar headgroups.

Choice between the two will therefore depend on availability, cost, the desirability or otherwise of using natural products (as are phospholipids) and, for pharmaceutical applications, regulatory considerations. While there is no consensus as to what NIOS is most suitable among those used, the phospholipid of choice is phosphatidyl choline (PC), which is the most readily available and which is the predominant lipid in natural cell membranes. In what follows, the assumption will be made that phosphatidyl choline of some sort is being used as the staple membrane lipid.

2. Synthetic or Natural?

Use of synthetic lipids enables one to work with pure well-characterized materials—usually single species, perhaps displaying special qualities that are unavailable with natural lipids. They are good for performing controlled studies, but the relationship between the results obtained and the behavior of natural membranes is not always a close one. Natural membranes, and the phospholipids derived from them, are in fact a mixture of different species, each molecule consisting of permutations of two fatty acids on the two distinct positions on the glycerol bridge, out of a total pool of 10 to 15 fatty acids varying in chain length, saturation, and in some cases position of double bonds along the chain. In natural membranes, a preference is usually displayed for "mixed" phospholipids—i.e., for those consisting of a combination of saturated and unsaturated chains within the same molecule. This mixture, both within and among phospholipid molecules, often gives rise to membranes with an integrity that cannot be matched by individual components.

3. Saturated or Unsaturated?

Membranes composed wholly of unsaturated lipids are fluid, highly permeable, susceptible to oxidation, and easily disrupted by contact with serum proteins. On the other hand, unsaturated lipids are easier to work with, being more soluble in organic solvents, hydrating more readily, and, since they are fluid at room temperature, allow packing defects to disperse more easily. Size modification of liposomes after they have been formed is also easier with unsaturated lipids.

In many cases, a highly stable membrane is not essential, particularly if the liposome is being used to carry materials within its lipophilic compartment (i.e., the membrane itself) rather than in the aqueous phase. Natural phosphatidyl choline extracted from egg yolk is widely used. If retention of water-soluble compounds is required, however, egg PC alone will not be sufficient, and inclusion of other materials (see next section) will be necessary. Saturated lipids form stable liposomes, provided they have been "annealed" to remove packing defects by incubation above the phase transition temperature (or some other form of energy input). Synthetic lipids such as dipalmitoyl PC are often used or, alternatively, hydrogenated soya PC, which is mainly a mixture of C16 and C18 fatty acid chains.

4. Inclusion of Cholesterol?

Cholesterol reduces the fluidity of membranes above the phase transition temperature, with a corresponding reduction in permeability to aqueous solutes. Consequently, inclusion of cholesterol into unsaturated membranes is often essential in order to achieve sufficient stability. Cholesterol can be incorporated up to a level of 50 mol %—i.e., at a PC:cholesterol ratio of 1:1, at which it displays its maximum stabilizing effect both *in vitro* and *in vivo*. On the other hand, cholesterol increases the fluidity of membranes below the phase transition temperature, so that its inclusion in saturated membranes, which are usually in the lower, gel phase at ambient temperature, may result in a reduction in stability.

For maximum stability, therefore, the choice exists between saturated PC liposomes, composed of synthetic or hydrogenated lipid, and unsaturated egg PC liposomes incorporating cholesterol. Development work usually proceeds evaluating these two approaches side by side.

5. Charged Lipids?

Charged lipids may be included in the membrane for a large number of reasons. Firstly, to increase the surface charge density to prevent close approach of liposomes and, hence, prevent aggregation, fusion, etc. Also, to push adjacent membranes apart within a multilamellar liposome to increase the internal aqueous volume. The most suitable phospholipid for this purpose is probably the negatively charged phosphatidyl glycerol. Other negative phospholipids such as PA or PS may be unsuitable because they cause aggregation in the presence of calcium. No natural positively-charged phospholipids exist, and inclusion of cationic lipids of any class gives rise to liposomes that bind strongly to proteins and other molecules. Incorporation of PG at a 10% mol ratio of total lipids is usually sufficient to reduce intermembrane interactions.

A second reason for including charged lipids is to introduce controlled instability into a membrane—for example, with PE, which has a partial negative charge at or above pH 7 but which forms non-bilayer aggregates at low pH. For this to happen, the fatty acyl chains of PE should be unsaturated. Other lipids with an ionizable functional group (e.g., cholesterol hemisuccinate) behave in the same way. In such applications, the charged lipid may be the predominant species in the membrane.

Finally, molecules such as PE or their derivatives may be included in a membrane because of their functional groups, which allow them to be conjugated with proteins or other molecules, and act as an anchor for binding these molecules to the liposome surface. Again, between 5 and 10% is a good starting point for the quantity of minor lipid included.

Unless the intention is deliberately to prepare an unstable membrane, care needs to be taken to match the phase transition temperatures of all the lipids (charged and uncharged) making up the membrane. Combination of lipids with widely differing T_cs will result in a membrane displaying phase separation, in which fluid and gel phases co-exist—leading to increased permeability. In general, all saturated or all unsaturated lipids should be employed.

6. Alternative Lipids?

If appropriate care is taken over conditions employed during preparation and storage, then liposomes can be obtained that display adequate stability. In cases where the material entrapped imposes conditions that depart from ideal, or where the application to which the liposomes are put requires them to survive a very aggressive environment, alternative lipids may be employed to overcome these problems.

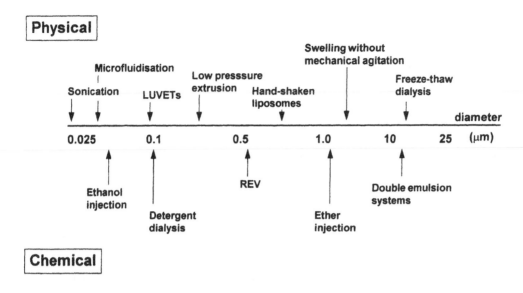

Figure 2 Methods of preparation in relation to size range of liposomes produced.

Tighter packing can be achieved by using lipids that have the opportunity for extra intermolecular hydrogen bonding (see Figure 1), as is the case for sphingomyelin, which has an amide linkage and a hydroxyl group at the interfacial region. Other phospholipid analogues where the ester bond is replaced by an amide also show this effect. It has also been suggested that such lipids may aid incorporation of protein into membranes. Sphingomyelin is the major component of sheep red cell membranes, and liposomes composed of sphingomyelin display increased resistance to disruption by bile salts.

A further effect of modification of the ester linkages is to confer resistance to phospholipases, which is accomplished effectively by replacement with ether linkages, replacement by a carbamoyl moiety, or introduction of a function on the α carbon of the fatty acid in the 2– position.

Not only are both sphingomyelin and ether-linked lipids more resistant to chemical hydrolysis, but they also share another property that may render membranes more stable—namely, the ability of the hydrocarbon chains of the outer and inner leaflets to interdigitate. In the case of sphingomyelin, and certain natural PEs, this occurs by virtue of marked inequality in the length of the two hydrocarbon chains. With ether-linked lipids, a difference in packing at the bridge region is presumably responsible.

Modifications can be made in the nature of the acyl chains of phospholipids (Figure 1). Use of branched chains should permit the possibility of fully saturated long-chain phospholipids with low transition temperatures. Fatty acids with diacetylenic groups have been employed that undergo intramolecular polymerization on exposure to UV light. Bis-diazo phospholipids can be converted between fluid and gel phases by exposure to light. Brominated phospholipids have been employed in liposomes as contrast agents, and liposomes composed of fluorinated phospholipids have been proposed as carriers for fluorinated compounds. They have also been demonstrated to display reduced leakage due to the extra barrier posed by the inner membrane core of fluorocarbon, which is neither hydrophilic nor conventionally lipophilic.

B. CONTROL OF LIPOSOME SIZE

Although liposomes can vary tremendously in size, most workers have settled on liposomes of a very narrow size range for study, round about the middle of the spectrum. Liposomes can be too large, tending to be very fragile mechanically, and may not handle well, getting stuck in tubing, filters or capillaries *in vivo*. Liposomes can also be too small, requiring input of large amounts of energy to achieve that size, and having a very small internal volume, so that the encapsulation efficiency is poor.

For most applications the ideal size is between 0.1 and 0.5µm, the lower end of the scale being amenable to sterile filtration. Unilamellar liposomes at the upper end of this scale can be produced by two-phase systems such as the REV method and at the lower end by detergent dialysis (see Figure 2). Mechanical methods of dispersion will give liposomes over the whole size range, depending on the extent of subdivision of the original dry lipid particles and the amount of energy put in during the process of swelling of the dry lipid in the presence of water. With the exception of detergent dialysis and extensive sonication or microfluidization, most preparation methods give rise to vesicle populations of fairly heterogeneous size distribution. The size range can be reduced by filtering out or breaking up the liposomes at the larger extreme and by removal of small vesicles by dialysis or centrifugation.

III. CONCLUSION

As can be seen from the above, although there is a tremendous choice in the methods of preparation, the lipids incorporated, and the conditions employed, in most cases, the liposomes one ends up using are identical in the majority of their characteristics and composition to those employed by everyone else. The chief difference is the material that is being incorporated into the liposomes, and this itself can have a tremendous influence on the properties of liposomes, for which allowances and adjustments have to be made. For this there are no rules laid down, for one compound behaves differently from another, and it is here, in overcoming the unexpected problems that interaction between membrane and entrapped material have introduced, that the major work in development of liposomes lies, usually through a process of trial and error.

Underlying it all, however, there are a number of rules of thumb and well-known phenomena that workers in the field have at the back of their minds when working out new systems, and I have tried here to finish by encapsulating these into a set of, if not rules, at least basic principles, which form the starting point for conceptualizing how liposomes work.

This summary is followed by a glossary, which describes in more detail the technical background to some of the terms commonly employed.

IV. SUMMARY OF THE BASIC PRINCIPLES OF LIPOSOMOLOGY

A. PREPARATION

1. *Spontaneous assembly* — Phospholipids aggregate spontaneously in aqueous media to form bilayer membranes, adopting the configuration of closed vesicles (liposomes) that may consist of single or multiple concentric lamellae.
2. *Dispersion of lipids* — Phospholipids may be dispersed in the dry form, as solutions in organic solvents or as complexes with surface-active agents.
3. *Fluidity* — Most effective dispersion takes place under conditions where the phospholipids of the membrane formed are in a fluid state—e.g., with unsaturated lipids at high temperatures.
4. *Physical energy* — Higher numbers of small or unilamellar liposomes are produced with increasing shear rate or energy imparted during dispersion.
5. *Water-soluble compounds* may be entrapped within the liposomes by inclusion in the aqueous medium during dispersion of the phospholipids. Conventional separation procedures can be used to remove unentrapped material.
6. *Lipid-soluble compounds* may be incorporated within the membrane by inclusion in the phospholipid mixture during dispersion; 100% incorporation may be achieved provided the ratio relative to phospholipid is not excessive.
7. *Post-assembly incorporation* — Materials may be introduced into liposomes after the membranes have been formed either by irreversible passage across the membrane, governed by pH, osmotic or chemical potential gradients (remote loading), or by controlled rupture and reformation of the vesicles, e.g., during freeze-drying or extrusion.
8. *Size* — Liposome size is critically dependent on the method of manufacture. Size may be reduced by mechanical methods—extrusion, microfluidization—and increased by methods inducing fusion—exposure to calcium, pH, freeze-thawing. Loss of aqueous contents is a risk with these procedures.

B. PROPERTIES OF LIPOSOMES

1. *Selective permeability* — The membranes formed allow water and certain small molecules to pass through unhindered but display limited permeability to most hydrophilic molecules.
2. *Lipid phases* — The phospholipids in the membrane can pack together to give one of two different phases—a low temperature gel phase and a high-temperature fluid or liquid-crystal phase.
3. *Transition temperature* — The temperature of transition (T_c) between the two phases is higher for phospholipids with long-chain, saturated fatty acids.

4. *Fluidity* and permeability of membranes increase with increasing temperature.
5. *Phase separation* — Irregularities in phospholipid packing increase at the phase transition temperature T_c itself, arising from co-existence of domains of the two phases within the same membrane and giving rise to increased membrane permeability and susceptibility to intermembrane fusion, vesicle aggregation, and so on at the T_c.
6. *Cholesterol* incorporated in the membrane (up to 50 mol %) decreases fluidity above the T_c, increases it below the T_c, and reduces the enthalpy of the phase transition itself.
7. *Charge* — Inclusion of charged phospholipids encourages intermembrane repulsion, thereby increasing the volume of internal compartments of MLVs and reducing the propensity of vesicles to aggregate.
8. *pH* — Inclusion of lipids with ionizable headgroups capable of forming hexagonal phases (e.g., as in PE) results in membranes that display instability at certain pH values, which can lead to leakage of contents, fusion, or rupture of liposomes. The stability of such membranes can be modulated by the presence of other membrane components such as proteins.
9. *Calcium* — Liposomes containing ionizable negatively-charged lipids (e.g., PA or PS) undergo aggregation and/or fusion in the presence of calcium and other divalent cations.

C. STABILITY AND STORAGE

1. *Leakage* — Retention of water-soluble contents is greater for high-molecular weight-charged compounds. Hydrophobic molecules incorporated into the membrane will remain entrapped indefinitely. Compounds that display both hydrophilicity and lipophilicity pass through membranes readily.
2. *Aggregation and fusion* — Close approach between membranes, leading to aggregation and fusion, can be prevented by introducing charge into the membranes and reducing the size of the liposomes. Conditions that lead to packing irregularities and increased fluidity increase the tendency for interaction between vesicles.
3. *Oxidation* leads to chain breakage, generation of toxic lipids, and introduction of membrane changes encouraging processes (1) and (2) above. It is entirely due to action of free radicals (and subsequently molecular oxygen) on unsaturated hydrocarbons and is avoided by elimination of one or all of these factors. Antioxidants (e.g., α-tocopherol) and metal ion chelators are beneficial.
4. *Hydrolysis* — De-esterification within phospholipid molecules brings about release of free fatty acids and lyso-derivatives, which render the membrane less stable. Hydrolysis is encouraged at extremes of pH, and by the presence of buffer ions, and is more pronounced in fluid, cholesterol-free membranes.
5. *Lyophilization* — Reduction of the extent of the above phenomena as time progresses may be achieved by freeze-drying. High concentrations of sugars such as trehalose or sucrose need to be present to avoid fusion and leakage, which may occur due to formation of ice crystals or phase changes within the membrane.
6. *Sterilization* — Liposomes may be autoclaved with minimal chemical damage or alteration of membrane characteristics but display elevated leakage rates during the process itself. Liposomes may be passed through a 0.2 μm sterile filter without damage, providing they are smaller than this pore size.

D. BIOLOGICAL BEHAVIOR
1. Proteins

1. *Adsorption* — Even in the absence of packing irregularities, liposomes of all charges adsorb proteins onto their surface from biological media, such as plasma.
2. *Tightness of packing* — Liposomes with a tight packing structure, such as those composed of saturated phospholipids, or combined with high levels of cholesterol, display reduced protein interactions, but adsorption is never completely eliminated.
3. *Stability* — Adsorption of proteins tends to exacerbate stability problems displayed by liposomes in the absence of proteins. However, in certain cases proteins increase stability, probably by masking packing defects or by facilitating their dispersal. Increased saturation and cholesterol content increases stability.
4. *Lipoproteins* — Liposomes with a high degree of fluidity may interact with lipoproteins to such an extent that phospholipids are abstracted from the liposome membrane, and the contents released.
5. *Specific interactions* with proteins possessing binding sites for phospholipid headgroups (e.g., anti-PC antibodies, C-reactive protein, phospholipases) occur in relation to the extent of packing irregularities in the membrane.

6. *Coating* of the liposome membrane by polyols compounds such as PEG or oligosaccharides prevents strong association with proteins; increase of stability and impermeability of the liposome membrane is thought to manifest via a mechanism known as steric compression.

2. Cells

7. *Non-specific uptake* by cells is principally through phagocytosis of liposomes opsonized by adsorption of surface protein.
8. *Lysosomes* — Liposomes are readily broken down in lysosomes after phagocytosis. Escape from the lysosome may be engineered by using metastable liposomes whose fusogenic potential is increased at low pH.
9. *Fusion* — A small proportion of liposomes may enter the cytoplasm by fusion directly with the plasma membrane, and this process can be enhanced by incorporating fusogenic lipids or viral fusogens into the membrane.
10. *Uptake by pinocytosis* can occur for small liposomes coated with ligands specific for cell surface receptors. Choice of the correct receptor-ligand interaction is critical in determining whether the liposomes are internalized or remain bound on the surface.
11. *Avoidance of uptake* — Interaction with cells can be reduced by adoption of measures that reduce protein adsorption (see above).

3. Whole Body

12. *Administration i.v.* — Liposomes introduced into the bloodstream are taken up rapidly by the RES, probably as a result of opsonization by non-specifically adsorbed proteins, leading to phagocytosis. Coating of liposomes with sugars recognized specifically by phagocytes reduces the clearance time.
13. *Avoidance of RES* — Liposomes coated with PEG or appropriate cerebrosides ("Stealth" liposomes) display markedly prolonged circulation times in the bloodstream as a result of avoidance of the RES through reduction in protein adsorption.
14. *Natural tissue distribution* — For as long as liposomes in the bloodstream avoid the RES, their distribution is determined by size; small liposomes pass through fenestrae into the liver and large liposomes become lodged in the microvasculature, e.g., of the lung. Intermediate-sized liposomes have longest circulation half-times.
15. *Targeting* — Liposomes administered i.v. bearing surface receptors for specific cells display increased localization in tissues containing such cells, providing they have access to the blood compartment.
16. *Lymphatic drainage* — Liposomes injected into peripheral tissues drain via the lymphatics into lymph nodes. Antigens incorporated in liposomes may gain access to antigen-processing cells and invoke a strong immune response.
17. *Cutaneous administration* — Liposomes applied topically may deliver material either to the dermis, or transdermally. The route followed is thought to be via water channels in the skin barrier, either associated with hair follicles or the junction between skin lipids and cells of the *stratum corneum.*
18. *Oral delivery* — Liposomes administered orally are most often broken down by digestive enzymes and gut surfactants; entrapment of materials inside liposomes may aid their dispersion within lipid micelles and consequent uptake into the systemic circulation.

V. GLOSSARY

A. QUANTIFICATION OF LIPID

Quantities of lipid are defined either in terms of weight or in terms of moles. The use of molarity is valuable in cases where interactions are determined according to a particular stoichiometry. Thus, it is useful to describe the ratio of cholesterol to phospholipid in a membrane in terms of molar ratio rather than weight ratio because the quantity of cholesterol that can be accommodated into a membrane appears to be defined by a one-to-one interaction between the two molecules.

In other cases, however, reference to liposomes containing a certain number of micromoles of lipid is distinctly unhelpful, because the point of comparison is rarely described in similar terms. Animals, cells, etc. are quantitated by weight or number; physical manipulations in the laboratory involve measurements of weight or volume, and, indeed, properties such as solubility or uptake by cells are determined by the total mass of material introduced, irrespective of molecular size. In aqueous media, phospholipids associate with each other and rarely act as individual molecules, so colligative properties or receptor interactions do not apply, and no contribution is made by osmolarity, tonicity, or ionic strength. Thus, while description of membrane composition in terms of molar ratio is very much to be encouraged, this author recommends against use of molarities in other contexts.

1. Entrapment

A number of criteria exist to characterize the entrapment of materials inside liposomes. An alternative term for entrapment is "capture," and the two words are often used interchangeably. In what follows, however, the author will employ the term capture to refer to conditions that prevail at the time of manufacture of liposomes, while entrapment will relate to the liposome suspension at any time during manufacture, purification, or further processing. There are no fixed conventions for the terms used to describe particular parameters, and the ones given below are not intended to be definitive. It is important to be aware of the confusions that can arise, however, and to be as precise as possible in description of the liposomes one is using.

2. Entrapped Volume

This is the most unambiguous parameter and refers to the total volume of liquid entrapped within a certain quantity of liposomes. It may be quoted either with reference to the total volume of suspension, or with reference to the weight of lipid. In the former case, a notion is conveyed very clearly of the proportion of total liquid that constitutes the internal compartment and helps to visualize the density of liposomes in suspension. In the latter case, (μl/mg) an indication of the size of the liposomes is given. SUVs have an entrapped volume:weight ratio of ~0.3, while LUVs of 0.2 μm diameter have a value of 10 for this parameter. The captured volume clearly refers to the entrapped volume at the time of manufacture and gives a measure of the efficiency of the process.

3. Capture Efficiency

This term is somewhat ambiguous. It usually refers to the proportion of starting material (i.e., aqueous volume or dissolved solute) that ends up inside liposomes after manufacture. However, whether this constitutes an efficient process or not depends on the quantity of lipid that is used. One can envisage a situation in which 90% of the liquid volume is entrapped inside liposomes (giving a capture efficiency of 90%), but if this is achieved by use of lipid concentrations that greatly exceed that of the solute, the process may be considered inefficient in terms of lipid usage. Sometimes it is found that the efficiency measured in terms of volume entrapped is different from that for the solute dissolved in that volume. In certain cases the solute may be concentrated within the liposome—perhaps as a result of interaction between the solute and membrane components, resulting in a high efficiency of entrapment, even if the entrapped volume is low. At other times the converse is true.

Since efficiency is a concept related to a process, rather that a condition, its use for entrapment within liposomes after further processing has no meaning.

4. Percentage Entrapped

This term refers to the quantity of solute inside liposomes in relation to the total solute present in solution at any given time. At the start of the manufacturing process, it will be identical to the capture efficiency. At other times it is indicative of the efficiency of the process used for purifying the liposomes free of unentrapped material or, if the liposomes have been stored for a period of time, of their stability and the rate of leakage of entrapped solute over that time period.

B. LIPIDS

1. Membrane-forming lipids

A wide range of amphiphiles with strongly contrasting structures for both lipid and polar portions is capable of forming membranes when mixed with water. Three parameters may be used as indicators of the propensity for assembly in lamellar phases:

1. The hydrophilic-lipophilic balance (HLB) ratio quantitates the bulk hydrophilic and lipophilic portions of the molecule relative to one another, and it is independent of the size of the molecule itself.
2. The critical packing parameter (CPP) is concerned more with the geometry of the two portions of the surfactant, and is expressed as

$$CPP = v/al$$

where l = length of lipid chain
a = cross-sectional area of the headgroup
v = total volume of the molecule
and identifies those molecules between 0.5 and 1.0 as having a cylindrical shape compatible with formation of lamellae, in contrast to values on either side of this region, representing conical structures more conducive to micellar or hexagonal phases.

3. Solubility of the monomer in aqueous phase determines the proportion of molecules that may exist in lamellar phase at a given concentration and can vary with molecular weight at a given HLB value.

Other considerations can come into play, however, relating to the way in which the molecules fit together geometrically when in close proximity, and packing may be facilitated by inclusion of other molecules—e.g., non-amphiphilic lipids—in the mixture.

The lipid portions of non-ionic vesicle-forming surfactants can range from sterols, to single, double, or triple long chain acyl or alkyl chains. Being non-ionic in nature, the hydrophilic portions of this class of compounds are restricted to linear or cyclic structures containing oxygens either as ether or hydroxyl groups.

Ionic amphiphiles that form vesicles tend to possess hydrophilic headgroups that are much more compact. Cholesterol hemisuccinate and cholesterol sulfate will both organize into lamellar structures, as will sodium oleate or dimethyl dioctadecyl ammonium salts. However, the most widely used compounds for this purpose, both in nature and in the laboratory, are phospholipids.

2. Phospholipids

Natural phospholipids have the general structure shown in Figure 3 in which two hydrocarbon chains are linked to a phosphate-containing polar headgroup. In phosphoglycerides or "glycerophosphatides," the linkage of fatty acids to headgroup is via a bridge region consisting of the three carbon glycerol. In sphingolipids, the lipid sphingosine forms one of the hydrocarbon chains and is linked directly to the phosphate. Phospholipids can possess fatty acids of different chain length and unsaturation and may have different hydrophilic species linked to the phosphate, according to which individual members of the phospholipid category are classified. (See Figure 4.)

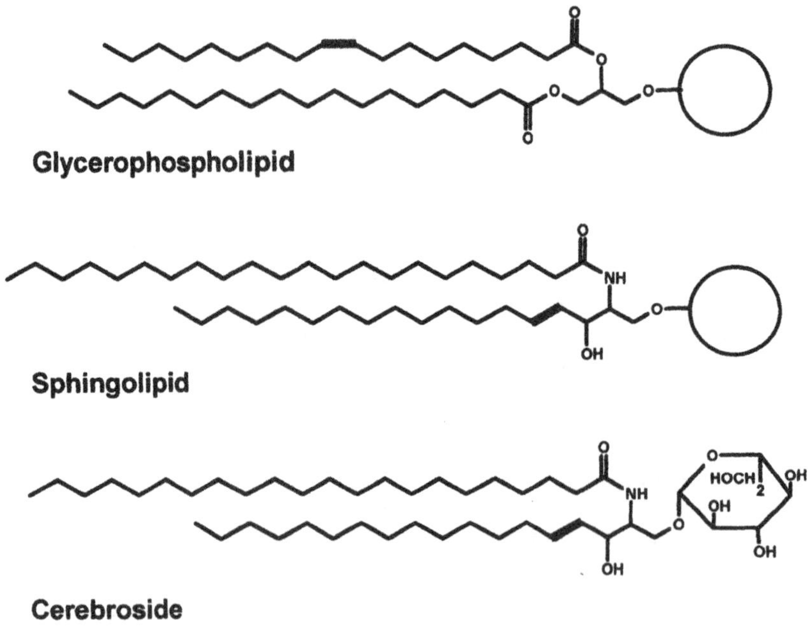

Glycerophospholipid

Sphingolipid

Cerebroside

Figure 3 Structures of representative natural phospholipids.

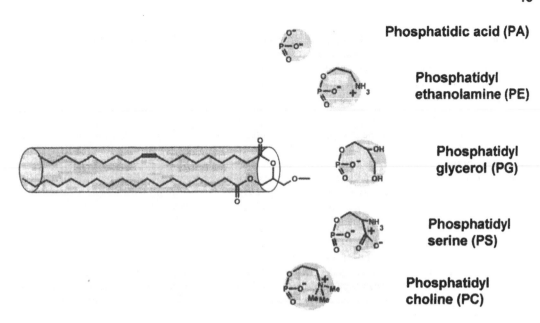

Phosphatidic acid (PA)

Phosphatidyl ethanolamine (PE)

Phosphatidyl glycerol (PG)

Phosphatidyl serine (PS)

Phosphatidyl choline (PC)

Figure 4 Structure of glycerophospholipids showing relative sizes of headgroup and lipid chains.

Phosphatidyl choline (PC)

PC is the predominant phospholipid found in natural membranes. The permanent positive charge on the choline of the headgroup counteracts the negative charge on the phosphate to give a neutral, very hydrophilic headgroup. In a membrane, interaction between the tertiary ammonium group and phosphates on adjacent molecules can contribute to the tightness of packing and help to disperse local fluctions in charge density.

Phosphatidyl ethanolamine (PE)

PE has a smaller headgroup than PC, and the presence of hydrogens directly attached to the nitrogen of ethanolamine permits interactions of adjacent molecules in the membrane by hydrogen bonding. At low or neutral pH, the amino group is protonated, giving a neutral molecule, which prefers to form hexagonal II phase inverted micelles to lamellar structures when above the main phase transition temperature. The presence of other lipids can stabilize the membrane so that this is prevented, and the ratio of lipids can be carefully arranged if so desired, such that the membrane converts from stable lamellar to non-lamellar with change of pH. Natural PEs tend to be more highly unsaturated than average and have fatty acids of longer and more assymetric chain lengths.

Phosphatidyl glycerol (PG)

PG possess a permanent negative charge over the normal physiological pH range. In addition to isolation direct from natural sources, it may be readily prepared semi-synthetically from other lipids by the action of phospholipase D in the presence of glycerol.

Cardiolipin (CL)

Identical to PG, except that a glycerophosphatide moiety is linked to both ends of the headgroup glycerol, to give a molecule with two negatively charged phosphates and four fatty acid chains. Cardiolipin is found in high proportion in mitochondria of heart tissue. The alternating phosphate-glycerol structure produces an entity that is similar antigenically to the sugar-phosphate backbone of DNA, and, for this reason, CL liposomes can be used diagnostically for detection of conditions such as SLE.

Phosphatidyl serine (PS)

Serine is linked to the phosphate via its hydroxyl group, leaving the carboxyl and amino functions both free and ionized to form a neutral zwitter ion. The net charge of the PS headgroup is therefore negative, as a result of the charge on the phosphate. Membranes containing PS show a marked sensitivity to calcium, which interacts directly with the carboxyl functions on the headgroups, causing PS molecules to aggregate within the membrane, resulting in a condensed phase separate from that of the bulk lipids.

Together with this phase separation goes the appearance of packing irregularities at phase boundaries. Calcium also causes bridging interactions between PS on membranes of different liposomes, so that aggregation of these liposomes, in which packing defects have been introduced, often results in fusion. However, it has been reported that the presence of PS in membranes helps to stabilize them during freeze-drying in the presence of sugars.

Phosphatidic acid (PA)

Absence of any substitution on the phosphate in PA confers a very strong negative charge to the molecule. Dispersions of PA alone in water have a pH of between 2 and 3, and rapid neutralization with acid can cause membrane reorganization, under the influence of electrostatic effects, to produce unilamellar vesicles. In a similar way to PS, addition of calcium can lead to aggregation and fusion, although higher concentrations of the divalent cation are usually required.

Sphingomyelin (SM)

SM is found to varying extents in the erythrocyte plasma membranes of a number of mammalian species and completely replaces PC in sheep red cells. It is also readily extracted from nervous tissue. It is a neutral molecule with the same phosphocholine headgroup as PC. SMs have hydrocarbon chains often markedly different in length and with a degree of unsaturation giving rise to T_cs between 20°C and 40°C. Membrane packing is tighter than for PC, by virtue of the extra hydrogen bonding made possible in the bridge region by the presence of the amide hydrogen, which participates in interaction between adjacent sphingomyelin molecules, and probably also with cholesterol.

Lyso-phosopholipids

Any of the lipids described above can lose a fatty acid chain, by either chemical or enzymatic hydrolysis, to give single chain amphiphiles. While they do not form membranes themselves, they are often present in membranes as impurities, either of the starting components, or as a result of degradation during storage. In high concentrations, lysophospholipids can disrupt membranes, and, indeed, they can be highly toxic for cells and whole organisms.

Membrane disruption with 1-PC only occurs when there is an imbalance in chains in the membrane relative to the headgroups. The action of phospholipase A, converting PC to 1-PC and fatty acid does not lead to perturbations until the fatty acid has been removed from the membrane (e.g., by incubating with albumin) whereupon increase in permeability, etc. is readily observed.

Cholesterol

Although cholesterol has a very different structure from the fatty acids of phospholipids, it is able to incorporate into phospholipid membranes very efficiently—up to a 1:1 molar ratio with PC without markedly affecting the dimensions of the membrane. At these levels, however, it has a profound effect on the order of the fatty acyl chains, increasing their rigidity for the first nine or so carbons from the carboxyl end, while permitting as much or greater freedom of motion for the remaining carbons in the chain. This may be expected, since the cholesterol molecule positions itself toward the outer half of the lipid portion of the membrane, with its polar hydroxyl group located at the level of the bridge region, where hydrogen bonding can take place.

Cholesterol reduces the net fluidity of membranes in the fluid phase above the main phase transition temperature but increases it in gel phase membranes below the T_c. Permeability to water-soluble solutes is affected accordingly. In addition, although cholesterol has virtually no effect on the temperature as the phase transition occurs, at high levels it reduces the enthalpy of the transition, which results in the discontinuities that occur in this region also being eliminated—further increasing the stability of the membrane as the temperature changes. A number of other natural sterols found as major membrane components in plants or fungi (e.g., sitosterol, stigmasterol, and ergosterol) display similar behavior.

Miscellaneous phospholipids

Other phospholipids found in small proportions in natural membranes are phosphatidyl inositol (PI), partially N-methylated forms of phosphatidyl ethanolamine, and a form of sphingomyelin in which the choline group is replaced by ethanolamine. Variants of some lipids (particularly PEs) are found in which the ester linkages joining the fatty acids to the glycerol bridge are replaced by ethers, to give compounds called plasmalogens. Platelet activating factor is a specific example. Phospholipids have been extracted from certain classes of halophilic bacteria in which a glycerol-linked headgroup is located at either end of a pair of long-chain di-carboxylic acids.

Another class of lipids often incorporated as minor constituents into liposomes are glycolipids, in particular gangliosides and cerebrosides. The lipophilic portion of these molecules is ceramide, the same species as that found in sphingomyelin; in the glycolipids, however, the phosphate-containing headgroup of sphingomyelin is replaced by single sugar molecules, such as galactose or glucose, or straight and branched chain oligosaccharides. Membranes composed of sphingomyelin in combination with a significant proportion of such glycolipids have been shown to display reduced protein interaction compared with conventional PC liposomes.

Cationic lipids

Lipids bearing a net positive charge are almost unknown in nature. However, liposomes can be prepared in which amphiphiles such as stearylamine or cetyl trimethyl ammonium bromide are incorporated in the membrane, where the same effects are noted, due to increased charge density and charge repulsion, as for negative liposomes. The liposomes can bind and absorb negatively charged drugs but also interact strongly with proteins and other anionic macromolecules. This property has been put to use in transfection of cells with DNA, in which liposomes composed of cationic lipid (such as DOTMA - dioleyl oxypropyl trimethyl ammonium chloride) and dioleyl PE form a complex with DNA that aids entry into the cell, probably by virtue of increased cell binding, coupled with fusogenicity of the lipids themselves (see Chapter 9 and 10). It is likely that, after binding with DNA to form a complex, the lipids are no longer in liposomal form.

Fatty acid chains

In mammals, the fatty acids of membrane phospholipids have even-numbered chains of 16 or greater carbons, up to C24. C14 (myristoyl) fatty acid species may be found in some plant or bacterial sources. Lauryl PCs are reported to be toxic, while shorter chain fatty acids do not tend to form stable liposomes on their own.

The fatty acids may be either saturated or unsaturated, with single unsaturations being located usually in the middle of the chain in the 9-position. The double bonds are usually all in the *cis*- configuration, which is the form that leads to maximum disruption in chain linearity and packing. The middle of the chain is also the position that leads to maximum disorder. Multiple unsaturations are spaced along the chain every three carbons so that none of the double bonds are conjugated with each other. In mammals, any additional double bonds are inserted toward the front end of the chain, while in plants, multiple unsaturations extend from the 9 position toward the tail end of the chain. Since the solubility of foreign lipids is increased in membranes as a result of disorder introduced by unsaturation, this difference in distribution of double bonds may have some practical significance. Lipids that prefer to locate themselves in the middle of the bilayer should incorporate better into mammalian PCs, while plant PCs could favor lipids that tend to locate themselves closer to the bridge region.

Plant lipids have lower proportions of saturated fatty acids than do those of egg yolk or mammalian sources. Because the headgroup of phospholipids is attached to the glycerol bridge at one end (the 3-position), the two fatty acid chains are located assymetrically within the molecule in the remaining two positions on the glycerol. In natural phospholipids, fatty acids in the 2- (central) position tend to have a higher degree of unsaturation, compared with the 1- (terminal) position, as indicated in Figure 4.

C. MEMBRANE BEHAVIOR
1. Lamellar phases

The lamellar configuration that phospholipids adopt actually constitutes a distinct physico-chemical phase, and this lamellar phase can adopt a number of different forms, the transition between which, for any given phospholipid species, depends on temperature, degree of hydration, solute effects, etc. The names given to these phases are a little imprecise and confusing. At low temperature there is a crystalline phase, "L_c", which is followed by a gel, or liquid-crystalline phase, "L_β", as the chains acquire a degree of mobility; thereupon a fluid phase, "L_α", forms where the fatty acid chains have completely melted and form disordered structures. Superimposed on this there are transitions that reflect changes in headgroup orientation (e.g., the pretransition, just before going from gel to fluid phase).

2. Transition temperature

The most important characteristic is this gel-fluid transition—generally referred to as the "main" transition. The temperature at which this occurs can vary from –60°C for di-myristoleyl PC to 80°C for di-tetracosanoyl PC. Plant and egg PCs have T_cs below 0°C, while mammalian PCs are considerably higher.

The determining factor in deciding the onset of the fluid phase is the ease with which disorder can be introduced into the hydrocarbon chains, so fatty acids of shorter chain length, those with unsaturations, particularly toward the center of the chain, polyunsaturated fatty acids, and *cis*-unsaturations (as opposed to *trans*-) are all factors which predispose to a low transition temperature.

3. Membrane dimensions

The membrane surface area per unit lipid in the gel phase is dependent almost entirely upon the headgroup and varies little in relation to the nature of the fatty acids or the temperature. The cross-sectional area of the hydrocarbon chains in the gel phase is usually smaller than that for the headgroup, so that when the headgroups have approached as closely as they can, the chains must tilt in order to achieve the tightest possible packing, instead of adopting an orientation perpendicular to the plane of the membrane. On increasing the temperature from the gel-fluid phase, the acyl chains become more disordered and bulky and are able to occupy the internal membrane space without the need for tilting. PE is one exception, where the headgroup is so small that there is no tilt observed even below the main transition. Typical membrane surface areas per molecule for gel phase are 0.4–0.6 nm^2, increasing to 0.75 nm^2 in the fluid phase. In contrast, the thickness of the bilayer decreases as the transition from gel to fluid phase is made— e.g., from 4.5 nm to 4 nm for distearyl PC.

4. Discontinuities

In the same way as for any other crystal structure, packing defects can be introduced during the formation of lipid lamellae. These can take the form of "point defects," where one single molecule is out of place or in a disordered non-regular conformation, or it may extend to fault lines where two whole sheets abut in a discontinuous manner. In the gel phase, where rotational and lateral mobility is severely affected, the phospholipid molecules do not have the opportunity to reorient themselves easily, and these packing defects can persist in membranes for long periods of time (the order of days).

In the fluid state, such discontinuities have the opportunity to form more readily but, by the same token, last for only short periods of time because of the increased mobility. Packing discontinuities introduced into the membrane in the gel state as a result of shear stress during formation may be dissipated by raising the temperature to the fluid state and then cooling, to give an essentially fault-free membrane. Irregularities in the gel state can be inferred by differential scanning calorimetry, since their presence causes an earlier onset of the main transition during the heating cycle.

Discontinuities can also be introduced as a result of chemical degradation of membrane lipids. The formation of lyso-phosphatides by hydrolysis and of truncated fatty acid chains as a result of oxidation reactions, introduces species, into the membrane which prefer to adopt non-bilayer states, and can disrupt packing within the lipid matrix.

5. Phase separation

Because of differences in physical dimensions of the two phases, major discontinuities exist at the phase transition temperature itself, when gel and fluid phases co-exist in the same membrane. For a single molecular species, this situation occurs at a single temperature, with a very narrow range. For liposomes composed of mixture lipids whose T_cs are close to each other, the membrane will have a single T_c intermediate between the individual components. It may even be lower than any of the individual T_cs— in the same way that a solute depresses the freezing point of a liquid. Such mixtures will also have a phase transition that is much broader than for single components, as is seen with phospholipids extracted from natural sources.

In the case of membrane components differing markedly in their transition temperatures, mixtures give rise to membranes in which one component forms domains of condensed (gel) material floating in a sea of fluid phase of the other components. This situation can persist stably over a wide temperature range, the breadth of this range being determined by the temperatures at which all components are in fluid phase and all components in gel phase. During this state of phase separation of the different components, numerous discontinuities are present at the interface between the solid domains and the surrounding fluid lipid.

6. Asymmetry

Natural cell membranes often display asymmetry with respect to the distribution of phospholipids of different headgroups on the inner and outer surfaces. A similar situation can arise in artificial membranes, and this can be brought about in several ways.

First, in the formation of liposomes of high curvature, packing constraints dictate that phospholipids with a large headgroup will preferentially be located on the outer surface, while those with smaller headgroups can be accommodated internally. SUVs composed of a mixture of PE and PC, for example, will have the majority of the PC located externally, while a higher proportion of PE will be on the inner face of the membrane. Even in liposomes that are quite large, such effects can be seen if the headgroups are also large—for example, with hydrophilic polymers such as PEG.

A second mechanism for introduction of asymmetry is via chemical modification procedures, which have access only to the outside of preformed liposomes. Alternatively, use of double emulsion preparation procedures, in which the inner and outer leaflets of unilamellar vesicles are formed from lipids provided by different pools, permits assymetry of acyl chains to be achieved.

Asymmetries across the membrane give rise to a further type of phase separation, as can happen with PS-containing membranes exposed to calcium, thereby inducing a change from fluid to condensed phase. Since calcium has access to the outer leaflet only, a situation must exist where different phases can be present in different halves of the membrane at the same time.

7. Aggregation

Aggregation occurs naturally between neutral planar phospholipid membranes as a result of weak interactions, such as van der Waals forces. Large, uncharged PC liposomes will display this phenomenon, therefore, and the effect can be minimized by increasing the surface curvature (i.e., reducing the size of the liposomes) and by incorporating charged lipids into the membrane, employing electrostatic repulsion to prevent close approach.

A second type of aggregation occurs over a period of time, even with small vesicles, as a result of the presence of discontinuities in the membrane. Such discontinuities can expose, or facilitate access to, some lipophilic areas within the membrane when the liposomes collide, and the opportunity for van der Waals interactions to operate and the exclusion of water may provide the driving force for the continuation of these associations.

8. Permeability

Flux of water across phospholipid membranes is surprisingly high and thought to occur as a result of formation of narrow water channels between the acyl chains spanning the region below the glycerol bridges. Permeability of most other hydrophilic molecules is much lower, so that liposomes membranes can display osmotic sensitivity in the presence of concentration differences between outer and inner compartments. Large molecules, such as proteins, cannot diffuse across membranes.

Permeability of small aqueous solutes is closely related to the presence of discontinuities in the membrane. For a liposome that has been formed free of packing defects in the gel phase, leakage of contents increases with temperature, is higher in the fluid phase, and is highest at the phase transition temperature, where irregularities at the phase boundaries create pores that are permanently open to allow molecules to pass through.

Molecules that can pass through membranes without difficulty are those that are both hydrophilic and lipophilic; such molecules do not need the formation of pores in the membrane to permit entry, since they are able to dissolve among the fatty acyl chains themselves. To be able to achieve this dual solubility characteristic, these molecules often possess an ionizable moiety, such as an amino group, which is charged in aqueous environments but loses this charge when it passes into lipids. Many drug molecules display this behavior, and advantage has been taken of it using a technique called "remote loading" in which the drug is allowed to pass through the membrane into a liposome in the uncharged form and is then held there as the cation, as a result of a lowered pH inside the vesicle.

9. Fusion

Fusion is the interaction of two separate membrane sheets such that they combine to form a single sheet in which membrane lipids are shared. Like aggregation and permeability phenomena, the propensity for membranes to fuse is very much linked with phase discontinuities and irregularities in packing. In fusion it goes one step further, since a necessary condition for lipids to rearrange so that their acyl chains come into contact is the ability to form non-lamellar lipid phases. Thus, while liposomes of normal composition can fuse to a small extent, if oxidation and hydrolysis processes go unchecked, liposomes that are specially designed to fuse usually contain high proportions of unsaturated PE, which form inverted hexagonal structures, or PS, in which the headgroups condense in the presence of calcium.

It should be noted that membranes with a tendency to fuse are often less stable in other respects (e.g., retention of contents prior to fusion). Furthermore, while these membranes fuse readily with each other, it has not been demonstrated as a general rule that they can fuse with other, normal, non-fusogenic membranes.

D. TYPES OF LIPOSOMES

A number of different subsets within the class of vesicles termed liposomes have been described and given various names relating to certain distinguishing characteristics such as size, "morphology," method of preparation, and chemical composition or behavior in biological systems. The different types of liposomes are discussed below under these headings.

1. Size

This was the first criterion ever used to characterize liposomes that gave rise to specific names, principally because this was the first feature that was easily manipulable and whose control brought about marked differences in behavior.

Small unilamellar vesicle (SUV)

This term refers to single-shelled vesicles produced as a result of high-intensity (probe) ultrasonication, and the abbreviation may thus also be considered to stand for "sonicated unilamellar vesicles." The liposomes prepared by this method are of the limit size, i.e., the smallest possible size that curvature of bilayer membranes will permit on steric grounds, and to this day, ultrasonication, together perhaps with certain high-pressure extrusion techniques and the alcohol injection method of Batzri and Korn, is the only method that is capable of giving a preparation of vesicles in this smallest size range. Because the SUVs approach the "limit size" in diameter, they are a population of liposomes more homogenous in size than liposomes prepared by other methods and have often been chosen for study for precisely this reason.

Subsequently, it has been realized that because of the high energy imparted by ultrasonication and the constraints in packing resulting from forcing the membrane to adopt such a high degree of curvature, SUVs are in fact a rather unusual type of liposome and demonstrate many properties atypical of liposomes in general. At the time when the term was coined, SUVs were being compared and contrasted with the only other type of liposomes extant, namely MLVs (multilamellar vesicles—see below), which were the result of dispersion of phospholipids in water without the aid of sonication.

Large unilamellar vesicle (LUV)

This term has been used to denote single-shelled vesicles of diameters greater than that of SUVs, but opinions differ as to what constitutes "large" in this context. The first methods developed to prepare such vesicles were calcium-induced fusion of liposomes composed of SUVs, and ether injection—i.e., introduction of ether solutions of PC into hot aqueous buffer to form large planar sheets of bilayer membrane that folded in on themselves. Liposomes produced by these methods were of the order of 0.5 μm in diameter. Other workers, however, have used the term LUV in reference to any unilamellar vesicle larger than an SUV; this usage is unfortunate and should be discouraged since it gives very little information about the actual size, which for liposomes may vary through several orders of magnitude from 25nm in diameter to 25 microns.

Intermediate-sized unilamellar vesicle (IUV)

This is a term that has not been widely adopted but whose use would help to identify liposomes within the 100 to 200-nm region between SUVs (25 nm) and LUVs (500 nm). Liposomes of this size are easily prepared by high-pressure extrusion or by detergent dialysis and are important in pharmaceutical applications since they fit into a size window that displays longer circulation times in the bloodstream, good stability, and ease of sterilization by membrane filtration.

Other types of large liposomes described in published works are "cell-size" liposomes, and "giant" liposomes, referring to vesicles of many microns in diameter.

2. Morphology
Multilamellar vesicle (MLV)

MLVs can be liposomes of any size that are composed of more than one bilayer membrane. Since even a liposome of just two bilayers is at least twice the size of an SUV, MLVs are readily distinguishable from SUVs in terms of size. MLVs are the type of liposome formed most easily, being obtainable simply by gentle manual shaking of dry phospholipids in water, and preparations thus formed are often called "hand-shaken" liposomes. The lamellarity of these MLVs depends on lipid composition among other factors, but it typically varies between 5 and 20 bilayers. Liposomes with lower numbers of lamellae are sometimes referred to as oligo-lamellar or pauci-lamellar liposomes, although acronyms have not been adopted for these terms.

Multivesicular liposome (MVL)
This type of liposome is bounded by an external bilayer membrane shell, but it has a very distinctive internal morphology, which arises as a result of the special method employed in the manufacture. A double emulsion is formed (water-in-oil) in which multiple aqueous droplets are suspended within single droplets of organic solvent, with phospholipids forming monolayers at both the external and internal oil-water interfaces. Removal of the organic solvent gives a particle composed of numerous distinct compartments distributed throughout the interior, separated from each other by single bilayer membranes. Topologically, each internal compartment is equivalent to every other, in contrast to the different compartments within conventional MLV, in which the separate aqueous compartments are all located concentrically within the vesicle.

The unusual structure of MVLs necessitates junctions in two or three dimensions in which three or four different membrane sheets come together, and to stabilize this configuration it appears that inclusion of neutral, non-bilayer-forming lipids in the membrane may be advantageous. The presence of internal membranes distributed as a network throughout MVLs may also serve to confer increased mechanical strength to the vesicle, while still maintaining a high volume:lipid ratio compared with MLVs. The multivesicular nature also indicates that, unlike LUVs, a single breach in the external membrane will not result in total release of the internal aqueous contents, giving rise to additional stability *in vitro* and *in vivo*.

Stable plurilamellar vesicle (SPLV)
Although this title could be considered to describe any oligolamellar vesicle with a tolerable shelf-life, the term was in fact coined by its makers to refer to liposomes manufactured by a special process that results in the entrapped solute being evenly distributed throughout the entire vesicle. This appears to be something that is not always achieved by conventional methods for preparation of MLVs, which give rise to osmotic differences between internal compartments that leave the intervening membranes in a stressed (and therefore unstable) condition. In the SPLV method, bath sonication during removal of solvent from a water-in-oil emulsion consisting of an ethereal PC solution relieves this stress.

3. Method of Preparation
Large unilamellar vesicle by extrusion technology (LUVET)
Extrusion of liposomes through porous membranes was developed as a method of modifying their size, liposomes being broken down as they passed through to give a population with an upper size limit closely approximating that of the pores of the membrane themselves. At relatively low pressures (100 psi), MLVs retain their multilamellar characteristics, while displaying a reduced-size heterogeneity. At higher pressures, however, the higher shear forces resulting from the greater pressure differential across the membrane filter result in reorganizations of the phospholipid bilayers giving rise to unilamellar vesicles, which are termed LUVETs. Repetition of the process several times again leads to a population with an upper size limit determined by the pore size of the membrane.

Reverse-phase evaporation vesicle (REV)
The key feature of the preparation method for this type of liposome is the formation of a water-in-oil dispersion (i.e., a reverse-phase emulsion) from which the organic phase (usually an ether) is evaporated off. The result is a gel consisting of aqueous vesicles bounded by a single monolayer of phospholipid. Mechanical agitation ruptures a proportion of the vesicles, and the phospholipid released provides the outer monolayer to convert the remaining vesicles into large unilamellar bilayer liposomes.

Dehydration-rehydration vesicle (DRV)
In this type of vesicle, a process of dehydration followed by rehydration has been employed to entrap material inside the liposomes. The starting point is a suspension of empty SUVs, to which the solute to be entrapped is added, such that the solute is outside the liposomes in the external medium. Lyophilization of the mixture, followed by subsequent re-addition of a limited volume of water brings about a reorganization of the lipid membranes such that after fusion they reform liposomes in which a considerable

proportion of the aqueous solute is now located within the vesicles. The liposomes obtained are somewhat larger than the original SUVs started with. Entrapments greater that 50% can be achieved. Because the energy to which the lipids are subjected is imparted in the absence of the solute (which is added only after formation of the SUVs), the method is good for the entrapment of sensitive molecules such as proteins.

4. Composition
Immunoliposomes
Contrary to expectation, this term has been used, not to describe liposomes designed to elicit an immune response but to describe liposomes that use immunological molecules, in particular, immunoglobulins for targeting purposes. In a similar way to any other protein, immunoglobulins may be attached to the liposome surface by covalent linkage through membrane lipids possessing functional groups. In certain circumstances, the proteins may be conjugated to free lipids, and the whole conjugate incorporated into the membrane during formation of the liposome.

With the development of PEG coatings for liposomes to increase circulation time in the blood (see "Stealth" liposomes below) the preferred method of linkage of ligands in general appears to be the terminal of the PEG chains, rather than the membrane surface, in order to permit maximum exposure of the ligand and optional access to its target.

Proteosomes
A term applied to lipid vesicles incorporating proteins in or on the outer membrane. The same consideration applies as for immunoliposomes above.

5. Function
Stealth
Although this is a registered trademark for a proprietary technology, it is also used in a wider context by researchers in the field to indicate any liposome that avoids uptake by the RES (and hence has a long circulation in the bloodstream) as a result of coating the liposomes surface with hydroxylated polymers. Two approaches have been identified to achieve this aim: in one case a combination of sphingomyelin and gangliosides (e.g., oligosaccharide-containing ceramides) are incorporated directly into the membrane; in the second approach, polyethylene glycol chains are conjugated to the surface. Optimal molecular weight of the chains is 2000.

Transfersomes
The term refers to a liposome of a particular composition that has been demonstrated to be capable of transferring aqueous contents across the skin. The important features of the liposomes are that the membrane contains a certain proportion of bile salt distributed among the phospholipid, which confers markedly increased flexibility on the membrane, and that the liposome adopts a configuration in which the internal volume is insufficient to inflate the external membrane fully, so that it is flaccid and readily deformable. Such liposomes are capable of being extruded through narrow channels without rupture and can be driven across the skin barrier under the influence of an osmotic potential difference.

Chapter 2

Chemistry of Ligand-Coupling to Liposomes

Francis Schuber

CONTENTS

I. INTRODUCTION

Numerous methods have been utilized in the last two decades to increase the interaction between liposomes and cells. In order, for example, to achieve the targeting of liposomes, which carry drugs or other bioactive molecules, it is indeed necessary that the vesicles bind to their target cells with a high selectivity and affinity. A fruitful approach consists of attaching to the surface of the liposomes various ligands that are recognized by receptors expressed by the target cells. Among the most widely used ligands are (monoclonal) antibodies directed against specific antigens of tumoral cells (immunoliposomes) and glycosidic residues (e.g., mannose or galactose), which interact with membrane-bound lectins that are present at the surface of certain cell types (macrophages, hepatocytes, metastases). Many other ligands have been attached to liposomes (for reviews see References 1 and 2), including proteins (plant lectins, glycoproteins, enzymes, hormones), particles such as LDL,[3] and small molecules such as peptides or ecto-enzyme inhibitors.[4,5] Another reason to conjugate a ligand to the surface of vesicles is the possibility of increasing their intracellular delivery; indeed in many cases the uptake of such targeted liposomes by cells is receptor-mediated (endocytosis and phagocytosis), i.e, a process generally much more efficient than the passive uptake of non-targeted liposomes. However, as discussed in this book, liposomes have gained a wide use as tools in many fields other than drug delivery. The elaboration of the chemistry of ligand-coupling to liposomes has important applications in immunochemistry, e.g., to increase the immunogenicity of proteins, haptens,[6,7] or peptides (see Chapters 6 and 12) in diagnostics[8] (Chapter 15) and imaging.[9] Finally, it should be noted that grafting the outer surface of liposomes with short chains of polyethylene glycols has a profound effect on the circulation time of liposomes *in vivo*, allowing them to better escape capture by the reticuloendothelial system (Chapter 11).

The aim of this chapter is to describe some basic techniques that allow ligands to attach to the surface of liposomes. The first section is devoted to methods that permit direct covalent or non- covalent (via the formation of a complex) coupling of ligands to preformed liposomes (Figure 1A,B). This approach is the most common and is also the best controlled from a chemical standpoint. It involves coupling procedures in aqueous media; this special type of chemistry has developed greatly during the last decade and is found

0-8493-4569-3/95/$0.00+$.50
© 1995 by CRC Press, Inc.

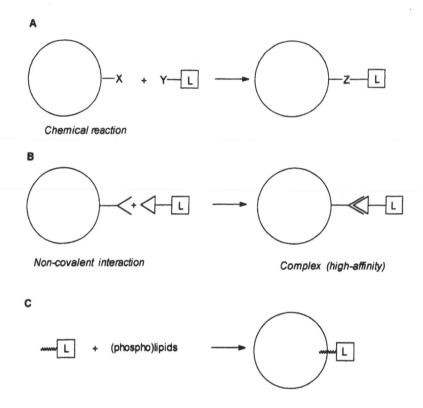

Figure 1 Different modes of attachment of ligands to liposomes. The coupling between a ligand and preformed liposomes can be *covalent* (**A**), by reaction between a function X introduced at the surface of the vesicle and a reactive function Y of the ligand, or (**B**) *non-covalent*, resulting from the formation of a high-affinity complex between a recognition element present at the surface of the liposome and the (modified) ligand. A ligand can also be incorporated into liposomes via a *hydrophobic anchor* (**C**).

in other domains such as the preparation of affinity chromatography supports,[10] the protein conjugation[11] and the coupling of haptens or antigens to carriers (proteins, polymers).[12-14] The second section deals with the incorporation of ligands that have been modified e.g., by hydrophobic anchors into liposomes, (Figure 1C). The chapter concludes with some useful techniques related to this field, such as the purification of the liposomes after ligand grafting. Several reviews have been published previously that deal mostly with the coupling of proteins to liposomes[1,2,15-17] and the application of targeted liposomes.[1,2,18-22]

II. METHODS FOR THE CONJUGATION OF LIGANDS TO PREFORMED LIPOSOMES

This approach implies a reaction in aqueous media between preformed liposomes, bearing at their surface appropriate reactive groups, and ligands. It is of importance that such chemical reactions can take place under experimental conditions (such as pH, osmolarity, temperature) that respect the integrity of the vesicles and also, if applicable, minimize the efflux of encapsulated material. Such constraints, and the fairly limited number of chemical functions susceptible to be present into liposomes, have somewhat restricted the options available. Essentially two types of techniques allow the coupling of ligands to preformed vesicles:

1. Direct covalent coupling between a ligand and (phospho)lipid incorporated into the bilayer of a preformed liposome (Figure 1A). In most cases the (phospho)lipid has been modified in order to make it react with chemical functions present, or introduced, in the ligand. The covalent coupling is the prevalent technique and has been used to graft many ligands, from small molecules to proteins. It has greatly benefited from the elaboration of sophisticated and efficient heterobifunctional reagents that permit simple and chemically unequivocal conjugation reactions, including in aqueous media.*

2. Formation of a stable non-covalent complex between the ligand and a molecule covalently linked at the surface of the liposome (Figure 1B). This technique, although less common, has been used with the couples protein A (bound to the vesicle)/antibody and avidin/biotin. It has the advantage to allow the preparation of "universal" liposomes, which can thereafter complex a large number of different ligands.

There are many advantages to these conjugation techniques. They allow the use of liposomes that have been prepared by any method, including those that favor high encapsulation yields and efficient sizing. Moreover they give predictable (phospho)lipid/ligand ratios and preclude the exposure of the ligands to deleterious conditions sometimes used in the preparation of liposomes (such as organic solvents, ultrasounds, etc....). Finally, the separation of the conjugated liposomes from the free ligands is generally easy (see Section IV. A).

In order to choose among the different coupling methods it is important to underline some other considerations. Overwhelmingly, the aim of coupling a ligand to the surface of a liposome is to favor its interaction with a receptor (recognition site) expressed at the surface of the target cell. The coupling strategy should therefore avoid masking, chemically or sterically, the elements of the ligand engaged in the recognition process. These requirements are very similar to those encountered in affinity chromatography. For that reason one should select specific chemical reactions that only involve functions that are distal to such elements. Moreover, and this is particularly important for low molecular weight ligands, one should pay attention to the critical distance between the ligand and the surface of the liposome. If this distance is too short it might disfavor, if not prevent, the formation of the recognition complex. For such cases, it is advisable to adopt techniques, such as the ones using heterobifunctional reagents, which allow the introduction of spacer arms. It is noteworthy that such considerations also apply to the coupling of high molecular weight ligands, such as proteins, to preformed liposomes. In this latter case, coupling yields can dramatically depend on the accessibility of the reactive group at the surface of the vesicle. Finally, the number of targeted liposome/cell interactions are of polydendate nature; i.e., they involve the formation of multiple complexes between one liposome (which carries, at its surface, a great number of ligands) and a cell. In some cases where, for selectivity and affinity reasons, one wishes to favor such multiple interactions it important to adjust the surface density of the ligands (occurrence of threshold phenomena). The best methods will be those that allow the strict control of such a parameter.

A. COVALENT COUPLING TECHNIQUES

The covalent coupling techniques rely on the reaction between a (phospho)lipid, or a derivative of a (phospho)lipid, incorportated in the bilayer of the liposome with a reactive group of the ligand. The most interesting derivatives that were first introduced in this domain are those formed between PE-NH$_2$** and heterobifunctional reagents such as N-succinimidyl-3-(2-pyridyldithio)propionate (SPDP) and N-succinimidyl-4-(p-maleimidophenyl)butyrate (SMPB) (Figures 2 and 3). Originally these derivatized phosphatidylethanolamines have been developed by Leserman et al. and Papahadjopoulos et al.[23] After their incorporation into liposomes, they react with ligands bearing thiol residues. Importantly, these reagents also introduce a spacer arm between the ligand and the liposome surface. Other methods involve the use of oxidized glycolipids and the coupling of the ligand by reductive amination.

1. Coupling of Ligands-SH to preformed liposomes

As mentioned, essentially two types of PE-NH$_2$ derivatives have been prepared by reaction, in organic media, of the amine function of the phospholipid and the activated (by a N-hydroxysuccinimide group) ester of a heterobifunctional reagent:

---

*This chapter refers only to the most recent coupling techniques that allow the formation of chemically defined linkages and compounds, or at least that allow unequivocal reactions in which the ligand reacts exclusively with the surface of the vesicles. This is in opposition to earlier methods using homobifunctional reagents (e.g., glutaraldehyde, diethylsuberimidate, etc...) or carbodiimides, which, in many cases in reactions also plagued by low yields, give complex and ill-defined (polymers) mixtures or which need site-modifications of proteins. The reader interested by these earlier approaches is refered to References 15 and 22.

**The terminology adopted in this chapter is the following: molecule-X is given when one wishes to highlight the reactive function X (such as -NH$_2$, -SH, -COOH, etc...). For example, PE-NH$_2$ indicates that the primary amine function of phosphatidylethanolamine is important and reacts with the activated ester of a heterobifunctional reagent.

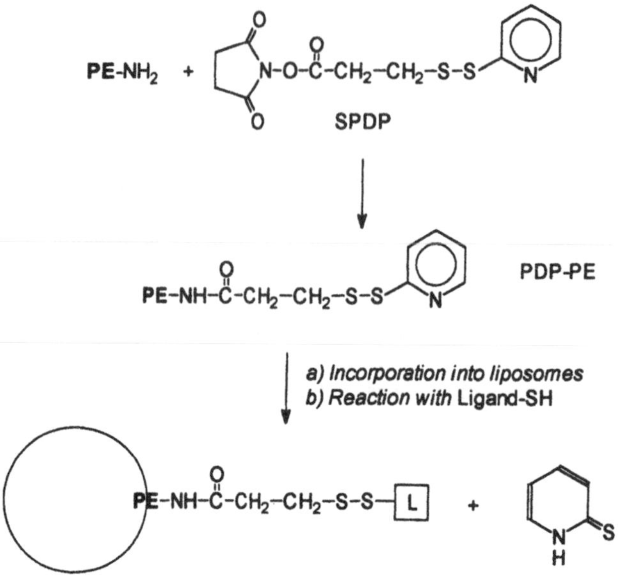

PE-NH$_2$ + [SPDP structure]

SPDP

PDP-PE

a) Incorporation into liposomes
b) Reaction with Ligand-SH

Figure 2 Covalent coupling of a ligand-SH to preformed liposomes containing PDP-PE. PDP-PE, obtained by reaction of PE-NH$_2$ with SPDP, is incorporated into the (phospho)lipid mixture used for the preparation of the vesicles. The dithiopyridine group allows an easy exchange reaction with ligand-SH and permits the coupling of the ligand via a new disulfide bond. This technique is well-suited to graft, e.g., proteins to the surface of liposomes.

PE-NH$_2$ + [SMPB structure]

SMPB

MPB-PE

b, c

Figure 3 Covalent coupling of a ligand-SH to preformed liposomes containing MPB-PE. MPB-PE, obtained by reaction of PE-NH$_2$ with SMPB, is incorporated into the (phospho)lipid mixture used for the preparation of the vesicles. Its maleimide residue reacts, under mild conditions of pH and temperature, with the thiol group of a ligand-SH yielding a stable thioether bond. This technique is useful to conjugate proteins and small molecular ligands to the surface of liposomes.

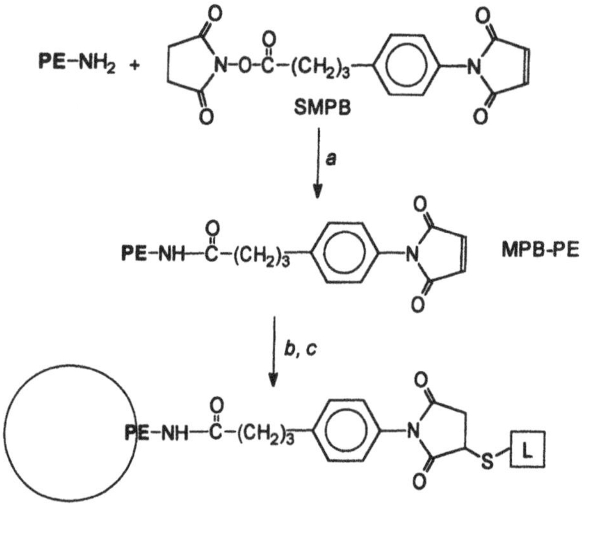

a - Reaction in organic solvent ; b - Incorporation into liposomes
c - Reaction with Ligand-SH.

PDP-PE (N-3-(2-pyridyldithio)propionylphosphatidylethanolamine) — obtained by reaction with SPDP[24], a reagent introduced by Carlsson et al., in 1978.[25] This derivative has a reactive pyridyl-disulfide bond (Figure 2).

MPB-PE (N-4-(p-maleimidophenyl)butyrylphosphatidylethanolamine) — obtained by reaction with SMPB[26]. This derivative bears a maleimide group, i.e. another thiol-reactive function (Figure 3).

These modified PE, via an amide linkage, can be added to the (phospho)lipid mixture used to make liposomes. When the ligand to be grafted is a protein, PDP-PE and MPB-PE are generally used in a 1–5 mol% range, whereas for small molecular weight ligands this proportion can be increased (e.g., up to 25 mol% of the total lipids). The vesicles thus prepared bear, at their surface, groups which react rapidly and in excellent yields with ligands possessing a free thiol function. Importantly, the coupling reaction is realized with the aqueous liposome preparation under mild experimental conditions (e.g., pH 6.5 under argon) that are fully compatible with the integrity of the vesicles and non-denaturing for protein ligands. An excellent review covering the practical aspects of these techniques has been published by Martin et al.[17]

In the case of PDP-PE, in the presence of an excess of ligand-SH, an exchange reaction will occur with the heterodisulfide bond and the conjugation of the ligand via a new disulfide bridge (Figure 2). This classical reaction, very specific for thiol functions, exploits the known reactivity of aromatic disulfides which, for example, is also used with the Ellman's reagent (see Section II.A.2). The release of an equivalent of 2-pyridinethiol, a chromophoric reaction product ($\varepsilon_{343\,nm} = 8080\ M^{-1}\cdot cm^{-1}$), can be used to quantify the conjugation process.[25] The chemical properties of 2-pyridinethiol, i.e., tautomerism, make this exchange reaction virtually irreversible. This approach has been mostly used to attach monoclonal antibodies to the surface of liposomes.[27] The free thiol group can be generated by transformation of the IgG into Fab' fragments. Other types of proteins, such as protein A, have also been coupled to the surface of liposomes by this technique.[28] Importantly, thiol functions can also be introduced into proteins, including native immunoglobulins, or other ligands with the help of thiolation agents (see Section II.A.2). One potential limit to the use of PDP-PE is the possible lability of a disulfide bridge (via a reductive process) under *in vivo* conditions, resulting in the release of the ligand from the liposome and obliterating the targeting process. The incidence of such an event seems relatively minor, at least with protein ligands that might conceal the disulfide bridge.

The maleimide group of MPB-PE is a good electrophile: incorporated into liposomes it reacts easily with the thiol function of a ligand-SH (Figure 3). The addition product is a thioether, which is chemically stable. This type of reaction is analogous to the classical modification of proteins, in aqueous media, with e.g., *N*-ethylmaleimide. It should be noted that if it is performed under pH conditions close to, or slightly below, neutrality, this reaction will be fairly specific for thiols; moreover, the hydrolysis of the maleimide group, which occurs under basic pH, remains limited. This technique has also been widely used to couple monoclonal antibodies to liposomes.[23] However, because of its versatility it has been extended to the conjugation of other proteins,[29] including lectins,[30] or to small molecular weight ligands such as glycosidic structures[31-33] and peptides.[34-36] It should be noted that SMPB is not the only heterobifunctional reagent available bearing a maleimide group: e.g., succinimidyl 6-(*N*-maleimido)-*n*-hexanoate can be also be used. The advantage of this latter molecule is that the immune response found against its spacer arm is less important than with SMPB.[37] With proteins, such as antibodies or streptavidin, up to 50% of MPB-PE incorporated into the liposomes were found to react;[38,39] with smaller ligands nearly quantitative coupling yields (100% of exposed MPB-PE) could be observed.[4,35]

A potential problem, which is often overlooked, is that the utilization of PDP-PE and MPB-PE in the preparation of liposomes results in the presence of thiol-reactive groups also on the inner surface of the vesicles and inner bilayers if multilamellar liposomes are used. When the ligand-SH is non-permeant, which is the case for proteins and polar molecules, these reactive groups remain intact after the coupling step; to our knowledge, the incidence of this, e.g., in the interaction of targeted liposomes with cells, has not been properly addressed. In our laboratory, after the reaction of ligand-SH with liposomal MPB-PE, we treat the conjugated vesicles with an excess of 2-mercaptoethanol in order to eliminate all the residual maleimide groups. A similar result can be obtained for the *externally* exposed groups with cysteine. These thiols, including the excess of ligand-SH, are then separated from the vesicles (see Section IV. A). Finally, when the presence of such thiol-reactive groups are detrimental to the preparation of liposomes, for example in the case they react with the molecules to be entrapped, it was suggested to react SPDP or SMPB with preformed vesicles containing PE.[17] It should be noted that many heterobifunctional reagents, which can be used in this domain, are now commercially available. Moreover, such reagents with *N*-hydroxysulfosuccinimide activated esters are also useful; they have a better water solubility and cross membranes less easily.

Other approaches use the reaction between PE-NH_2, or fatty acids derivatized with amines (e.g., N^2-palmitoyl-L-lysine methyl ester), and incorporated in preformed liposomes, with other maleimide-containing heterobifunctional reagents.[40,41] Similarly, Goundhalkar et al., have synthesized PDP-SA, a derivative obtained by reaction between stearylamine-NH_2 and SPDP.[42] This compound has been used similarly to PDP-PE to attach antibodies to liposomes.

An important notion is the influence of the grafted ligands on the liposome stability. This parameter depends on the nature of the ligand and on the number of conjugated ligands per surface unit.[43,44] For example, it was mentioned that a concentration of MPB-PE superior to 2.5 mol % in liposomes, after coupling to certain antibodies, dramatically increased the efflux of the encapsulated molecules.[43] It is therefore advisable to systematically verify the incidence of the coupling procedure on the liposome integrity (see Section IV. B). On the other hand, coupling of proteins-SH to the surface of liposomes is sometimes accompanied by aggregation phenomena;[17] such vesicles can nevertheless be sized by extrusion techniques through polycarbonate filters.[38]

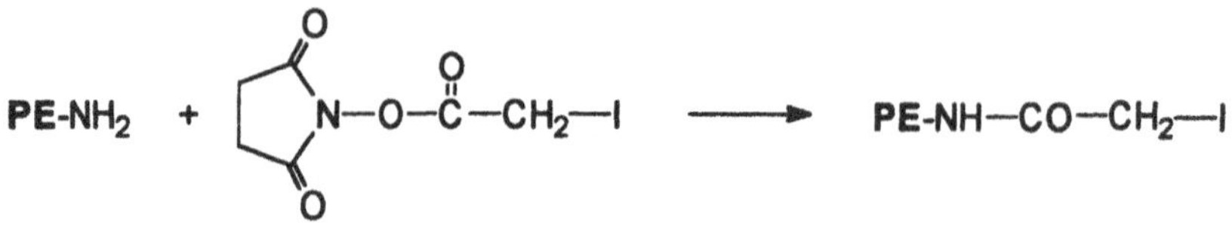

PE-NH$_2$ + [N-succinimidyl structure] N—O—$\overset{\overset{\text{O}}{\|}}{\text{C}}$—CH$_2$—I $\longrightarrow$ PE-NH—CO—CH$_2$—I

Figure 4 Synthesis of a iodoacetylated derivative of phosphatidylethanolamine. This derivative can be incorporated into liposomes; it allows the coupling a ligand by a substitution reaction with nucleophilic functions (such as -SH).

Other derivatives of phospholipids have been synthesized in order to react, once incorporated into liposomes, with ligands-SH. Thus reaction of PE-NH$_2$ with N-succinimidyliodoacetate yields iodoacetyl-PE[45,46] (Figure 4); this derivative can undergo a substitution reaction with a ligand-SH to give a stable sulfide bond. It seems, however, at least in the case of proteins-SH, that the coupling yields are somewhat low; this is probably due to steric hindrance problems because of the shortness of the spacer-arm. In agreement with this interpretation, Kinsky et al. have observed that the iodoacetyl group reacts efficiently, with a thiolated IgG, provided that this reactive group is linked to PE through a longer spacer-arm.[46] Recently Lasch et al. have used Traut's reagent (see Section II. A.2) to transform PE-NH$_2$, incorporated in liposomes, into a SH-bearing derivative of phosphatidylethanolamine (PE-SH).[47] After formation of a mixed disulfide bond with Ellman's reagent (see Section II. A.2) the liposomes could then be coupled to proteins-SH via an exchange reaction similar to the one with PDP-PE (see above).

2. Thiolation reactions

As mentioned above, the coupling reactions between ligands-SH and the different PE-derivatives can involve thiol functions that are intrinsic to the ligand. For example, immunoglobulins can be transformed into Fab' fragments that possess reactive thiols distal to the CDR (antigen combining site). However in some cases, e.g., when intrinsic thiols of some proteins do not react readily with PDP-PE or MPB-PE[48], it might be of importance to be able to introduce extrinsic thiol functions into ligands. To this end, one can utilize so-called thiolation agents, which can also be used in aqueous media. The best known is SPDP (Figure 5A), which reacts, via its active ester part, with amines (e.g., in proteins with the ε-NH$_2$ of a lysine) to yield an amide derivative bearing at the extremity of the spacer-arm a protected thiol function.[25] The free thiol can be obtained by reduction of the pyridyldisulfide bridge, for example by dithiothreitol (DTT).[25] The excess of reducing agent, which under the experimental conditions used does not attack the disulfide bridges of proteins, must then be eliminated before the thiolated ligand reacts with the liposomes. In the case of proteins-SH this is easily performed by dialysis or gel filtration (see Section IV. A). However, these techniques might not be easily applicable to small molecular weight ligands; alternatively therefore the thiol function can also be freed by reduction with NaBH$_4$.[33]

N-succinimidyl-S-acetylthioacetate (SATA), another heterobifunctional reagent used to generate extrinsic thiol groups in proteins destined to be coupled to liposomes, has been introduced more recently by Derksen and Scherphof[49] (Figure 5B). This molecule also carries an active ester that reacts, e.g., with accessible amine functions of the ligands, and therefore grafts a thioacetyl group.[50] The free -SH function can be released by reaction with an excess of hydroxylamine (Figure 5B); this procedure does not interfere with the coupling step of the thiolated ligand with, for example, liposomal MPB-PE. The main advantage of SATA is therefore to circumvent the separation step of the ligand-SH with other reagents and to allow a "one-pot" conjugation method. The use of SATA with ligands other than proteins is, however, not without some pitfalls. In our laboratory we have shown that this reagent, in both organic and aqueous media, can also act as an acetylating agent for hydroxyl and amine functions present in molecules to be thiolated.[31,51] Such side-reactions, which because of their extent cannot be neglected, are attributable to the chemical reactivity of the thioacetate moiety of SATA. If O-acetyl groups can be eliminated during the hydroxylamine treatment step, this is not the case for the N-acetyl ones. In conclusion, SATA should also be considered as an N-acetylating reagent and used as a thiolation agent only if this secondary reaction has a limited incidence on the ligand (decreased net charge, aggregation). The respective merits of SPDP and SATA for the thiolation of antibodies with the aim to prepare immunoliposomes have been discussed by Schwendener et al.[41]

Other thiolation agents have been used in the present context, such as 2-(S-acetyl)mercapto-succinic anhydride (SAMSA)[46] or 2-iminothiolane (Traut's reagent).[52] The experimental conditions for the use of SAMSA are similar to those for SPDP; however, SAMSA presents the potential disadvantage to

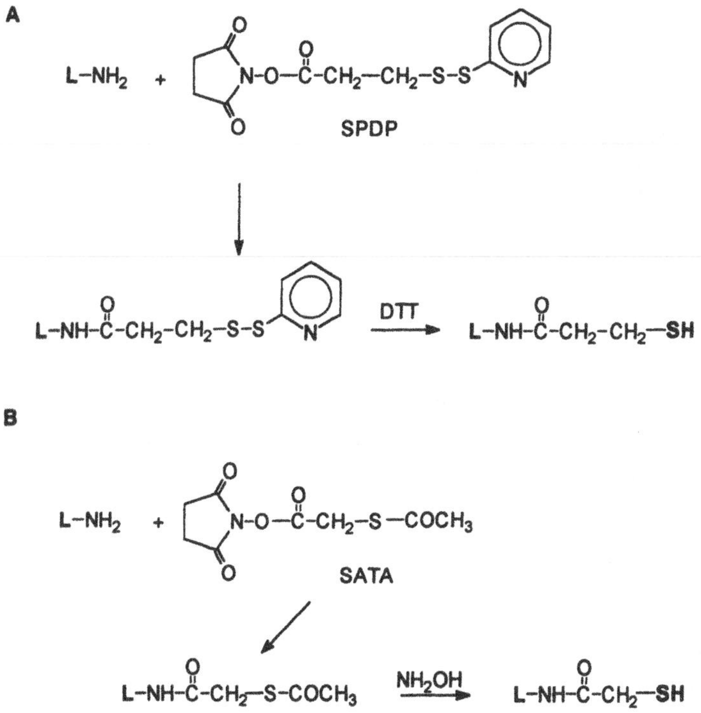

Figure 5 Thiolation reactions of ligand-NH$_2$. Ligand-NH$_2$, after reaction with a thiolation agent such as SPDP (**A**) or SATA (**B**), is transformed into ligand-SH, respectively, by reduction of the disulfide bond with dithiothreitol (A) or by reaction of the thioacetyl moiety with hydroxylamine (B). In the first case, the excess of DTT must be separated from ligand-SH before performing the coupling with the liposomes.

introduce supplementary carboxylic groups into the modified ligand. Traut's reagent, which reacts with the amine functions of ligands, is generally used under fairly basic conditions (pH 9), which are not always well tolerated by proteins or other pH-sensitive ligands.

The introduction of extrinsic thiol residues into ligands can be quantified by use of Ellman's reagent, 5,5′-dithio-*bis*-(2-nitrobenzoate), or DTNB. However, sometimes underestimated, the reaction of DTNB with thiols (Figure 6) is very sensitive to experimental conditions, especially to the pH value.[53] For example, 2-nitro-5-thiobenzoic acid, the chromophore formed after the exchange reaction has a $\varepsilon_{412\,nm} = 1.36 \; 10^4 \, M^{-1}\cdot cm^{-1}$ at pH 8.0. It should be borne in mind that, as any chemical modification, thiolation can sometimes result in the partial or total loss of the recognition sites of a ligand (for proteins, see[54]). Interestingly, the same Ellman's reagent can also be used to determine the concentration of maleimide groups present at the surface of a liposome. To this end, liposomes (containing, for example, MPB-PE) are first treated with a 2- to 2.5-fold excess of cysteine, the residual thiol is then measured with DNTB.

3. Coupling of ligands-NH$_2$ to preformed liposomes

Essentially two types of methods have been developed to attach ligands-NH$_2$ to preformed liposomes: (1) formation of a Schiff base between an aldehyde function generated at the surface of liposomes and a ligand-NH$_2$, and (2) formation of an amide bond between a carboxylic function present at the liposome surface and a ligand-NH$_2$.

Schiff base formation — Vicinal hydroxyl functions present in phosphatidylglycerol (PG) or in sugar residues can be easily transformed into aldehydes by oxidative cleavage with sodium periodate (Figure 7). This reaction allows to generate aldehydes at the surface of preformed liposomes* containing PG or glycolipids (e.g., lactoceramide, galactocerebrosides, gangliosides, phosphatidylinositol, etc...). These

---

*The reactions should be performed at pH 8.4 to limit the leakage of liposomes. Acidic pHs such as 5.5 can also be used; in this case, however, the oxidant concentration should be kept under 10 mM.[17]

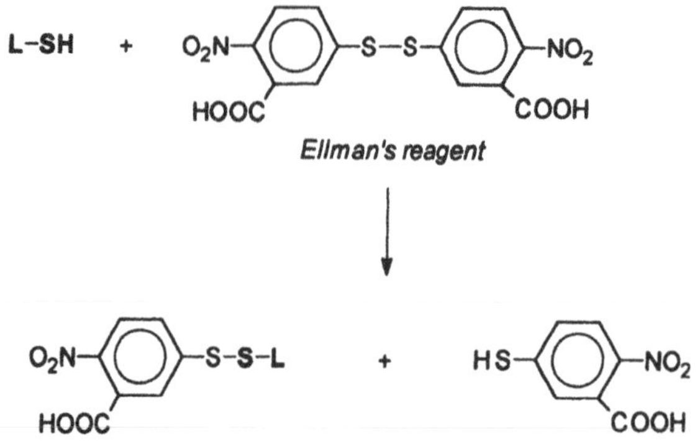

Figure 6 Quantification of thiols with Ellman's reagent. Reaction of ligand-SH with Ellman's reagent results, by an exchange process, in the formation of a highly chromophoric product: 2-nitro-5-thiobenzoic acid, which is measured spectrophotometrically.

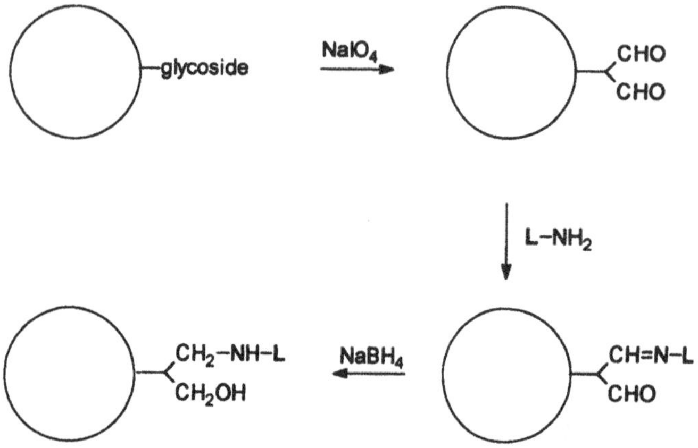

Figure 7 Coupling of ligand-NH_2 to preformed liposomes bearing aldehyde functions. In the first step, aldehydes are generated, by oxidative cleavage with periodate, from glycosidic residues (glycolipids) present at the surface of the vesicles. In the second step, a reductive amination is performed in the presence of ligand-NH_2 (formation of a Schiff base), followed by a reduction with sodium borohydrate.

aldehydes (liposomes-CHO) react easily with primary amine functions of ligands-NH_2, such as antibodies or other proteins, to yield a Schiff base (liposome-CH = N-ligand).[55-59] These linkages being somewhat labile (hydrolysis), it is preferable, from a chemical point of view, to reduce them into stable amines (liposome-CH_2-NH-ligand) with sodium borohydride ($NaBH_4$) or better, with sodium cyanoborohydride ($NaBH_3CN$). This latter reducing agent is more selective for Schiff bases and does not normally reduce aldehydes or, more important, disulfide bond in proteins; moreover, it can be used in water at neutral pH. One of the first applications of this method has been the coupling of horseradish peroxidase to the surface of a liposome.[60] It is remarkable that all that conjugation chemistry can take place, under controlled conditions, without altering too much the integrity of the vesicles.

Schiff bases can also be formed at the surface of liposomes by generating the aldehyde functions in the ligand, such as a glycoprotein. The ligand-CHO is made to react with amines present at the surface of preformed liposomes, originating from phospholipids (PE-NH_2 or phosphatidyserine-NH_2) or from fatty acids bearing an amine (stearylamine) or hydrazides (lauryl-CONH-NH_2) function.[59]

Amide bond formation — From PE-NH_2 one can synthesize phospholipids possessing a carboxylic function (i.e., PE-COOH): for example, reaction of succinic anhydride with PE results in the formation of N-succinyl-PE (Figure 8).[61] The size of the spacer-arm can be modulated by reacting PE-NH_2 with diacids or anhydrides of various lengths. Kinsky et al.[62] have prepared N-hydroxysuccinimide derivatives (activated esters) of these PE-COOH and incorporated them into liposomes. These molecules, after reaction with a ligand-NH_2 form conjugates via a stable amide bond (Figure 8). Unfortunately the yield of such coupling procedures are strongly diminished by the spontaneous hydrolysis of the active ester moiety in water. Small molecular weight molecules (haptens-NH_2) have been coupled, according to similar procedures, to N-succinyl-PE or to cholesterol-hemisuccinate activated by N-hydroxysuccinimide.[61] An alternative approach, which gives better coupling yields, consists first in the activation, by a water-soluble carbodiimide at a relatively acidic pH, of the carboxylic function present at the surface of the liposome. After rapid elimination of the excess reagent, the pH is then increased and the ligand-NH_2 added to form the amide bond.[63] Finally, it should be mentioned that a similar technique has been used to covalently graft sugar residues to the surface of liposomes. Thus a PE-COOH derivative, obtained by

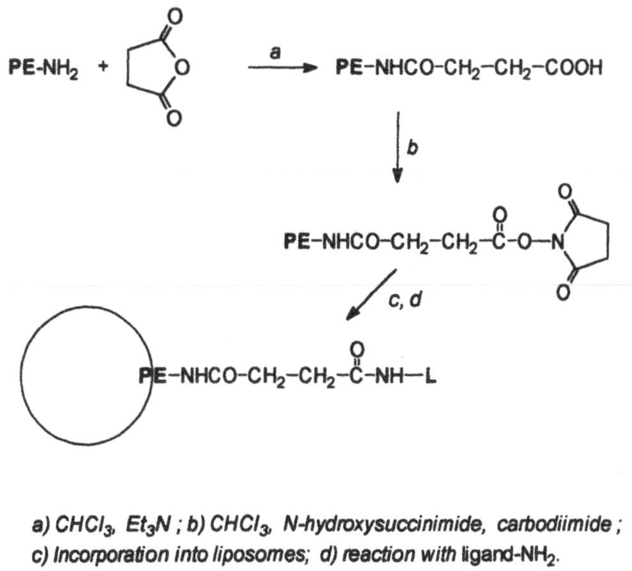

Figure 8 Coupling of a ligand-NH$_2$ to preformed liposomes containing carboxylic functions (here as an activated ester). Steps **a** and **b** consist in the transformation of PE-NH$_2$ into a homologous phospholipid PE-COOH whose carboxylic function can be activated, e.g., by *N*-hydroxysuccinimide. After incorporation into a liposome, the activated ester can react with a ligand-NH$_2$ to form a stable amide bond. This technique is less used because of the relative lability of the ester in water.

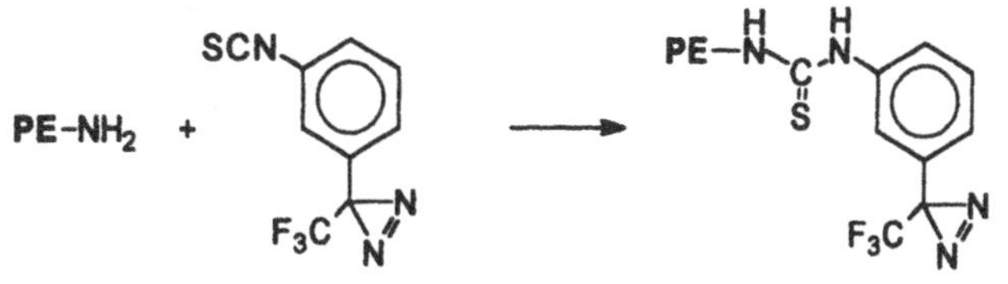

a) CHCl$_3$, Et$_3$N ; b) CHCl$_3$, N-hydroxysuccinimide, carbodiimide ; c) Incorporation into liposomes; d) reaction with ligand-NH$_2$.

Figure 9 PE-diazirine, a photoactivable derivative of PE. The molecule was produced by reaction of PE-NH$_2$ with 3-(trifluoromethyl)-3-(*m*-isothiocyanophenyl)diazirine.[67] Upon UV treatment, the molecule, incorporated into liposomes, decomposes into nitrogen and a carbene that initiates ligand coupling.

reaction of PE-NH$_2$ with glutaric anhydride, was coupled, by means of a carbodiimide after incorporation into liposomes, to a glycoside-NH$_2$ (*p*-amino-phenyl-α-D-mannopyranoside).[64]

4. Miscellaneous covalent coupling techniques

A few techniques have been developed that permit the covalent coupling of proteins or peptides to the surface of liposomes and that do not require the use of specifically functionalized ligands such as protein-SH. Accordingly, Snyder and Vannier[65,66] have used the well known reactivity of aryl diazonium groups with aromatic residues such as tyrosine (diazoic coupling reaction). N-(*p*-aminophenyl)stearylamide is incorporated into liposomes and after diazotization with HNO$_2$, at low temperature at acidic pH, is rapidly reacted with the ligand (protein). Although these reactions seem somewhat drastic, the authors suggest that they have only a marginal impact on liposome stability. Recently, a potentially attractive procedure to prepare proteoliposomes has been published by Sänger et al.[67] It is based on a photoactivated coupling between a PE-diazirine (Figure 9), incorporated into liposomes, and proteins (e.g., streptavidin) to be immobilized. The diazirine undergoes, at $\lambda \geq 320$ nm, a photodecomposition into a highly reactive carbene intermediate that reacts readily by insertion into chemical bonds of the target protein.

B. FORMATION OF STABLE NON-COVALENT COMPLEXES

As mentioned in the Introduction, an alternate mode to covalent attachment of a ligand to the surface of liposomes consists in the formation of a stable non-covalent complex (such as "sandwich" type) between the ligand and a complexing entity bound to the vesicle. This approach allows the preparation of "universal" liposomes that in turn can complex different types of ligands or lead to homogeneous populations of liposome each one carrying a different ligand. Essentially two techniques have been developed based on the following interactions: protein A/immunoglobulins and (stept)avidin/biotin.

1. Protein A/antibodies complexes

Protein A, isolated from *Staphylococcus aureus* (M$_r$ 42 kDa), is capable of specifically binding, with excellent affinities (Ka = 4–20$\cdot$10^7 M^{-1}), the (constant) Fc part of a large number of different IgGs. By

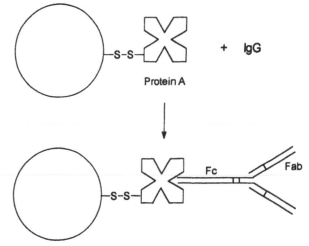

Figure 10 Association of a IgG to a lipo-some with protein A coupled at its surface. Protein A attached to the liposome, e.g., according to the technique given in Figure 2, can complex the invariant moiety (Fc part) of an antibody. This procedure allows, with one type of liposomes, to prepare a series of immunoliposomes that differ by the antibody carried by the vesicle.

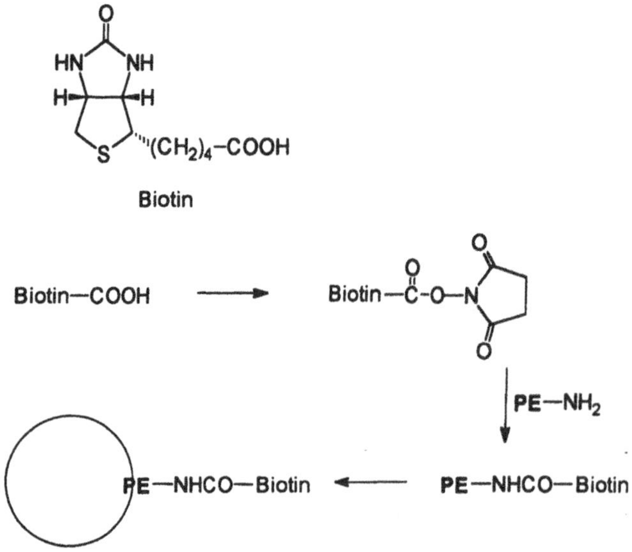

Figure 11 Preparation of bio-tinylated phosphatidylethanolamine. The carboxylic function of biotin is first reacted with N-hydroxysuc-cinimide, followed by reaction with PE-NH$_2$. The PE derivative can be mixed with the (phospho)lipids to make liposomes.

coupling this protein, which has 4 binding sites, to the surface of liposomes, Leserman and co-workers have introduced the notion of universal liposomes, i.e. these preformed liposomes can be mixed with a solution of a given antibody to yield, after formation of the complex, a targeted immunoliposome.[27] The interest of protein A resides also in the fact that, since it binds the antibodies at a distal part from the antigen combining site, the immunoliposome/antigen interaction is not disturbed (Figure 10). However, not all classes of IgGs, including certain monoclonal antibodies, are recognized by protein A thus limiting its use. Such a situation could in principle be remedied by using protein G instead. The coupling of protein A to liposomes was performed according to the classical procedures outlined above: the protein was first thiolated with SPDP and attached to liposomes containing PDP-PE according to the thiol exchange reaction.[48] The different applications of this technique have been discussed by Machy and Leserman.[1]

2. (Strept)avidin/biotin complexes

Biotin is a low molecular weight molecule (MW = 244; Figure 11) that presents an extraordinarily high affinity (Ka $\approx 10^{15}$ M^{-1}) for proteins such as avidin (a glycoprotein isolated from egg white) and streptavidin (isolated from *Streptomyces*). These two proteins of, 67 and 60 kDa molecular weight respectively, exhibit four independent binding sites for biotin.[68] The (strept)avidin/biotin interaction has been exploited in numerous applications in biochemistry, cell labeling, affinity chromatography, and immunochemical techniques.[69-71] Streptavidin is generally preferred because of its pI value close to 7 and its absence of glycosidation, which limits non-specific interactions. With regard to the use of the (strept)avidin/biotin couple for targeting, several approaches can be followed to realize a stable complex between a ligand and the surface of a liposome.

1. Biotin is coupled covalently to the surface of a liposome and to a ligand (e.g., an antibody), and the complex is realized by bridging the two biotins by (strept)avidin. The carboxylic function of biotin can be easily activated, e.g., via esterification with N-hydroxysuccinimide, and reacted with the amine function of PE (Figure 11), of PS or of any ligand-NH$_2$ (for example an ε-amino lysyl residue of an antibody), to yield, after the formation of an amide bond, a biotinylated derivative.[72-74] It should be noted that many biotin conjugates, such as PE-biotin, are now commercially available. Moreover the biotinylation of antibodies, under controlled conditions, was found to leave the antigen binding site essentially intact.[75] PE-biotin, after its incorporation into liposomes is added to a biotinylated ligand in the presence of (strept)avidin; because of its multivalency, the protein will bridge ("sandwich" type complex) the two entities and realize the heterologous complex at the surface of the vesicle.[76] Understandably, this procedure needs well controlled experimental conditions; it can easily lead to the formation of homologous complexes and to the aggregation of the liposomes.[72] To circumvent this, Schott et al.[77] have first incubated the liposomes-biotin in the presence of avidin. After elimination of the excess of protein by gel filtration, the resulting vesicles (i.e., liposome-biotin-avidin) could thereafter be incubated with biotinylated IgGs to realize the final complex. Noteworthy in this context is the study by Hashimoto et al.[46] on the incidence of the spacer-arm length, between the liposome surface and the conjugated biotin, on its recognition by free avidin.

2. (Strept)avidin is covalently linked to the surface of liposomes and the complex is realized with biotinylated ligands. It is relatively easy to biotinylate a protein (ligand-NH$_2$). A complementary approach consists in the preparation of liposomes bearing (strept)avidin at their surface. This can be realized by using one of the methods described above, which allows the attachment of proteins to the surface of vesicles. In the most recent approach, avidin was first thiolated with SPDP, then coupled to liposomes containing MPB-PE.[78] This procedure is not hampered by the drawbacks mentioned above for the other method and the final complex, in the presence of the ligand-biotin, forms readily under mild conditions. The same technique was used to target liposomes to cells expressing the nerve growth factor (NGF) receptor; in this case NGF-biotin was complexed to liposomes-streptavidin.[79]

III. ATTACHMENT OF LIGANDS TO LIPOSOMES THROUGH HYDROPHOBIC INTERACTIONS

As opposed to the previous procedures that allow conjugation reactions, in aqueous media, between ligands and preformed liposomes, one can also envisage the direct anchoring of ligands in the bilayer of the vesicles using hydrophobic interactions. To this end, ligands of interest are first chemically coupled to hydrophobic moieties (such as fatty acids, cholesterol, or phospholipids) and then added to the (phospho)lipid mixture that is used to prepare the liposomes (Figure 1C). An alternate approach consists of anchoring ligands, rendered hydrophobic, to preformed liposomes. Such methods have been applied to proteins (such as monoclonal antibodies) and to low molecular weight ligands. The potential interests to this hydrophobic anchoring are the possibility to attach different types of ligands to a single liposome and to avoid the exposure of the vesicles, and their contents, to chemical reagents. On the negative side is the inevitable loss of encapsulated molecules, from preformed liposomes, during the insertion step of the hydrophobic ligand. Related to this problem some recent liposome preparation techniques developed by Gregoriadis et al.[80] might be useful. Importantly, the same approach allows the incorporation into the bilayers of liposomes of prodrugs[81-83] and other biological molecules that have been rendered hydrophobic, e.g., PE-muramyl tripeptide[84] or oligoribonucleosides.[85]

A. ACYLATION OF PROTEINS AND LIPOSOME PREPARATION

The techniques available to graft hydrophobic anchors to proteins are, from a chemical point of view, very similar to the one described above for protein conjugation (e.g., thiolation and biotinylation reactions). Here, an amine function of the protein is reacted with an acylating derivative of a fatty acid (Figure 12). Thus, antibodies have been rendered hydrophobic by reaction with palmitic acid N-hydroxysuccinic ester[86] or with palmitoyl chloride.[87,88] These methods must, however, be carefully controlled, especially the acylating agent/protein molar ratio. Huang et al.[86] have shown that the introduction of 3 to 4 palmitoyl chains per IgG has a limited impact on the antigen recognition and binding properties of the antibody (Ka decreased by factor 3 to 4).

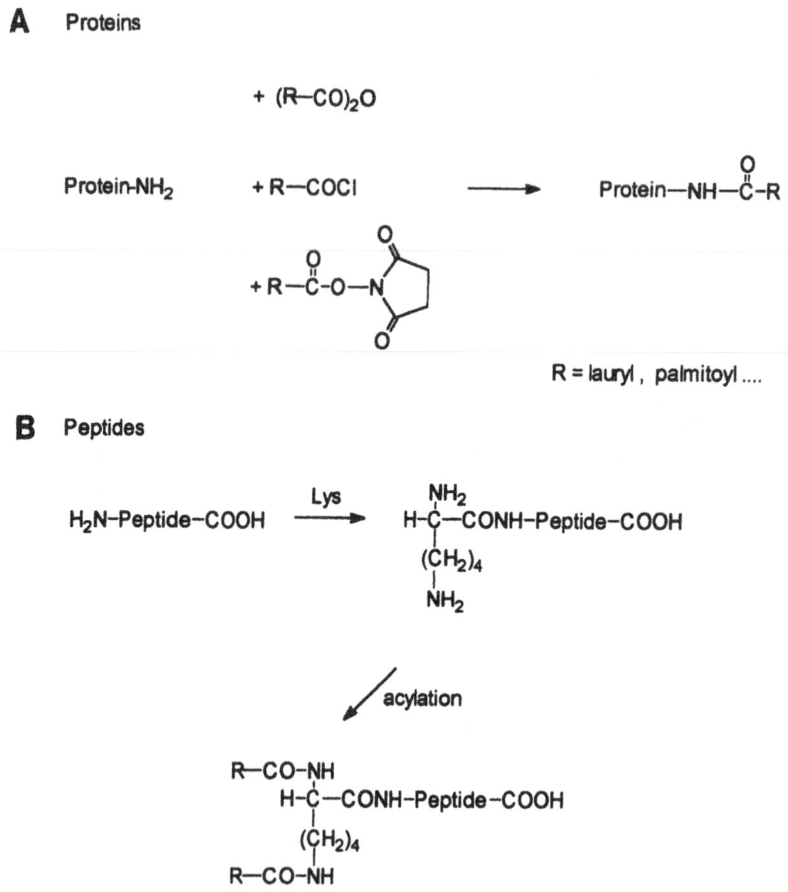

A Proteins

$$+ (R-CO)_2O$$

Protein-NH$_2$ $+ R-COCl$ $\longrightarrow$ Protein$-$NH$-\overset{\overset{\displaystyle O}{\|}}{C}$-R

$+ R-\overset{\overset{\displaystyle O}{\|}}{C}-O-N$

R = lauryl , palmitoyl ....

B Peptides

H$_2$N-Peptide-COOH $\xrightarrow{\text{Lys}}$ H-$\overset{\displaystyle NH_2}{\underset{\displaystyle (CH_2)_4}{\underset{\displaystyle NH_2}{C}}}$-CONH-Peptide-COOH

acylation

R$-$CO$-$NH
H-$\overset{}{\underset{\displaystyle (CH_2)_4}{C}}$-CONH-Peptide-COOH
R$-$CO$-$NH

Figure 12 Fatty acid grafting to proteins and to peptides. Hydrophilic molecules to be attached to liposomes (see Figure 1C) are first conjugated to hydrophobic anchors. This can be realized by reacting ligand-NH$_2$, e.g. protein-NH$_2$, with derivatives of fatty acids such as anhydrides, acyl chlorides, or activated esters (**A**). An example is given for the hydrophobic anchoring of a peptide (**B**). The peptide is first extended, at its N-terminal part, by a lysyl residue; the two extraneous amine functions thus introduced can then be acylated as above, providing an anchoring through two fatty acid chains.

According to similar chemical approaches, Fab' fragments, previously citraconylated, have also been coupled to PE-NH$_2$ with a carbodiimide.[89] More recently, several authors have described the use of phosphatidylethanolamine modified with glutaric anhydride (see Section II.A.3.); the analogue obtained, which bears a free carboxylic function (i.e., PE-COOH), can be directly conjugated, by use of a carbodiimide (see Footnote 1), with amine residues of soluble proteins.[90,91] The proteins that have been rendered hydrophobic are then incorporated into liposomes. This technique, which has been optimized by Huang et al., has yielded particularly efficient immunoliposomes.[92]

The incorporation of ligand associated to a hydrophobic anchor in liposomes is performed in most cases by two types of detergent-dialysis methods: (1) the derivatized protein and the (phospho)lipids are mixed in the presence of the detergent, the latter is then eliminated by dialysis to form liposomes. The fatty acid chain, by insertion into the forming bilayers, will permit the anchoring of the protein.[86] (2) the derivatized protein is added to preformed liposomes (SUV or REV) in the presence of an optimal concentration of detergent, followed by a dialysis step.[93,94]

Generally the detergents used belong to the high-CMC class (e.g., deoxycholate or β-octylglucoside). β-Octylglucoside can be rapidly eliminated during the dialysis step by adsorption on SM2-beads.[93,95] It should be noted that these procedures require an optimization of the detergent/(phospho)lipids and proteins/(phospho)lipids ratios.[93,96] The conjugation of antibodies by these techniques allowed the preparation of classical and pH-sensitive immunoliposomes.[94]

The same detergent-elimination methods (by dialysis or gel filtration) can also be used to reassociate to the bilayers of liposomes, naturally occurring hydrophobic proteins (such as membrane-bound (glyco)proteins).[97-99] One can thus prepare immunosomes, such as rabies immunosomes,[100,101] i.e., liposomes carrying at their surface immunogenic proteins or liposomes presenting cellular recognition determinants, e.g., CD4-liposomes.[102]

B. ATTACHMENT OF SMALL MOLECULES

A large variety of relatively low molecular weight molecules have also been attached to liposomes through derivatization with hydrophobic anchors (see the Introduction of this section). In the majority of the cases, these derivatives have been incorporated into liposomes by simple addition to the initial (phospho)lipid mixture used to make the vesicles. Following is a limited and representative number of examples; in this field each ligand presents a chemical challenge of its own and necessitates an adapted, and sometimes more complex, chemistry.

Peptides — Several peptides, acylated with a fatty acid derivative (see Section III.A.), have been prepared, such as: (1) a nonapeptide of myelin basic protein, allowing the targeting of liposomes toward lymphocytes responsible for an autoimmune disease,[103] (2) peptides incorporated into liposomes to increase their immunogenicity,[104] and (3) tuftsine, a small peptide, which is bound by a receptor at the surface of macrophages and allows vesicle targeting toward these cell types.[105] The chemistry of peptides allows some special strategies, such as the introduction of extraneous lysyl residues at the N-termini (Figure 12B) and therefore the coupling of two hydrophobic tails.[106,107]

Methotrexate — Methotrexate has been coupled, through its free carboxylic functions to PE-NH$_2$ by use of a carbodiimide in the presence of N-hydroxysuccinimide.[108] Incorporated into liposomes, it was used as a antiproliferative agent.[109] Interestingly, liposomes that carry methotrexate on the surface do not seem to interact with the folate carrier (which is responsible for methotrexate uptake by cells).[110] This might be due, in this particular case, to the absence of a spacer arm between the molecule and the vesicle surface.

Neoglycolipids — Because of the paucity of natural glycolipids and the limits imposed by their chemical structures, several groups have elaborated synthetic methods for preparing glycosidic molecules with the aim to target vesicles to cell expressing lectins at their surface.[111,112] Several approaches exist, depending on the nature of the hydrophobic anchor, i.e., cholesterol or (phospho)lipids. Merck (U.S.A.) has developed a comprehensive strategy allowing the coupling of S-glycosides (such as mannose (Figure 13), galactose, ...), via a spacer arm, to the 3β-OH group of cholesterol,[113] (see also[114]). More complex molecules, such as triantennary structures, have also been synthesized by coupling glucidic moieties (e.g., mannose[115,116]) on the amine functions of a lys-lys dipeptide or on the hydroxyl functions of tris(hydroxymethyl) aminomethane (e.g., galactose[117,118]); in the examples cited the final hydrophobic anchor remains cholesterol, which allows the incorporation of these structures into liposomes. Another procedure, somewhat more limited in its extent, employs natural disaccharides (such as lactose) to prepare synthetic glycolipids; the reaction consists of a reductive amination of the aldehyde function of the disaccharide with PE-NH$_2$ in the presence of NaBH$_3$CN[119] (see section II.A.3). More recently, Haensler and Schuber[33] have synthesized neogalactolipids (Figure 14), which could be incorporated into liposomes by mixing directly with (phospho)lipids.

IV. ADDITIONAL TECHNIQUES

On several occasions it is of importance to be able to determine the quantity of ligand coupled to the surface of liposomes. A prerequisite, is therefore, after the preparation of the vesicles and the coupling

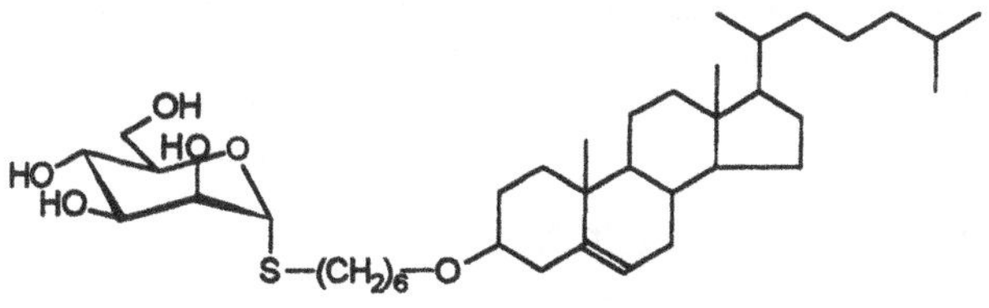

Figure 13 1-Thio-α-ᴅ-mannose coupled to cholesterol: a hydrophobic anchor. This molecule is a representative example of a hydrophobic derivative of a sugar residue, obtained by chemical synthesis. It allows the attachment of glycosides, via a spacer-arm, to the surface of liposomes.

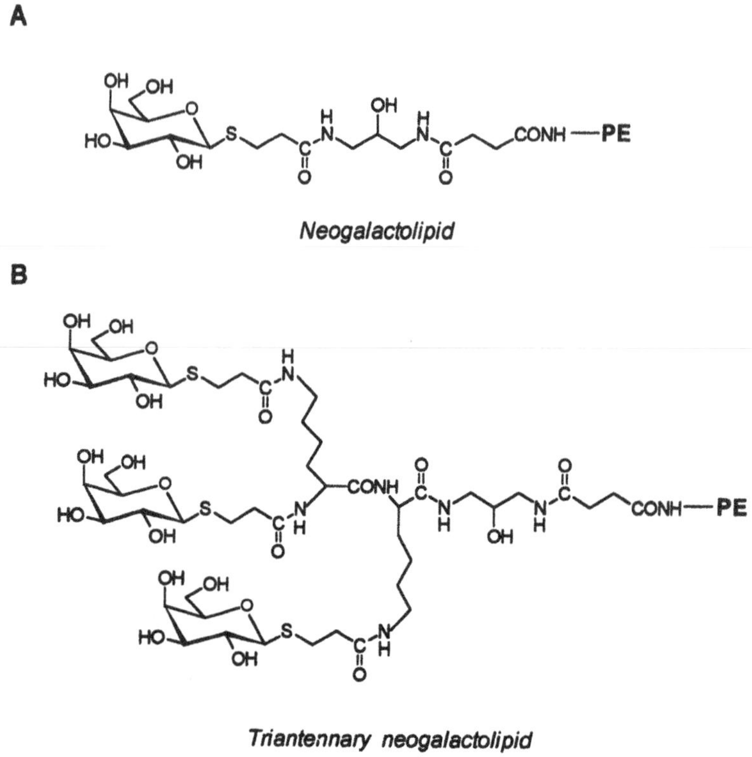

A

Neogalactolipid

B

Triantennary neogalactolipid

Figure 14 Neogalactolipids: (A) Derivative of 1-thio-β-D-galactose coupled to PE-NH$_2$; (B) Structure of a triantennary neogalactolipid. Examples of synthetic glycosidic structures for targeting liposomes to cells expressing (gal)lectins at their surface.

step, to be able to separate the conjugated liposomes from the excess free ligand. Equally crucial is the possibility of control if the coupling techniques respect the integrity of the vesicles and do not provoke an efflux of entrapped material.

A. SEPARATION OF CONJUGATED LIPOSOMES FROM FREE LIGANDS

After coupling low molecular weight ligands to the surface of preformed liposomes, the classical methods in use to separate, for example, entrapped molecules from the excess free ones are perfectly appropriate. Thus, one can dialyze or ultra-filtrate the liposomal suspension or, by a more expedient procedure, use gel filtration (on Sephadex G-75, for example[120]). However, the separation of conjugated liposomes from free protein ligands is more elaborate and requires more controls. Gel filtration on Sepharose 4B, for example, can also be used, but problems might be encountered due to the adsorption of the vesicles to the gel and/or to their failure to separate liposomes from protein aggregates formed spontaneously (e.g., IgGs) or resulting from the coupling procedures.[15] In control experiments, it is therefore important to assess the separation of coupled liposomes from free proteins. With respect to this problem, the method developed by Heath et al.[56] is particularly adapted. It consists of loading, in a centrifuge tube, the liposomal mixture under isotonic cushions of different metrizamide or ficoll concentrations. Due to centrifugal forces, the liposomes, because of their buoyancy, float through the cushions and can be collected, concentrated, at the upper interface, whereas the free proteins remain at the bottom of the tube. This technique is particularly well-suited to larger vesicles, and in the case of smaller liposomes such as SUV, much higher centrifugal forces must be used (e.g., 16 h at $4.10^5 \times g$).[15]

B. VERIFICATION OF LIPOSOME INTEGRITY

A simple means to verify if the coupling procedures used with preformed liposomes respect the integrity of the vesicles consists of measuring the retention of the encapsulated molecules. To this end several markers can be used, such as radiolabeled sucrose or inulin.[15] One of the best method uses 5(6)-carboxyfluorescein; entrapped at high concentrations (≈ 50 mM) the fluorescence of this dye is almost extinct (quenching). The technique consists therefore of measuring the fluorescence extinction coefficient before and after the conjugation step.[120] The use of precise equations allows one to quantify the leakage of this molecule from vesicles, and to therefore appreciate the incidence of the chosen coupling procedure on vesicle integrity.

REFERENCES

1. Machy, P. and Leserman, L.D., *Liposomes in Cell Biology and Pharmacology,* John Libbey, London, 1987.
2. Gregoriadis, G., Ed., *Liposome Technology,* 2nd Edition, Vol. III, CRC Press, Boca Raton, FL, 1992.
3. Vidal, M., Sainte-Marie, J., Philippot, J.R., and Bienvenue, A., LDL-mediated targeting of liposomes to leukemic lymphocytes in vitro, *EMBO J.,* 4, 2461, 1985.
4. Salord, J., Tarnus, C., Muller, C.D., and Schuber, F., Targeting of liposomes by covalent coupling with ecto-NAD+ glycohydrolase ligands, *Biochim. Biophys. Acta,* 886, 64, 1986.
5. Salord, J. and Schuber, F., In vitro drug delivery mediated by ecto-NAD+ glycohydrolase ligand-targeted liposomes, *Biochim. Biophys. Acta,* 971, 197, 1988.
6. Gregoriadis, G., Immunological adjuvants: a role for liposomes, *Immunol. Today,* 11, 89, 1990.
7. van Rooijen, N., Liposomes as carriers and immunoadjuvant of vaccine antigens, in *Bacterial Vaccines,* Alan R. Liss, New York, 255, 1990.
8. Martin, F.J. and Kung, V., Use of liposomes as agglutination-enhancement agents in diagnostic tests, *Methods Enzymol.,* 149, 200, 1987.
9. Gabizon, A., Price, D.C., Huberty, J., Bresalier, R.S., and Papahadjopoulos, D., Effect of liposome composition and other factors on the targeting of liposomes to experimental tumors: biodistribution and imaging studies, *Cancer Res.,* 50, 6371, 1990.
10. Dean, P.D.G., Johnson, W.S., and Middle, F.A, Eds., *Affinity Chromatography: a Practical Approach,* IRL Press, Oxford, U.K., 1985.
11. Wong, S.S., *Chemistry of Protein Conjugation and Cross-linking,* CRC Press, Boca Raton, FL, 1991.
12. Erlanger, B.F., The preparation of antigenic hapten-carrier conjugates: a survey, *Methods Enzymol.,* 70, 85, 1980.
13. Blair, A.H. and Ghose, T.I., Linkage of cytotoxic agents to immunoglobulins, *J. Immunol. Methods,* 59, 129, 1983.
14. Muller, S., Peptide-carrier conjugation, in *Synthetic Polypeptides as Antigens,* Van Regenmortel, M.H.V., Briand, J.P., Muller, S., and Plaué, S., Eds., Elsevier, Amsterdam, 95, 1988.
15. Heath, T.D. and Martin, F.J., The development and application of protein-liposome conjugation techniques, *Chem. Phys. Lipids,* 40, 347, 1986.
16. Heath, T.D., Interaction of liposomes with cells, *Methods Enzymol.,* 149, 111, 1987.
17. Martin, F.J., Heath, T.D., and New, R.R.C., Covalent attachment of proteins to liposomes, in *Liposomes: a Practical Approach,* New, R.R.C., Ed., IRL Press, Oxford, U.K., 163, 1990.
18. Toonen, P.A.H.M. and Crommelin, D.J.A., Immunoglobulins as targeting agents for liposome encapsulated drugs, *Pharm. Weekblad Sci. Ed.,* 5, 269, 1983.
19. Poznansky, M.K. and Juliano, R.L., Biological approaches to the controlled delivery of drugs: a critical review, *Pharmacol. Reviews,* 36, 277, 1984.
20. Weinstein, J.N. and Leserman, L.D., Liposomes as drug carriers in cancer chemotherapy, *Pharmacol. Ther.,* 24, 207, 1984.
21. Gregoriadis, G., Ed. *Liposome Technology,* Vol III, CRC Press, Boca Raton, FL, 1984.
22. Connor, J., Sullivan, S., and Huang, L., Monoclonal antibody and liposomes, *Pharmacol. Ther.,* 28, 341, 1985.
23. Papahadjopoulos, D., Heath, T., Bragman, K., and Matthay, K., New methodology for liposome targeting to specific cells, *Ann. N.Y. Acad. Sci.,* 446, 341, 1985.
24. Barbet, J., Machy, P., and Leserman, L.D., Monoclonal antibody covalently coupled to liposomes: specific targeting to cells, *J. Supramol. Struct. Cell. Biochem.,* 16, 243, 1981.
25. Carlsson, J., Drevin, H., and Axen, R., Protein thiolation and reversible protein-protein conjugation, *Biochem. J.,* 173, 723, 1978.
26. Martin, F.J. and Papahadjopoulos, D., Irreversible coupling of immunoglobulin fragments to preformed vesicles. An improved method for liposome targeting, *J. Biol. Chem.,* 257, 286, 1982.

27. Leserman, L.D., Machy, P., and Barbet, J., Cell-specific drug transfer from liposomes bearing monoclonal antibodies, *Nature,* 293, 226, 1981.

28. Leserman, L.D., Barbet, J., Kourislsky, F., and Weinstein, J.N., Targeting to cells of fluorescent liposomes covalently coupled with monoclonal antibody or protein A, *Nature,* 288, 602, 1980.

29. Shek, P.S. and Heath, T.D., Immune response mediated by liposome associated protein antigens, *Immunology,* 50, 101, 1983.

30. Hutchinson, F.J. and Jones, M.N., Lectin-mediated targeting of liposomes to a model surface. An ELISA method, *FEBS Lett.,* 234, 493, 1988.

31. Haensler, J. and Schuber, F., Preparation of neo-galactosylated liposomes and their interaction with mouse peritoneal macrophages, *Biochim. Biophys. Acta,* 946, 95, 1988.

32. Muller, C.D. and Schuber, F., Neo-mannosylated liposomes. Synthesis and interaction with mouse Kupffer cells and resident peritoneal macrophages, *Biochim. Biophys. Acta,* 986, 97, 105, 1989.

33. Haensler, J. and Schuber, F., Influence of the galactosyl ligand on the interaction of galactosylated liposomes with mouse peritoneal macrophages, *Glycoconjugate J.,* 8, 116, 1991.

34. Krowka, J., Stites, D., Debs, R., Larsen, C., Fedor, J., Brunette, E., and N. Düzgüneş, Lymphocyte proliferative responses to soluble and liposome conjugated envelop peptides of HIV-1, *J. Immunol.,* 144, 2535, 1990.

35. Frisch, B., Muller, S., Briand, J.P., Van Regenmortel, M.H.V., and Schuber, F., Parameters affecting the immunogenicity of a liposome-associated synthetic hexapeptide, *Eur. J. Immunol.,* 21, 185, 1991.

36. Friede, M., Muller, S., Briand, J.P., Van Regenmortel, M.H.V., and Schuber, F., Induction of immune response against a short synthetic peptide antigen coupled to small neutral liposomes containing monophosphoryl lipid A, *Mol. Immunol.,* 30, 539, 1993.

37. Peeters, J.M., Hazendonk, T.G., Beuvery, E.C., and Tesser, G.I., Comparison of four bifunctional reagents for coupling peptides to proteins and the effect of the three moieties on the immunogenicity of the conjugates, *J. Immunol. Methods,* 120, 133, 1989.

38. Loughrey, H.C., Wong, K.F., Choi, L.S., Cullis, P.R., and Bally, M.B., Protein-liposome conjugates with defined size distributions, *Biochim. Biophys. Acta,* 1028, 73, 1990.

39. Bragman, K.S., Heath, T.D., and Papahadjopoulos, D., Cytotoxicity of antibody-directed liposomes that recognize two receptors on K-562 cells, *J. Natl. Cancer Inst.,* 73, 723, 1984.

40. Schott, H., Hess, W., Hengartner, H., and Schwendener, R.A., Syntheses of lipophilic amino and carboxyl components for the functionalization of liposomes, *Biochim. Biophys. Acta,* 943, 53, 1988.

41. Schwendener, R.A., Trüb, T., Schott, H., Langhals, H., Barth, R.F., Groscurth, P., and Hengartner, H., Comparative studies of the preparation of immunoliposomes with the use of two bifunctional coupling agents and investigation of in vitro immunoliposome-target cell binding by cytofluorometry and electron microscopy, *Biochim. Biophys. Acta,* 1026, 69, 1990.

42. Goundalkar, A., Ghose, T., and Mezei, M., Covalent binding of antibodies to liposomes using a novel lipid derivative, *J. Pharm.Pharmacol.,* 36, 465, 1984.

43. Bredehorst, R., Ligler, F.S., Kusterbeck, A.W., Chang, E.L., Gaber, B.P., and Vogel, C.W., Effect of covalent attachment of immunoglobulin fragments on liposomal integrity, *Biochemistry,* 25, 5693, 1986.

44. Francis, S.E., Lyle, I.G., and Jones, M.N., The effect of surface-bound protein on the permeability of proteoliposomes, *Biochim. Biophys. Acta,* 1062, 117, 1991.

45. Wolff, B. and Gregoriadis, G., The use of monoclonal anti-Thy1 IgG1 for the targeting of liposomes to AKR-A cells in vitro and in vivo, *Biochim. Biophys. Acta,* 802, 259, 1984.

46. Hashimoto, K., Loader, J.E., and Kinsky, S.C., Iodoacetylated and biotinylated liposomes: effect of spacer length on sulfhydryl ligand binding and avidin precipitability, *Biochim. Biophys. Acta,* 856, 556, 1986.

47. Lasch, J., Niedermann, G., Bogdanov, A.A., and Torchilin, V.P., Thiolation of preformed liposomes with iminothiolane, *FEBS Lett.,* 214, 13, 1987.

48. Leserman, L.D., Machy, P., and Barbet, J., Covalent coupling of monoclonal antibodies and protein A to liposomes: specific interaction with cells in vivo and in vitro, in *Liposome Technology,* Gregoriadis, G., Ed., CRC Press, Boca Raton, FL, Vol. III, 29, 1984.

49. Derksen, J.T.P. and Scherphof, G.L., An improved method for covalent coupling of proteins to liposomes, *Biochim. Biophys. Acta,* 814, 151, 1985.

50. Duncan, R.J.S., Weston, P.D., and Wrigglesworth, R., A new reagent which may be used to introduce sulfhydryl groups into proteins, and its use in the preparation of conjugates for immunoassay, *Anal. Biochem.,* 132, 68, 1983.

51. Haensler, J., Mise au point de liposomes néo-galactosylés pour le transport de molécules bioactives vers des cibles cellulaires spécifiques. Etude de systèmes modèles *in vitro* et *in vivo*. Ph.D. thesis, Université Louis Pasteur, Strasbourg, France, 1990.

52. Traut, R.R., Bollen, A.A., Sun, T.T., Hershey, J.W.B., Sundberg, J., and Pierce, L.R., Methyl 4-mercaptobutyrimidate as a cleavable cross-linking reagent and its application to the *Escherichia coli* 30S ribosome, *Biochemistry*, 12, 3266, 1973.

53. Riddles, P.W., Blakeley, R.L., and Zerner, B., Ellman's reagent: 5,5'-dithiobis(2-nitrobenzoic acid) — A reexamination, *Anal. Biochem.*, 94, 75, 1979.

54. Heath, T.D., Montgomery, J.A., Piper, J.R., and Papahadjopoulos, D., Antibody-targeted liposomes: increase in specific toxicity of methotrexate-gamma-aspartate, *Proc. Natl. Acad. Sci. USA*, 80, 1377, 1983.

55. Urdal, D.L. and Hakomori, S., Tumor-associated ganglio-*n*-triosylceramide. Target for antibody-dependent, avidin-mediated drug killing of tumor cells, *J. Biol. Chem.*, 255, 10509, 1980.

56. Heath, T.D., Macher, B.A., and Papahadjopoulos, D., Covalent attachment of immunoglobulins to liposomes via glycosphingolipids, *Biochim. Biophys. Acta*, 640, 66, 1981.

57. Torchilin, V.P., Klibanov, A.L., and Smirnov, V.N., Phosphatidylinositol may serve as the hydrophobic anchor for immobilization of protein on liposome surface, *FEBS Lett.*, 138, 117, 1982.

58. Heath, T.D., Martin, F.J., and Maches B.A., Association of ganglioside-protein conjugates into cell and Sendai virus. Requirement for the HN subunit in viral fusion, *Exp. Cell Res.*, 149, 163, 1983.

59. Chua, M.-M., Fan, S.-T., and Karush, F., Attachment of immunoglobulin to liposomal membrane via protein carbohydrate, *Biochim. Biophys. Acta*, 800, 291, 1984.

60. Heath, T.D., Robertson, D., Birbeck, M.S.C., and Davies, A.J.S., Covalent attachment of horseradish peroxidase to the outer surface of liposomes, *Biochim. Biophys. Acta*, 599, 42, 1980.

61. Kinsky, S.C., Loader, J.E., and Benson, A.L., An alternative procedure for the preparation of immunogenic liposomal model membranes, *J. Immunol. Methods*, 65, 295, 1983.

62. Kinsky, S.C., Hashimoto, K., Loader, J.E., and Benson, A.L., Synthesis of *N*-hydroxysuccinimide esters of phosphatidylethanolamine and some properties of liposomes containing these derivatives, *Biochim. Biophys. Acta*, 769, 543, 1984.

63. Kung, V.T. and Redemann, C.T., Synthesis of carboxyacyl derivatives of phosphatidyl-ethanolamine and use as an efficient method for conjugation of protein to liposomes, *Biochim. Biophys. Acta*, 862, 435, 1986.

64. Weissig, V., Lasch, J., and Gregoriadis, G., Covalent coupling of sugars to liposomes, *Biochim. Biophys. Acta*, 1003, 54, 1989.

65. Snyder, S.L. and Vannier, W.E., Immunologic response to protein immobilized on the surface of liposomes via covalent azo-bonding, *Biochim. Biophys. Acta*, 772, 288, 1984.

66. Vannier, W.E. and Snyder, S.L., Antibody responses to liposome/associated antigen, *Immunol. Lett.*, 19, 59, 1988.

67. Sänger, M., Borle, F., Heller, M., and Sigrist, H., Light-induced coupling of aqueous-soluble proteins formed from carbene-generating phospholipids, *Bioconjugate Chem.*, 3, 308, 1992.

68. Green, N.M., Avidin, *Adv. Protein Chem.*, 29, 85, 1975.

69. Wilchek, M. and Bayer, E.A., Introduction to avidin-biotin technology, *Methods Enzymol.*, 184, 5, 1990.

70. Bayer, E.A. and Wilchek, M., Application of avidin-biotin technology to affinity-based separations, *J. Chromatogr.*, 510, 3, 1990.

71. Suter, M., Butler, J.E., and Peterman, J.H., The immunochemistry of sandwich ELISAs-III. The stoichiometry and efficacy of the protein-avidin-biotin capture (PABC) system, *Mol. Immunol.*, 26, 221, 1989.

72. Loughrey, H., Bally, M.B., and Cullis, P.R., A non-covalent method of attaching antibodies to liposomes, *Biochim. Biophys. Acta*, 901, 157, 1987.

73. Bayer, E.A., and Wilcheck, M., Methodology involved in biotin-conjugated phospholipids, glycolipids, and gangliosides. In Liposome Technology, Gregoriadis, G., Ed., CRC Press, Boca Raton, FL, Vol III, 127, 1984.

74. Rivnay, B., Bayer, E.A., and Wilcheck, M., Use of avidin-biotin technology for liposome targeting, *Methods Enzymol.*, 149, 119, 1987.

75. Bayer, E.A., Rivnay, B., and Skutelsky, E., On the mode of liposome-cell interactions. Biotin-conjugated lipids as ultrastructural probes, *Biochim. Biophys. Acta*, 550, 464, 1987.

76. Trubetskoy, V.S., Berdichevsky, V.R., Efremov, E.E., and Torchilin, V.P., On the possibility of the unification of drug targeting systems. Studies with liposome transport to the mixtures of target antigens, *Biochem. Pharmacol.*, 36, 839, 1987.

77. Schott, H., Leitner, B., Schwendener, R.A., and Hengartner, H., Chromatography of functionalized liposomes and their components, *J. Chromatogr.*, 441, 115, 1988.

78. Loughrey, H.C., Choi, L.S., Cullis, P.R., and Bally, M.B., Optimized procedures for the coupling of proteins to liposomes, *J. Immunol. Methods,* 132, 25, 1990.

79. Rosenberg, M.B., Breakefield, X.O., and Hawrot, E., Targeting of liposomes to cells bearing Nerve Growth Factor receptors mediated by biotinylated Nerve Growth Factor, *J. Neurochem.*, 48, 865, 1987.

80. Gregoriadis, G., Garcon, N., da Silva, H., and Sternberg, B., Coupling of ligands to liposomes independently of solute entrapment: observation on the formed vesicles, *Biochim. Biophys. Acta,* 1147, 185, 1993.

81. Rosenmeyer, H., Ahlers, M., Schmidt, B., and Seela, F., A nucleolipid with antiviral acylguanosine as head-group. Synthesis and liposome formation, *Angew. Chem. Int. Ed. Engl.*, 24, 501, 1985.

82. Schwendener, R.A., Supersaxo, A., Rubas, W., Weder, H.G., Hartman, H.R., Schott, H., Ziegler, A., and Hengartner, H., 5′-O-Palmitoyl- and 3′,5′-O-dipalmitoyl-5-fluoro-2′-deoxyuridine. Novel lipophilic analogues of 5′-fluoro-2′-deoxyuridine: synthesis, incorporation into liposomes and preliminary biological results, *Biochem. Biophs. Res. Commun.*, 126, 660, 1985.

83. Kinsky, S.C., Loader, J.E., and Hashimoto, K., Inhibition of cell proliferation by putative metabolites and non-degradable analogs of methotrexate-gamma-dimyristoylphosphatidylethanolamine, *Biochim. Biophys. Acta,* 917, 211, 1987.

84. Fidler, I.J., Immunomodulation of macrophages for cancer and antiviral therapy, in *Site Specific Drug Delivery,* Tomlinson, E. and Davis, S.S., Eds., Wiley, New York, 111, 1986.

85. Oberhauser, B. and Wagner, E., Effective incorporation of 2′-O-methyl-oligoribonucleotides into liposomes and enhanced cell association through modification with thiocholesterol, *Nucleic Acid Res.*, 20, 533, 1992.

86. Huang, A., Tsao, Y.S., Kennel, S.J., and Huang, L., Characterization of antibody covalently coupled to liposomes, *Biochim. Biophys. Acta,* 716, 140, 1982.

87. Torchilin, V.P. and Klibanov, A.L., Preliminary "hydrophobization" of hydrophylic proteins increases its binding with liposomes, *Bioorg. Khim.*, 6, 791, 1981.

88. Torchilin, V.P., Immobilization of specific proteins on liposome surface: systems for drug targeting, in *Liposome Technology*, Gregoriadis, G., Ed., CRC Press, Boca Raton, FL, Vol. III, 94, 1984.

89. Jansons, V.K., Preparation and analysis of antibody-targeted liposomes, in *Liposome Technology,* Gregoriadis, G., Ed., CRC Press, Boca Raton, FL, Vol. III, 63, 1984.

90. Weissig, V., Lasch, J., Klibanov, A.L., and Torchilin, V.P., A new hydrophobic anchor for the attachment of proteins to liposomal membranes, *FEBS Lett.*, 202, 86, 1986.

91. Bogdanov, A.A., Jr., Klibanov, A.L., and Torchilin, V.P., Protein immobilization on the surface of liposomes via carbodiimide activation in the presence of N-hydroxysulfosuccinimide, *FEBS Lett.*, 231, 381, 1988.

92. Holmberg, E., Maruyama, K., Litzinger, D.C., Wright, S., Davis, M., Kabalka, G.W., Kennel, S.J., and Huang, L., Highly efficient immunoliposomes prepared with a method which is compatible with various lipid compositions, *Biochem. Biophys. Res. Commun.*, 165, 1272, 1989.

93. Huang, L., Connor,J., and Wang, C.Y., pH-sensitive immunoliposomes, *Methods Enzymol.*, 149, 88, 1985.

94. Connor, J. and Huang, L., Efficient cytoplasmic delivery of a fluorescent dye by pH-sensitive immunoliposomes, *J. Cell Biol.*, 101, 582, 1985.

95. Phillipot, J., Mustaftschiev, S., and Liautard, J.P., A very mild method allowing the encapsulation of very high amounts of macromolecules into very large (1000 nm) unilamellar liposomes, *Biochem. Biophys. Acta,* 734, 137, 1983.

96. Huang, L., Huang, A., and Kennel, S.J., Coupling of antibodies with liposomes, in *Liposome Technology,* Gregoriadis, G., Ed., CRC Press, Boca Raton, FL, Vol. III, 51, 1984.

97. Eytan, G.D., Use of liposomes for reconstitution of biological functions, *Biochim. Biophys. Acta,* 694, 185, 1982.

98. Etemadi, A.-H., Functional and orientational features of protein molecules in reconstituted lipid membranes, *Adv. Lipid Res.*, 21, 281, 1985.

99. Madden, T.D., Current concept in membrane protein reconstitution, *Chem. Phys. Lipid,* 40, 207, 1986.

100. Perrin, P., Thibodeau, L., Dauget, C., Fritsch, A., and Sureau, P., Amplification des propriétés immunogènes de la glycoprotéine rabique par ancrage sur de liposomes préformés, *Ann. Virol. (Inst. Pasteur)*, 135E, 183, 1984.

101. Perrin, P., Thibodeau, L. and Sureau, P., Rabies immunosome/subunit vaccine/structure and immunogenicity. Pre- and post-exposure protection studies, *Vaccine*, 3, 325, 1985.

102. Cudd, A., Noonan, C.A., Tosi, P.-F., Melnick, J.L., and Nicolau, C., Specific interaction of CD4-bearing liposomes with HIV-infected cells, *J. Acqu. Immun. Defic. Syndr.*, 3, 109, 1990.

103. Boggs, J.M., Goundalkar, A., Doganoglu, F., Samji, N., Kurantsin-Mills, J., and Koshy, K.M., Antigen targeted liposome-encapsulated methotrexate specifically kills lymphocytes sensitized to the nonapeptide of myelin basic protein, *J. Neuroimmunol.*, 17, 35, 1987.

104. Lowell, G.H., Smith, L.F., Seid, R.C., and Zollinger, W.D., Peptides bound to proteosomes via hydrophobic feet become highly immunogenic without adjuvants, *J. Exp. Med.*, 167, 658, 1988.

105. Guru, P.Y., Agrawal, A.K., Singha, U.K., Singhal, A., and Gupta, C.M., Drug targeting in leishmania donovani infections using tuftsin-bearing liposomes as drug vehicles, *FEBS Lett.*, 245, 204, 1989.

106. Hopp, T.P., Immunogenicity of a synthetic HBsAg peptide: enhancement by conjugation to a fatty acid carrier, *Mol. Immunol.*, 21, 13, 1984.

107. Brynestad, K., Babbitt, B., Huang, L., and Rouse, B.T., Influence of peptide acylation, liposome incorporation, and synthetic immunomodulators on the immunogenicity of a 1-23 peptide of glycoprotein D of herpes simplex virus: implications for subunit vaccines, *J. Virol.*, 64, 680, 1990.

108. Hashimoto, K., Loader, J.E., and Kinsky, S.C., Synthesis and characterization of methotrexate-dimyristoylphosphorylethanolamine derivatives and the glycerophophorylethanolamine analogs, *Biochim. Biophys. Acta*, 816, 163, 1985.

109. Hashimoto, K., Loader, J.E., Knight, M.S., and Kinsky, S.C., Inhibition of cell proliferation and dihydrofolate reductase by liposomes containing methotrexate-dimyristoylphophatidylethanolamine derivatives and by the glycerophosphorylethanolamine analogs, *Biochim. Biophys. Acta*, 816, 169, 1985.

110. Kinsky, S.C., Hashimoto, K., Loader, J.E., Knight, M.S., and Fernandes, D.J., Effect of liposomes sensitized with methotrexate-gamma-dimyristoylphosphatidylethanolamine on cells that are resistant to methotrexate, *Biochim. Biophys. Acta*, 885, 129, 1986.

111. Barratt, G. and Schuber, F., Targeting liposomes with mannose-terminated ligands, in *Liposome Technology*, Gregoriadis, G., Ed., 2nd Edition, CRC Press, Boca Raton, FL, Vol. III, 199, 1992.

112. van Berkel, T.J.C., Kruijt, K.J., Spanjer, H.J., Kempen, H.J.M., and Scherphof, G.L., Targeting of liposomes with tris-galactoside-terminated cholesterol, in *Liposome Technology*, Gregoriadis, G., Ed., 2nd Edition, CRC Press, Boca Raton, FL, Vol. III, 219, 1992.

113. Ponpipom, M.M., Shen, T.Y., Baldeschwieler, J.D., and Wu, P.-S., Modification of liposome surface properties by synthetic glycolipids, in *Liposome Technology*, Gregoriadis, G., Ed., Vol III, CRC Press, Boca Raton, FL, 95, 1984.

114. Slama, J. and Rando, R.R., Lectin-mediated aggregation of liposomes containing glycolipids with variable hydrophilic spacer arms, *Biochemistry*, 19, 4595, 1980.

115. Ponpipom, M.M., Bugianesi, R.L., Robbins, J.C., Doebber, T.W., and Shen, T.Y., Cell-specific ligands for selective drug delivery to tissues and organs, *J. Med. Chem.*, 24, 1388, 1981.

116. Doebber, T.W., Wu, M.S., Bugianesi, R.L., Ponpipom, M.M., Furbisch, F.S., Barranger, J.A., Brady, R.O., and Shen, T.Y., Enhanced macrophage uptake of synthetically glycosylated human placental beta-glucocerebrosidase, *J. Biol. Chem.*, 257, 2193, 1982.

117. Kempen, H.J.M., Hoes, C. van Boom, J.H., Spanjer, H.H., Langendoen, A., and van Berkel, T.J.C., A water-soluble cholesteryl-containing trisgalactoside: synthesis, properties, and use in directing lipid-containing particles to the liver, *J. Med. Chem.*, 27, 1306, 1984.

118. Spanjer, H.H., van Berkel, T.J.C., Scherphof, G.L., and Kempen, H.J.M., The effect of a water-soluble tris-galactoside terminated cholesterol derivative on the in vivo fate of small unilamellar vesicles, *Biochim. Biophys. Acta*, 816, 396, 1985.

119. Bachhawat, B.K., Das, P.K., and Gosh, P., Preparation of glycoside-bearing liposomes for targeting, in *Liposome Technology*, Gregoriadis, G., Ed., CRC Press, Boca Raton, FL, Vol III, 117, 1984.

120. Szoka, F. and Papahadjopoulos, D., Comparative properties and methods of preparation of lipid vesicles-liposomes, *Annu. Rev. Biophys. Bioeng.*, 9, 467, 1980.

121. Barbet, J., Machy, P., Truneh, A., and Leserman, L.D., Weak acid-induced release of liposome-encapsulated carboxyfluorescein, *Biochim. Biophys. Acta*, 772, 347, 1984.

Chapter 3

Non-Phospholipid Molecules and Modified Phospholipid Components of Liposomes

Jean R. Philippot, Pierre G. Milhaud, Christophe O. Puyal, and Donald F.H. Wallach

CONTENTS

I. INTRODUCTION

Liposomes are analogs of biological cell membranes in that they are vesicular lipid membrane structures enclosing a volume of water. Interest in liposomes has soared in recent years, particularly since introduction of the idea that they might serve as carriers for drugs and enzymes.[1-3] In the first 4 months of 1993, over 170 full papers containing the word "liposome" (out of a total of more than 12,500) and more than 22 U.S. patents (out of a total of more than 446) have been published or issued. At least five companies are involved predominantly in liposome technology and several pharmaceutical and cosmetic companies have liposome divisions.

The membranes of liposomes are composed of amphiphiles, molecules with a hydrophilic head group and a hydrophobic tail (e.g., fatty acids, fatty alcohols, phospholipids, and glycolipids). Until recently, liposome technology has been concerned mostly with vesicles composed of phospholipids, predominantly lecithin,[3] and phospholipid liposomes (PL) continue to be the focus of most publications and patents. However, whereas PL are suited for certain pharmaceutical applications, PL technology has been beset by problems that hinder its general commercial or industrial application. For example, phospholipids turn over rapidly *in vivo* and are unstable *in vitro*. In addition, phospholipids are expensive to purify or synthesize and the manufacture of PL is difficult and costly to scale up.

For these reasons different technologies have been developed to make liposomes more useful. We will treat two types of liposomes:

1. The liposomes made of non-phospholipid, "membrane mimetic" amphiphiles,[4] which are molecules that have a hydrophilic head group attached to a hydrophobic "tail" and include long-chain fatty acids, long-chain alcohols and derivatives, long-chain amines, and polyol sphingo- and glycerolipids.

0-8493-4569-3/95/$0.00+$.50

2. The liposomes mostly made of phospholipids, but containing a few percent of additives (non-biological lipids or substituted lipids) which enhance particular properties of the vesicle.

Thus we will consider additives used to make polymerized PL, fusogenic PL (pH-sensitive or by extension fusogenic-peptide PL), cationic PL, and "Stealth®" PL.

A. MEMBRANE-MIMETIC AMPHIPHILES

The study of membrane-mimetic amphiphiles extends back to the first decade of this century. Experiments using physical and chemical methods[4-11] have shown that such molecules assume preferred arrays in the presence of water. Formation of these arrays, which include micelles, monolayers, and bimolecular layers is driven by the need for the polar head groups, which may be ionogenic or not, to associate with water, and the need of the apolar hydrophobic tails to be excluded from water.[12] Exactly which type of structure is assumed depends upon the nature of the amphiphile, its concentration, the presence of other amphiphiles, temperature, and presence of salt and other solutes in the aqueous phase.

Membrane-mimetic amphiphiles include molecules that are insoluble in water but can take up water, and molecules that have appreciable solubility in water under limiting conditions. Amphiphiles in the first group do not form molecularly disperse solutions in water but may swell considerably with water to form lamellar phases. The amphiphiles in the second category can, at some temperatures, form solutions of dispersed monomers in water and often undergo the following sequence as the concentration in water is increased: *Monomeric solution → micellar solution.*[4,11,13] The manufacture of non-phospholipid liposomes, as here described, depends on the manipulation of environmental variables (e.g., temperature, hydration) in an appropriate temporal sequence so as to cause non-phospholipid amphiphiles to form liposomal structures.

II. NON-PHOSPHOLIPID LIPOSOMES (NPL)

We will describe the major non-phospholipid constituents encountered in the literature before focusing on a family of liposomes called Novasomes®.

A. MAJOR NON-PHOSPHOLIPID MEMBRANE-MIMETIC AMPHIPHILES

Gebicki and Hicks[14,15] and Hicks and Gebicki[16] demonstrated the formation of water-containing vesicles — 'UFAsomes' — enclosed by bilayers of oleic acid. Hargreaves and Deamer[17] demonstrated that a number of long-chain soaps would form vesicles in the presence of nonionic surfactants.

Handjani and associates[18-24] and others[25-28] have made lipid vesicles called "Niosomes" from single- (or two-) tailed ether (or ester) derivatives of polyglycerol (or polyoxyethylene). These liposomes are being produced on a large scale by L'Oreal, France for use in cosmetic products. Different combinations of lipid moieties and polar head group joined by an ether or ester bonds have been tested for Niosome stability and permeability (Table 1). The physico-chemical properties of some nonionic, non-phospholipid liposome of the Niosome type have been reviewed.[28]

Kunitake and associates[29-37] have proven formation of water-containing vesicles limited by pseudobilayers — monolayers composed of two-headed ammonium amphiphiles. Kano et al.[38] demonstrated formation of bilayer vesicles with sonicated double-tailed cationic surfactants. Murakami et al.[39,40] formed stable single-compartment vesicles with one or more bilayer walls composed of cationic or zwitterionic two-chain amphiphiles involving amino acid residues. Shigami et al.[41] and Schenk et al.[42] have prepared multibilayer vesicles from two-tailed sucrose fatty acid esters.

Kaler et al.[43] have demonstrated that appropriate aqueous mixtures of single-tailed cationic and anionic surfactants spontaneously form single-walled vesicles, presumably via production of salts that then act as double-tailed zwitterionic surfactants. Wallach and associates[44-63] have developed methods for the manufacture of paucilamellar, non-phospholipid liposomes, called Novasomes®, that can be formed from a variety of membrane-mimetic amphiphiles, as well as from certain phospholipids. These liposomes, made by MicroVesicular Systems, Inc., Nashua, NH, have two or more membranes surrounding an amorphous core, each membrane being composed of amphiphile molecules in bilayer array. The core accounts for most of the vesicle volume, providing a high carrying capacity for water-soluble and water-immiscible substances.

B. MAJOR CONSTITUENTS OF NOVASOMES®

The non-phospholipid liposomes made by Micro Vesicular Systems, Inc. are engineered for particular applications. Appropriate non-phospholipid liposomes are assembled using a series of membrane "mod-

Table 1   **Existence and properties of several types of Niosomes[1]**

Hydrocarbon chain[2]	Bond	Head Groups[3,4]	
		[oxyethylene]$_n$	[glycerol]$_n$
C_{12}	-O	n = 4 (+)	n = 2 (+)
	-CO-O		n = 1 (0)
C_{14}	-O		n = 2 (++)
C_{16}	-O		n = 2 (+++)
			n = 3 (+++)
	-CO-O		n = 2 (+++)
C_{18}	-O	Brij (0)	n = 2 (++)
			n = 2 (0)
	-CO-O	Tween (0)	n = 1 (0)

[1]From Reference 28

[2]Saturated or unsaturated chain

[3]Homogeneous condensate (n), statistical condensate (n)

[4](0) no vesicle, (+) vesicle highly permeable, (++) medium permeability, (+++) low permeability

ules", each module imparting desired characteristics to the liposomes (Figure 1). The modular approach provides great breadth and flexibility to membrane design. It also allows the combination of numerous amphiphiles with each other, or with phospholipids, sphingolipids, or both, to yield membrane hybrids with novel properties.

1. Major Structural Modules

Table 2 lists some of the major structural membrane amphiphiles currently used in this non-phospholipid liposome vesicle design. In any given liposome type these molecules account for more than 50% of the membrane lipid.

The fatty alcohols (Table 2, #1) form bilayer vesicles only in association with a modulating module (Table 3) and an ionogenic module (Table 4).

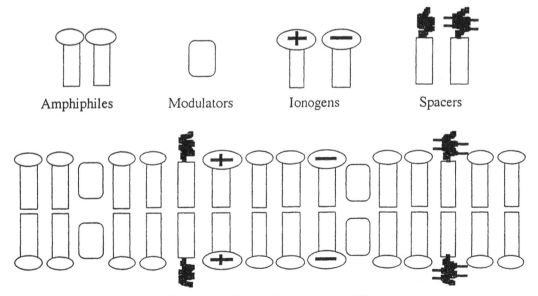

Amphiphiles   Modulators   Ionogens   Spacers

A Novasome™ vesicle membrane bilayer

Figure 1   Schematic representation of membrane modules of the Novasome® bilayer.

Table 2   Principal wall-forming materials

Classification	Hydrocarbon Chain	Bond	Head Group
1. Fatty alcohol	C_{12}-C_{20} (0-1 unsaturation)		-OH
2. Fatty acid	C_{12}-C_{20} (0-1 unsaturation)		-COOH
3. Ethoxylated fatty alcohol	C_{12}-C_{20} (0-1 unsaturation)	-O	-$(CH_2CH_2O)_{2.5}H$
4. Glycol ester of fatty acid	C_{12}-C_{20} (0-1 unsaturation)	-CO-O	-CH_2CH_2OH
5. Ethoxylated fatty acid	C_{12}-C_{20} (0-1 unsaturation)	-CO-O	-$(CH_2CH_2O)_{2-8}H$
6. Glycerol fatty acid monoester	C_{12}-C_{20} (0-1 unsaturation)	-CO-O	-$CH_2CHOHCH_2OH$
7. Glycerol fatty acid diester	C_{12}-C_{18} (1 unsaturation)	-CO-O	-CH-CHO
	C_{12}-C_{18} (1 unsaturation)	-CO-O	-CH_2
8. Ethoxylated glycerol fatty acid ester	C_{16}-C_{18} (0 unsaturation)	-CO-O	-$CH_2CHOHCH_2O(CH_2CH_2O)_9H$
9. Fatty acid diethanolamide	C_{12}-C_{20} (0-2 unsaturations)	-CO-N	-$(CH_2CH_2OH)_2$
10. Fatty acid dimethyl amide	C_{12}-C_{20} (0-2 unsaturations)	-CO-N	-$(CH_3)_2$
11. Fatty acid sarcosinates	C_{12}-C_{18} (0 unsaturation)	-CO-N | CH_3	-CH_2-COOH
12. "Alkyd"	C_{10} (0 unsaturation)	-O-CO	⬡-COO⁻
13. "Alkyd"	C_{12}-C_{18} (0-4 unsaturations)	-CO-O	-CH_2
	C_{12}-C_{18} (0-4 unsaturations)	-CO-O	-$(CH_2)_2$-O-CO-⬡(-COO⁻)(-COO⁻)

The fatty acids (Table 2, #2) serve as structural modules for acid pH-sensitive non-phospholipid liposomes. They form stable bilayers only when they are partly ionized (at bulk pH levels between 6 and 8, corresponding to membrane pH levels near 4-6) and in association with a nonionic amphiphile, such as a fatty alcohol, and a modulator module (Table 3). Full ionization of the carboxyl groups causes disruption of the bilayer, as does full protonation.

The ethoxylated fatty alcohols (Table 2, #3) and fatty acids (Table 2, #5) with 2-5 oxyethylene residues can form bilayers in the absence of ionogenic residues, but stable vesicles require a modulating module (Table 4). Polyoxyethylene (8-9) glycerol fatty acid monoesters (Table 2, #8) can form bilayer vesicles in the absence of ionogenic or modulator modules.

Fatty acid glycol esters (Table 2, #4), fatty acid glycerol monoesters (Table 2, #6), and fatty acid glycerol diesters (Table 2, #7) form satisfactory lipid bilayer vesicles when appropriately combined with a modulator and ionogenic module.

Table 3   Membrane modulators

Classification	Apolar Residue	Bond	Head Group
1. Cholesterol	Δ^5 cholestene 3-beta		-OH
2. Phytosterol (Beta sitosterol)	Δ^5 stigmastene 3-beta		-OH
3. Ethoxylated cholesterol	Δ^5 cholestene 3-beta	-O	-$(CH_2CH_2O)_nH$
4. Cholesterol phthalate	Δ^5 cholestene 3 beta	-O	-CO-⬡-COOH
5. Cholesterol hemisuccinate	Δ^5 cholestene 3 beta	-O	-CO-CH_2-CH_2-COOH
6. Thiocholesterol	Δ^5 cholestene 3 beta		-SH

Table 4 **Ionogenic modules**

Classification	Hydrocarbon Chain	Bond	Head Group
Anionic			
1. Fatty acid	C_{12}-C_{20} (0-1 unsaturation)		-COO⁻
2. Fatty acid sarcosinates	C_{12}-C_{18} (0 unsaturation)	-CO-N ⎸ CH₃	-CH₂-COO⁻
3. Diacyl phosphate	C_{16}-C_{18} (unsaturation)	-O⟍	O⁻ ⎸ P=O
	C_{16}-C_{18} (0 unsaturation)	-O⟋	
Zwitterionic			
4. Betaines	C_{12}-C_{18} (0-1 unsaturation)	-CH3 ⎸ -N⁺ ⎸ -CH₃	-CH₂-COO⁻
Cationic			
5.	C_{12}-C_{18} (0 unsaturation)		-N⁺H₃
6.	C_{16}-C_{18} (1 unsaturation)	-CO-O	-CH₂
	C_{16}-C_{18} (0 unsaturation)	-CO-O	-CH O ⎸ ‖ CH₂-O-P-CH₂-CH₂NH₃⁺ ⎸ O⁻
7.	C_{16}-C_{18} (0 unsaturation)		-N⁺(CH₃)₃
8.	C_{16} (0 unsaturation)		—N⁺⟨⬡⟩
9.	C_{16}-C_{18} (0 unsaturation)	⟍	CH₃ ⎸ N⁺ ⎸ CH₃
	C_{16}-C_{18} (0 unsaturation)	⟋	

Some fatty acid diethanolamides (Table 2, #9) form bilayer vesicles without ionogenic or modulator modules. The fatty acid N,N-dimethylacylamides (Table 2, #10) form vesicles without auxiliary amphiphiles and have been used to form liposomes for the delivery of highly water-insoluble fungicides.[54] Whereas most vesicle formers are amphiphiles with the polar end substantially hydrophilic (e.g., OH, $(CH_2CH_2O)_nH$, $CH_2CHOHCH_2OH$, COOH/COO⁻), N,N-dimethylamides lack a strongly hydrophilic head group. The head group instead consists of two methyl residues on the amide nitrogen, with the oxygen on the neighboring carbon having the potential for hydrogen bonding. Molecular modeling suggests that hydrogen bonding between water and the CO groups of adjacent N,N-dimethylamide molecules would favor interdigitation of the hydrocarbon chains, rather than the end-to-end configuration bound in typical lamellar phase structures.

The "alkyd" amphiphiles (Table 2, #12 and #13) were designed for the encapsulation of alkyd resin oils within non-phospholipid liposomes to create water-borne oil paints.[62] Unsaturated fatty acids are conjugated with polyols such as glycerol or pentaerythritol and further with an anhydride such as phthalic or trimellitic anhydrides, yielding anionic, "monomeric" alkyds. Oils at low pH, these molecules form vesicles at neutral pH and above, and can then be used to efficiently encapsulate conventional liquid alkyd resin in liposomes suspended in aqueous media. Preparations with 50% solids can be readily achieved. When painted out and dried, the "wall alkyds" copolymerize with the passenger alkyd to form coherent alkyd films.

2. Modulator Modules

The modular approach used in the design of non-phospholipid liposomes relies on the use of sterol modulator molecules (Table 3). Cholesterol can intercalate between amphiphile hydrocarbon chains in bilayers, and thereby allows the intermixing of different acyl chains without phase segregation and broadens the range of temperature within which the crystalline → liquid-crystalline transition occurs.[64]

It appears that cholesterol interacts strongly with the polar termini of the chains, leaving them less free to change conformation than the more disordered long-chain segments.[65] Presumably, phytosterols and ethoxylated sterols (Table 3) impart similar effects. Additional flexibility can be achieved by the use of ionogenic sterols such as cholesterol phthalate (Table 3, #4) or cholesterol hemisuccinate (Table 3, #5).

Thiocholesterol (Table 3, #6) is used as an attachment site for the coupling of proteins to the liposome surface via thiolated, bifunctional reagents.[49]

3 . Ionogenic Modules

Table 4 lists the major ionogenic membrane amphiphiles currently used in our non-phospholipid liposomes. They can be used in conjunction with any of the nonionic modules listed in Table 2 in proportions ranging from 0.05 to 10 moles percent.

Of the anionic modules, the fatty acids and sarcosinates (Table 4, #1 and #2) must be used at bulk pH levels of less than 7 (membrane pH greater than 5). Diacylphosphates (Table 4, #3) can be used over a broad pH range. The betaines (Table 4, #4) are zwitterionic at pH levels above neutrality, but become cationic at bulk pH levels below neutrality.

Of the cationic molecules, the fatty amines (Table 4, #5) are ionogenic up to about pH 10. They have been used for the membrane anchoring of proteins via bifunctional coupling reagents. Phosphatidylethanolamine (Table 4, #6), can be incorporated in small proportions for the same purpose. Single alkylated (Table 4, #7), and double alkylated (Table 4, #8) quaternary amines are cationic over the entire pH range. Cetyl pyridinium cation (Table 4, #9) is ionized over most of the pH range. It is widely used in oral hygiene products and therefore suitable for the preparation of non-phospholipid liposomes for dental care or oral care. Double chain quaternary amines are extensively used in hair and skin care preparations, and several types of non-phospholipid liposomes positively charged with such amphiphiles are in cosmetic use.

4. Spacer Modules

Spacer modules are employed to widen the interlamellar space, reducing the number of lamellae, and to make the external bilayer less accessible than in the case with compact head groups. Ethoxylated amphiphiles (Table 5, #1 and #3), propoxylated amphiphiles (Table 5, #2), and saccharide derivatives of fatty amines and fatty acids (Table 5, #4 and #5) allow the placement, atop the outermost bilayer surface, of bulky residues acting as steric barriers to regulate interaction of the vesicle surface with, for example, viruses and phagocytic cells.

5. Chain Matching

Provided there are no steric or other interferences between head groups, the most stable membranes are produced by optimal matching of the chain length of the membrane modules. However, the action of sterols allows assembly of membranes from modules of varying chain lengths, thereby providing some control over permeability.

III. MOLECULES MODIFYING PHOSPHOLIPID LIPOSOME MEMBRANES

In the last 10 years, extensive work has been directed to improve liposomes as drug carriers. The aim has been to develop a vehicle that is stable in physiological fluids (plasma, lymph), able to reach and bind to its target cells, and then to cause internalization of the drug. Several approaches incorporating or coupling of non-phospholipid or modified lipid molecules to the PL membranes have been used to enhance liposome stability and improve drug internalization. We will consider these molecules in this

Table 5 Spacer modules

Classification	Apolar Residue	Bond	Head Group
1. Ethoxylated fatty alcohol	C_{12}-C_{20} (0-1 unsaturation)	-O	-$(CH_2CH_2O)_{10\text{-}50}H$
2. Propoxylated fatty alcohol	C_{12}-C_{20} (0-1 unsaturation)	-O	-$(CH_2CHCH_3O)_{10\text{-}50}H$
3. Ethoxylated glycerol fatty acid ester	C_{16}-C_{18} (0 unsaturation)	-CO-O	-$CH_2CHOHCHO(CH_2CH_2O)_9H$
4. Aldosamides	C_{12}-C_{20} (0-1 unsaturation)	-NH-CO	-$(CHOH)_nCH_2OH$
5. Hexosamides	C_{12}-C_{20} (0-1 unsaturation)	-CO-NH	-Hexose

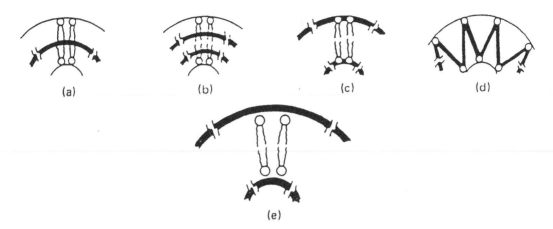

Figure 2 Possible lipid crosslinkages in polymerized liposomes.

section, excluding molecules used to target cell membrane receptors (e.g., Tf, LDL, mAb...) since these do not fall into the scope of this chapter.

A. MEMBRANE STABILIZERS

These are molecules incorporated in PL walls to increase the liposome lifetime *in vivo* and thereby extend the bioavailability of PL-associated drugs. Two principal techniques have been employed to manufacture such PL. One involves crosslinking phospholipid molecules to form an impermeable lipid vesicle (polymerized PL); the other involves coating the outermost surface of PL with naturally occurring polysaccharides or polyoxyethylenes absorbed or covalently bound to membrane lipid, in order to reduce the uptake of liposomes by the reticuloendothelial system.

1. Polymerized PL

In search of PL more stable and suitable for oral administration of drugs, several research groups[67-72] have formed and characterized polymerized PL made from synthetic, polymerizable derivatives of phosphatidylcholine or polymerizable surfactants. Regen[73] has reported on most of the molecules that have been used in this approach and describes how the polymeric backbone develops in the membrane bilayer (Figure 2). The polymeric phospholipids, mainly phosphatidylcholine, belong to four classes (Table 6). Polymerization is initiated either by chemical reagents or by UV irradiation. The most common strategy used to prepare polymerized PL has been first to assemble the vesicle from suitable, polymerizable lipids and then to carry out the polymerization reaction.[74] Polymerizable surfactants are used in the same way — 48 polymerizable surfactant molecules have been used to make liposomes.[73] Most authors find that polymerization increases the stability of liposomes,[73-76] but Bonte et al. find that such liposomes are cleared from the bloodstream more rapidly than normal PL.[77]

2. PL Stabilization by Polysaccharides

It is well known that certain sugar residues allow a selective binding of surface glycoproteins to various cells[78-80] and play an important role in biological recognitions. Such molecules, for example galactose or asialoglycoprotein, have been used to target liver cells and deliver drugs.[81-83] In addition, naturally occurring polysaccharides are required to maintain the shape and rigidity of bacterial and plant cell membranes, and coating of liposome surfaces with such polysaccharides might make liposomes more stable.

The stabilization of liposomes by simple polysaccharides[84] was first examined in 1972 and has been further studied recently.[78,85] The polysaccharides used were mannan, pullulan, levan, dextran, amylose, and phosphorylated mannan bound to the PL surface[86] either by simple interactions (with risks of desorption), or by chemical conjugation to cholesterol[86] or fatty acids.[87] The data suggest that such polysaccharide-coated PL are more stable in the circulation then uncoated PL.

Recently, new PL formulations containing anionic lipids such as monosialoganglioside GM1 or phosphatidylinositol,[78,85,88] (so-called "Stealth®" PL), were shown to exhibit extended persistence in the circulation, presumably due to steric stabilization of the liposomal surface. This type of liposome is described in Chapter 11 and will not be further discussed here. Some stabilization of PL in the circulation

Table 6   **Major polymerizable phosphatidylcholines**

Diacetylenic PC	$H_3C-(CH_2)_n-CH=C-C\equiv C-(CH_2)_m-COO-CH_2$
	$H_3C-(CH_2)_n-CH=C-C\equiv C-(CH_2)_m-COO-CH$
	$CH_2-O-P-O-(CH_2)_2-N^+(CH_3)_3$
Dienic PC	$H_3C-(CH_2)_n-CH=CH-CH=CH-COO-CH_2$
	$H_3C-(CH_2)_n-CH=CH-CH=CH-COO-CH$
	$CH_2-O-P-O-(CH_2)_2-N^+(CH_3)_3$
Methacrylate-derivatized PC	$H_2C=\overset{CH_3}{\underset{|}{C}}-COO-(CH_2)_{11}-COO-CH_2$
	$H_3C-(CH_2)_{14}-COO-CH$
	$CH_2-O-P-O-(CH_2)_2-N^+(CH_3)_3$
Thiol-derivatized PC	$H_3C-(CH_2)_n-\overset{SH}{\underset{|}{CH}}-COO-CH_2$ $\overset{SH}{\underset{|}{}}$
	$CH_2-COO-CH-(CH_2)_n-H_3C$
	$N^+(CH_3)_3-(CH_2)_2-O-P-O-CH_2$

has also been achieved by coupling polyoxyethylene groups[89,90] or uronic acid derivatives[87] to phospholipids or fatty acids, respectively.

B. INDUCERS OF MEMBRANE FUSION

Three types of methods have been used to introduce macromolecules into mammalian cells:

1. Facilitation of macromolecule uptake by polycations or nascent calcium phosphate,
2. Direct injection using microneedles, and
3. Internalization by fusion between liposomal vesicles and cell membranes.[91]

Since ordinary PL do not fuse spontaneously, specific strategies have been developed to make them fusogenic, or otherwise suitable for delivery of molecules and genes directly into the cytoplasm.

1. Molecules Creating Fusogenic Liposomes

pH-sensitive liposomes that are endocytosed and fuse with endosomal membranes at the acid pH of the endosomes were first developed by Yatvin et al.[92] Later, more subtle mechanisms of pH-dependent liposomal fusion were developed by Huang and associates.[93-94] The PL used in these studies contain dioleoylphosphatidylethanolamine which remains in a bilayer array at neutral pH, but undergoes a membrane-destabilizing phase change at acidic pH.[95] This approach is treated in Chapter 13.

In fact, membrane fusion is the process by which enveloped viruses enter cells.[96-97] Viruses such as influenza virus, Semliki Forest virus, and vesicular stomatitis virus enter cells via endocytosis — pH-sensitive release of fusion proteins from envelop glycoprotein precursors by endosomal proteases. Incorporation of such fusion protein precursors into liposomal membranes leads to pH-dependent fusion with endosomal membranes.[98-103] Viruses such as the paramyxoviruses and human or simian immunodeficiency viruses possess envelope glycoproteins that are fusogenic at neutral pH and can therefore fuse with the plasma membrane of target cells. Plasma membrane fusogenic activity can be induced by incorporating the

Table 7 **Amino acid sequences of peptides used to induce fusion**

1. Peptides from N-terminal segments of virus envelope glycoprotein

Virus type	Protein type	No. of a.a.	Sequence	Ref.
Influenza Virus	Hemag-	7 *	G-L-F-G-A-I-C	107
	glutinin	10 *	G-L-F-G-A-I-A-G-F-C	107
		15	G-L-F-A-G-F-I-E-N-G-W-M-I-D-G	108
		17 *	G-L-F-G-A-I-A-G-F-I-E-N-G-W-E-G-C	107
		20	G-F-F-G-A-I-A-G-F-L-E-G-G-W-E-G-M-I-A-G	99,100
		20	G-L-F-G-A-I-A-G-F-I-E-N-G-W-E-G-M-I-D-G	101
		20	G-L-F-G-A-I-A-G-F-I-E-G-G-W-T-G-M-I-D-G	102
		23 *	G-L-F-E-A-I-A-G-F-I-E-N-G-W-E-G-M-I-D-G-G-G-C	103
Vesicular Stomatis	G protein	7 *	K-F-T-I-V-F-C	107
Virus (VSV)		11 *	K-F-T-I-V-F-P-H-N-G-C	107
Sendaï Virus	F protein	8 *	F-F-G-A-V-I-G-C	107
Simian	gp32	12	G-V-F-V-L-G-F-L-G-F-L-A	109
Immunodeficiency		16	G-V-F-V-L-G-F-L-G-F-L-A-T-A-G-S	109
Virus (SIV)		24	G-V-F-V-L-G-F-L-G-F-L-A-T-A-G-S-A-M-G-A-A-S-L-T	109
Human	gp41	23	A-V-G-I-G-A-L-F-L-G-F-L-G-A-A-G-S-T-M-G-A-R-S	110
Immunodeficiency		23	A-I-G-I-G-A-M-F-L-G-F-L-G-A-A-G-S-T-M-G-A-A-S	110
Virus (HIV)				

2. Peptides from other proteins

Protein	No. of a.a.	Sequence	Ref.
Albumin	78	Serum albumin fragment comprising amino acids 307-385	111
Melittin	26	G-I-G-A-V-L-K-V-L-T-T-G-L-P-A-L-I-S-W-I-K-R-K-R-Q-Q	112
Gramicidin S	10	-V-O-L-F-P-V-O-L-F-P-	113

3. Model synthetic peptides

Peptide	No. of a.a.	Sequence	Ref.
GALA	30	W-E-A-A-L-A-E-A-L-A-E-A-L-A-E-H-L-A-E-A-L-A-E-A-L-E-A-L-A-A	114
		poly (K-Aib″-L-Aib)	115
	12	(L-A-X#-L)₃	116
		poly (K)	117
		poly (D)	118
		poly (R)	121
	51	C-(K-K-L-L)₅-K-K-N-P-G-(L-L-K-K)₆-Y	119
	12 •	palmitoyl-(L-A-R-L)₃	120
	12 •	L-A-R-L-W-(a-aminomyristic acid)-R-L-L-A-R-L	120

*These peptides contain an additional cysteine at C-terminus for use in coupling the peptides to liposomes or other carrier
″Aib = 2 amino isobutyric acid
#X = Arg or Lys
•Lipid-bound peptides

fusogenic virus proteins of such viruses into the liposomal membrane.[104-107] Good internalization has been achieved using liposomes containing paramyxovirus (Sendai virus) fusion protein.[104,105]

Because fusion proteins are immunogenic *in vivo*,[106] the possibility of using fusogenic peptides has been actively explored.[99-103,107-121] Table 7 presents peptides that have been shown to exhibit fusogenic

Table 8 Cationic lipids

Lipid		Ref.
DDA	$[CH_3-(CH_2)_{11}]_2-N^+-(CH_3)_2] \cdot Br^-$	132
CTAB	$[CH_3-(CH_2)_{15}-N^+-(CH_3)_3] \cdot Br^-$	133
DEBDA[OH]	$\left[(CH_3)_3-C-CH_2-C-(CH_3)_2 - \bigcirc - [O[CH_2]_2]_2 - \overset{\overset{CH_2-\bigcirc}{|}}{\underset{CH_3}{N^+-CH_3}} \right] \cdot OH^-$	118
DOTMA	$\left[\begin{array}{l} CH_3-(CH_2)_7-CH=CH-(CH_2)_8-O-CH_2 \\ \qquad\qquad\qquad\qquad\qquad\qquad | \\ CH_3-(CH_2)_7-CH=CH-(CH_2)_8-O-CH-CH_2-N^+-(CH_3)_3 \end{array} \right] \cdot OH^-$	130 / 131
DOTAP	$\left[\begin{array}{l} CH_3-(CH_2)_7-CH=CH-(CH_2)_7-CO-O-CH_2 \\ \qquad\qquad\qquad\qquad\qquad\qquad\quad | \\ CH_3-(CH_2)_7-CH=CH-(CH_2)_7-CO-O-CH-CH_2-N^+-(CH_3)_3 \end{array} \right] \cdot OH^-$	135
DOTB	$\left[\begin{array}{l} CH_3-(CH_2)_7-CH=CH-(CH_2)_7-CO-O-CH_2 \\ \qquad\qquad\qquad\qquad\qquad\qquad\quad | \\ CH_3-(CH_2)_7-CH=CH-(CH_2)_7-CO-O-CH-CH_2-CO-O-(CH_2)_2-N^+-(CH_3)_3 \end{array} \right] \cdot OH^-$	135
DOSC	$\left[\begin{array}{l} CH_3-(CH_2)_7-CH=CH-(CH_2)_7-CO-O-CH_2 \\ \qquad\qquad\qquad\qquad\qquad\qquad\quad | \\ CH_3-(CH_2)_7-CH=CH-(CH_2)_7-CO-O-CH-CH_2-CO-O-(CH_2)_2-CO-O-(CH_2)_2-N^+-(CH_3)_3 \end{array} \right] \cdot OH^-$	135
Cho TB	$[Chol-O-CO-(CH_2)_3-N^+-(CH_3)_3] \cdot OH^-$	135
Chol SC	$[Chol-O-CO-(CH_2)_2-COO-(CH_2)_2-N^+-(CH_3)_3] \cdot OH^-$	135
DC-Chol	$[Chol-O-CO-NH^+-(CH_2)_2-N^+H^+-(CH_3)_2] \cdot OH^-$	134
DOGS	$\left[\begin{array}{l} CH_3-(CH_2)_{15}-CH_2 \\ \qquad\qquad\quad | \\ \qquad\qquad\quad N-CO-CH_2-NH-CO-CH-(CH_2)_3-NH_2^+-(CH_2)_3-NH_3^+ \\ \qquad\qquad\quad | \qquad\qquad\qquad\qquad | \\ CH_3-(CH_2)_{15}-CH_2 \qquad\qquad\quad NH_2^+-(CH_2)_3-NH_3^+ \end{array} \right] \cdot 4\ OH^-$	136
DPPES	$\left[\begin{array}{l} CH_3-(CH_2)_{14}CO-O-CH_2 \qquad\quad O \\ \qquad\qquad\qquad\quad | \qquad\qquad\quad || \\ \qquad\qquad\qquad\quad CH-CH_2-O-P-O-(CH_2)_2-NH-CO-CH-(CH_2)_3-NH_2^+-(CH_2)_3-NH_3^+ \\ \qquad\qquad\qquad\quad | \qquad\qquad\quad || \qquad\qquad\qquad\qquad\qquad | \\ CH_3-(CH_2)_{14}CO-O-CH_2 \qquad\quad O \qquad\qquad\qquad\qquad NH_2^+-(CH_2)_3-NH_3^+ \end{array} \right] \cdot 3\ OH^-$	136

activity. Most of these have been tested in solution, but a few (asterisks in Table 7) have been tested after coupling to PL. The data indicate that fusogenic peptides, even when added in aqueous solution, must be inserted into the target membrane to achieve fusion.[122]

2. Reagents Promoting Liposomal Transfection (Lipofection)

Animal cells bear a negative surface charge due to extracellulary oriented carboxyl groups of plasma membrane glycoproteins and glycosphingolipids. It has been shown that despite this negative surface charge, polynucleic acids, DNA or RNA, and polyanions enter cells[123] via the endocytic pathway.[128] Presumably most of these polynucleic acids are destroyed within the lysosomes, but a few molecules reach the cytoplasm.[138]

It has been proposed that appropriate, positively charged polymers might adhere to the polyanionic plasma membrane and might thereby promote the internalization of any DNA (also proteins and drugs) bound to such polycations, and experiments with polyornithine[125] and polylysine[124-127] have demonstrated the feasibility of this approach. In these experiments the polycation acts not only as a vector for DNA transfer, but also protects the DNA against nucleolytic enzymes.

Work on gene transfer[129-131,133-137] has recently turned to the use of positively charged liposomes as vehicles for DNA transfection. Table 8 shows some of the cationic amphiphiles that have been incorporated into liposomal membranes to achieve gene transfer using liposomes. Relatively high efficiencies of gene transfer are achieved by lipofection techniques, even though the DNA is not encapsulated but carried on the positively charged surface of the cationic liposomes. Very recently Zhu et al.[137] attained systemic gene expression after intravenous injection of a cytomegalovirus-chloramphenicol plasmid bound to DOTMA: dioleoylphosphatidylethanolamine liposomes. A high liposome:DNA ratio of 8:1 was necessary to achieve such *in vivo* gene transfer. It is presumed that liposomal membrane in some way facilitates transfer of the DNA into the cytoplasm and that the electrostatic interaction between the liposomal surface and the absorbed DNA protect the latter against degradation. This topic is treated more extensively in Chapters 9 and 10.

IV. SUMMARY

This is an overview of the recent progress made in liposome technology that we classed according to their lipid constitution. We distinguished two main classes of liposomes: the non-phospholipid liposomes, made with non-phospholipid amphiphile molecules, and the phospholipid liposomes, bearers of a very specific function thanks to the addition of lipid or non-lipid molecules to their membranes.

In the analysis of the specificities of non-phospholipid liposomes we begin by an overall presentation of major molecules used to produce these kinds of vesicles. Then, we detail the role of each membrane constituent on liposome shape and fate for a category of such new liposomes (Novasomes®).

In the second part of the paper we compute major molecules used to make long-life phospholipid liposomes, thanks to membrane stabilizers, or to induce membrane fusion. Applications are reviewed in the corresponding specialized chapters.

ACKOWLEDGMENTS

This work was supported by grants from the l'Agence Nationale de la Recherche sur le SIDA (ANRS), l'Association pour la Recherche sur le Cancer (ARC), and le Centre National de la Recherche Scientifique (CNRS).

REFERENCES

1. Gregoriadis, G., Drug entrapment in liposomes, *FEBS Lett.*, 36, 292, 1973.
2. Gregoriadis, G., Enzyme entrapment in liposomes, *Methods Enzymol.*, 44, 218, 1976.
3. Ostro, M. J., *Liposomes,* Marcel Dekker, New York, 1983.
4. Fendler, J., *Membrane Mimetic Chemistry*, John Wiley & Sons, New York, 1982.
5. Schmitt, F. O., Bear, R. S., and Clark, G. L., X-ray diffraction studies on nerve, *Radiology*, 25, 131, 1935.
6. Schmitt, F. O. and Bear, R. S., The ultrastructure of the nerve axon sheath, *Biol. Rev.*,14, 27, 1939.
7. Bear, R. S., Palmer, K. J., and Schmitt, F. O., X-Ray diffraction studies of nerve lipids, *J. Cell Comp. Physiol.*, 17, 355, 1941.
8. Davson, H. and Danielli, J. F., *The Permeability of Natural Membranes,* Cambridge University Press, Cambridge, U.K., 1943.

9. Luzatti, V., X-ray diffraction studies of lipid water systems, in *Biological Membranes*, Chapman, D. and Wallach, D. F. H., Eds., Academic Press, New York, 1968.

10. Shipley, G. G., Recent X-ray diffraction studies of biological membranes and membrane components, in *Biological Membranes*, Vol. 2, Chapman, D. and Wallach, D. F. H., Eds., Academic Press, New York, 1973.

11. Small, D. M., Handbook of lipid research, Vol. 4, *Physical Chemistry of Lipids from Alkanes to Phospholipids*, Plenum Press, New York, 1986.

12. Tanford, C., *The Hydrophobic Effect*, John Wiley & Sons, New York, 1980.

13. Myers, D., *Surfactant Science and Technology*, VCH Publishers, New York, 1988.

14. Gebicki, J. M. and Hicks, M., Ufasomes are stable particles surrounded by unsaturated fatty acid membranes, *Nature*, 243, 232, 1973.

15. Gebicki, J. M. and Hicks, M., Preparation and properties of vesicles enclosed by fatty acid membranes, *Chem. Phys. Lipids*, 16, 142, 1976.

16. Hicks, M. and Gebicki, J. M., A quantitative relationship between permeability and the degree of peroxidation in ufasome membranes, *Biochem. Biophys. Res. Commun.*, 80, 704, 1978.

17. Hargreaves, W. R. and Deamer, D. W., Liposomes from ionic, single chain amphiphiles, *Biochemistry*, 17, 3759, 1978.

18. Vanlerberghe, G., Handjani-Vila, M. R., and Ribier, A., Les "niosomes", une nouvelle famille de vésicule, à base d'amphiphile non-ioniques, *Colloq. Nationaux CNRS*, 938, 303, 1978.

19. Handjani-Vila, R. M., Ribier, A., and Vanlerberghe, G., Les niosomes, in *Les Liposomes*, Puisieux, F. and Delattre, J., Technique et Documentation Lavoisier, Paris, 1985.

20. Azmin, M. N., Florence, A. T., Hadjani-Vila, R. M., Stuart, J. F., Vanlerberghe, G., and Whittaker, J. S., The effect of non ionic surfactant vesicle (niosome) entrapment on the absorption and distribution of methotrexate in mice, *J. Pharm. Pharmacol.*, 37, 237, 1985.

21. Azmin, M. N., Florence, A. T., Hadjani-Vila, R. M., Stuart, J. F., Vanlerberghe, G., and Whittaker, J. S., The effect of niosomes and polysorbate 80 on the metabolism and excretion of methotrexate in the mouse, *J. Microencapsulation*, 3, 95, 1985.

22. Handjani, R., Ribier, A., Vanlerberghe, G., Handjani-Vila, R. M., Preparation of More Stable Niosomes Useful in Preparation of Cosmetic Creams, Pharmaceutical Products, etc., and Obtained With Non-Ionic Lipid Phase and an Aqueous Phase, French Patent 2,597,346, 1987.

23. Vanlerberghe, G. and Handjani-Vila, R. M., Aqueous Dispersions of Lipid Spheres and Compositions and Contents of Same, U.S. Patent 4,772,471, 1988.

24. Griat, J., Handjani-Vila, R. M., Ribier, A., Vanlerberghe, G., and Zabotto, A., Cosmetic and Pharmaceutical Preparations Containing Niosomes and a Water-Soluble Polyamide, and a Process for Preparing These Compositions; Encapsulated Aqueous Phase in Spherule Formed From a Non-Ionic Amphiphilic Lipid; External Phase as Aqueous Poly-B-Alanine, U.S. Patent, 4,830,857, 1989.

25. Rogerson, A., Cummings, J., and Florence, A.T., Adriamycine-loaded niosomes. Drug entrapment stability and release, *J. Microencapsul.*, 4, 321, 1987

26. Rogerson, A., Cummings, J., Wilmott, N., and Florence, A. T., The distribution of doxorubicin in mice following administration in niosomes, *J. Pharm. Pharmacol.*, 40, 337, 1988.

27. Kerr, D. J., Rogerson, A., Morrison, G. J., Florence, A. T., and Kaye, S. B., Antitumor activity and pharmacokinetics of niosome-encapsulated adriamycin in monolayer, spheroid and xenograft, *Br. J. Cancer*, 58, 432, 1988.

28. Florence, A.T., Nonionic surfactant vesicles: preparation and characterization, in *Liposome Technology*, Vol. 1, Gregoriadis, G., Ed., CRC Press, Boca Raton, FL, 1992.

29. Kunitake, T. and Okahata, Y., A totally synthetic bilayer membrane, *J. Am. Chem. Soc.*, 99, 3860, 1977.

30. Kunitake, T., Okahata, Y., Tamaki, K., Kumamura, F., and Takayanagi, M., Formation of the bilayer membrane from a series of quaternary ammonium salts, *Chem. Lett.*, 387, 1977.

31. Kunitake, T., Nakashima, N., Hayashida, S., and Yonemori, K., Chiral, synthetic bilayer membranes, *Chem. Lett.*, 1413, 1979.

32. Okahata, Y. and Kunitake T., Formulation of a stable monolayer membrane and related structures in dilute aqueous solutions from two-headed ammonium amphiphiles, *J. Am. Chem. Soc.*, 101, 5231, 1979.

33. Okahata, Y., Ihara, H., Shimomura, M., Tawaki, S., and Kunitake, T., Formation of disk-like aggregates from single chain phosphocholine amphiphiles in water, *Chem. Lett.*, 1169, 1980.

34. Kunitake, T. and Okahata, Y., Formation of stable bilayer assemblies in dilute aqueous solution from ammonium amphiphiles with the diphenylazomethine segment, *J. Am. Chem. Soc.*, 102, 549, 1980.
35. Kunitake, T., Nakashima, N., Takarabe, K., Nagai, M., Tsuge, A., and Yanagi, H., Vesicles of polymeric bilayer and monolayer membranes, *J. Am. Chem. Soc.*, 103, 5945, 1981.
36. Kunitake, T., Okahata, Y., Shimomura, M., Yasunami, S., and Takarabe, K., Formation of stable bilayer assemblies in water from single-chain amphiphiles. Relationship between the amphiphile structure and the aggregate morphology, *J. Am. Chem. Soc.*, 103, 5401, 1981.
37. Kunitake T., Organization and functions of synthetic bilayers, *Ann. NY. Acad. Sci.*, 471, 70, 1986.
38. Kano, K., Romero, A., Djermouni, B., Ache, H. J., and Fendler, J. H., Characterization of surfactant vesicles as membrane mimetic agents. Temperature-dependent changes of the turbidity, viscosity, fluorescence polarization of 2-methylanthracene, and positron annihilation in sonicated dioctadecyldimethylammonium chloride, *J. Am. Chem. Soc.*, 101, 4030, 1979.
39. Murakami, Y., Nakano, A., and Fukuya, K., Stable single-compartment vesicles with zwitterionic amphiphile involving an amino acid residue, *J. Am. Chem. Soc.*, 102, 4253, 1980.
40. Murakami, Y., Nakano, A., and Ikeda, H., Preparation of stable single-compartment vesicles with cationic and zwitterionic amphiphiles involving amino acid residues, *J. Org. Chem.*, 47, 2137, 1982.
41. Ishigami, Y. and Machida, H., Vesicles from sucrose fatty acid esters, *J. Am. Oil Chem. Soc.*, 66, 599, 1989.
42. Schenk, P., Ausborn, M., Bendas, F., Nuhn, P., Arndt, D., and Meyer, H. W., The preparation and characterization of lipid vesicles containing esters of sucrose and fatty acids, *J. Microencapsulation*, 6, 95, 1989.
43. Kaler, E. W., Murthy, K., Rodriguez, B. E., and Zasadzinski, J. A., Spontaneous vesicle formation in aqueous mixtures of single-tailed surfactants, *Science*, 245, 1371, 1989.
44. Wallach, D. F. H., Method of Producing High-Aqueous Volume Multilamellar Vesicles, U.S. Patent 4,855,090, 1989.
45. Wallach, D. F. H. and Yiournas, C., Method and Apparatus for Producing Lipid Vesicles, U.S. Patent 4,895,452, 1990.
46. Wallach, D. F. H., Paucilamellar Lipid Vesicles, U.S. Patent 4,911,928, 1990.
47. Wallach, D. F. H., Lipid Vesicles Formed of Surfactants and Steroids, U.S. Patent 4,197,951, 1990.
48. Wallach, D. F. H. and Yiournas, C., Method and Apparatus for Producing Lipid Vesicles, U.S. Patent 5,013,497, 1990.
49. Wallach, D. F. H., Protein Coupling to Lipid Vesicles, U.S. Patent 5,000,960, 1991.
50. Wallach, D. F. H., Removing Oil From Surfaces With Liposomal Cleaner, U.S. Patent 5,019,174, 1991.
51. Wallach, D. F. H., Paucilamellar Lipid Vesicles Using Charge Localized Single Chain, Nonphospholipid Surfactants, U.S. Patent 5,032,457, 1991.
52. Tabibi, E. and Wallach, D. F. H., Theoretical consideration of lipid vesicle formation by Novamix. Interphex-USA, Proc. of 1991 Technical Program, 61, 1991.
53. Tabibi, E., Sakura, J. D., Mathur, R., Wallach, D. F .H., Schulteis, D. T., and Ostrom, J. K., The delivery of agricultural fungicides in paucilamellar amphiphile vesicles, in *Pesticide Formulations and Application Systems*, Vol. 12, ASTM STP 1146, B. N. Devisetty, D. G. Chasin, and P. D. Berger, American Society for Testing and Materials, Philadelphia, 1991, 155.
54. Wallach, D. F. H., Encapsulation of Parasiticides, U.S. Patent 5,019,392, 1991.
55. Wallach, D. F. H., Encapsulation of Ionophore Growth Factors, U.S. Patent 5,019,392, 1991.
56. Wallach, D. F. H., Hybrid Paucilamellar Lipid Vesicles, U.S. Patent 5,234,767, 1992.
57. Wallach, D. F. H., Reinforced Paucilamellar Vesicles, U.S. Patent 5,104,736, 1992.
58. Wallach, D. F. H., Paucilamellar Lipid Vesicles, U.S. Patent 5,147,723, 1992.
59. Wallach, D. F. H., Method of Making Oil Filled Paucilamellar Lipid Vesicles, U.S. Patent 5,160,669, 1992.
60. Wallach, D. F. H., Mathur, R. M., Reziniak, G. J. M., and Tranchant, J. F., Some properties of N-acyl sarcosinate lipid vesicles, *J. Cosmet. Chem.*, 43, 113, 1992.
61. Wallach, D. F. H. and Philippot, J., New type of lipid vesicle: Novasome™, in *Liposome Technology*, 2nd ed., Gregoriadis, G., Ed., CRC Press, Boca Raton, FL, 1992.
62. Wallach, D. F. H., Mathur, R., Chang, A. C., and Tabibi, E., Lipid Vesicles Having an Alkyd as Wall-Forming Material, U.S. Patent 5,164,191, 1992.
63. Vandergriff. K., Wallach, D. F. H., and Winslow, R. K., Encapsulation of hemoglobin in non-phospholipid vesicles, *Biomat. Art. Cells. Immob. Tech.*, in press, 1994.

64. Wallach, D. F. H., *Receptor Molecular Biology*, Marcel Dekker, New York, 1987.
65. Scott, H. L. and Kalaskar, S., Lipid chains and cholesterol in model membranes, *Biochemistry*, 28, 3687, 1989.
66. Mitchell, J., Tiddy, G. J. T., Waring, L., Bostock, T., and McDonald, M. P., Phase behavior of polyoxyethylene surfactants with water, *J. Chem. Soc. Faraday Trans.*, 79, 975, 1983.
67. Regen, S. L., Czech, B., and Singh, A., *J. Am. Chem.Soc.*, 102, 6638, 1980.
68. Hupfer, H., Hupfer, B., Koch, H., and Ringsdorf, H., *Angew. Chem. Int. Ed. Engl.*, 19, 938, 1980.
69. Johnston, D. S., Sanghera, S., Pons, M., and Chapman, D., Phospholipid polymers—synthesis and spectral characteristics, *Biochim. Biophys. Acta*, 602, 57, 1980.
70. O'Brien, D. F. and Whitesides, T. H., *J. Polym. Lett. Ed.*, 19, 95, 1980.
71. Wagner, N., Dose, H., Koch, H., and Ringsdorf, H., Incorporation of ATP synthetase into long-term stable liposomes of polymerizable synthetic sulfolipid, *FEBS Lett.*, 132, 313, 1981.
72. Juliano, R. L., Hsu, M. J., Regen, S. L., and Sing, M., Photopolymerized phospholipid vesicles. Stability and retention of hydrophilic and hydrophobic marker substances, *Biochim. Biophys. Acta*, 770, 109, 1984.
73. Regen, S. L., Polymerized liposomes, in *Liposomes from Biophysics to Therapeutics*, Ostro, M., Ed., Marcel Dekker, New York, 1987, chap 3.
74. Freeman, F. J. and Chapman, D., in *Liposomes as Drug Carriers*, Gregoriadis, G., Ed., John Wiley & Sons, Chichester, 1988, chap 59.
75. Machy, P. and Leserman, L., *Les Liposomes en Biologie Cellulaire et Pharmacologie*, Inserm, Paris, 1987, 18.
76. Johnston, D. S. and Chapman, D., Polymerized liposomes and vesicles, in *Liposome Technology*, Vol. 1, Gregoriadis, G., Ed., CRC Press, Boca Raton, FL, 1984, chap 9.
77. Bonté, F., Hsu, M. J., Papp, A., Wu, K., Regan, S. L., and Juliano, R. L., Interactions of polymerizable phosphatidylcholine vesicles with blood components: relevance to biocompatibility, *Biochim. Biophys. Acta*, 900, 1, 1987.
78. Allen, T.M. and Chonn, A., Large unilamellar liposomes with low uptake into the reticuloendothelial system, *FEBS Lett.*, 223, 42, 1987.
79. Ashwell, G. and Morell, A. G., The role of surface carbohydrates in the hepatic recognition and transport of circulating glycoproteins, *Adv. Enzymol.*, 41, 99, 1974.
80. Machy, P. and Leserman, L., Pilotage de drogues et de molécules biologiques: approches thérapeutiques, in *Les Liposomes en Biologie Cellulaire et Pharmacologie*, Inserm, Paris, 1987, chap 2.
81. Fiume, L., Busi, C., Mattiolo, A., Balboni, P. G., Barbanti-Bronado, G., and Wieland, Th., Hepatocyte targeting of antiviral drugs coupled to galactosyl-terminating glycoproteins, in *Targeting of Drugs*, Gregoriadis, G., Senior, J., and Trouet, A., Eds., Plenum Press, New York, 1982, 1–18.
82. Fiume, L., Mattiolo, A., Balboni, P. G.,Tognon, M., Barbanti-Bronado, G., De Vries, J., and Wieland, Th., Enhanced inhibition of virus DNA synthesis in hepatocytes by trifluorothymidine coupled to asialofetuin, *FEBS Lett.*, 103, 47, 1979.
83. Fiume, L., Mattiolo, A., Busi, C., Balboni, P. G., Barbanti-Bronado, G., De Vries, J., Altman, R., and Wieland, Th., Selective inhibition of Ectromelia virus DNA synthesis in hepatocytes by adenine-9-b-D-arabinofuranoside (ara A) and adenine-9-qb-D-arabinofuranoside 5'-monophosphate (ara-AMP) coupled to asialofetuin, *FEBS Lett.*, 116, 185, 1980.
84. Brooks, D. E. and Seaman, G. V. F., Electroviscous effect in dextran-erythrocyte suspensions, *Nature*, 238, 251, 1972.
85. Allen, T. M. and Papahadjopoulos, D., Sterically stabilized ("Stealth") liposomes: pharamacokinetic and therapeutic advantages, in *Liposomes Technology*, Vol. 3 2nd ed., Gregoriadis, G., Ed., CRC Press, Boca Raton, FL, 1993, chap. 5.
86. Sato, T. and Sunamoto, J., Recent aspects in the use of liposomes in biotechnology and medicine, *Prog. Lipid Res.*, 31, 345, 1992.
87. Yukihiro, N., Naoto, O., Fumiaki, I., Toshiyuki, S., and Shoji, O., Liposomal modification with uronate, which endows liposomes with long circulation in vivo, reduces the uptake of liposomes by J774 cells in vitro, *Life Sci.*, 50, 1773, 1992.
88. Gabizon, A. and Papahadjopoulos, D., Liposome formulations with prolonged circulation time in blood and enhanced uptake by tumors, *Proc. Natl. Acad. Sci. U.S.A.*, 85, 6949, 1988.

89. Papahadjopoulos, D., Allen, T. M., Gabizon, A., Mayhew, E., Matthay, K., Huang, S. K., Lee, K. D., Woodle, M. C., Lasic, D. D., Redemann, C., and Martin, F. J., Sterically stabilized liposomes: improvements in pharmacokinetics and antitumor therapeutic efficacy, *Proc. Natl. Acad. Sci. U.S.A.*, 88, 11460, 1991.

90. Woodle, M. C. and Lasic, D. D., Sterically stabilized liposomes, *Biochim. Biophys. Acta*, 1113, 171, 1992.

91. Kulka, R. G. and Loyter, A., The use of fusion methods for the microinjection of animals cells, in *Current Topics Membranes Transport*, 12, 365, 1979.

92. Yatvin, M. B., Kreutz, W., Horwitz, B. A., and Shinitzky, M., pH-sensitive liposomes: possible clinical implications, *Science*, 210, 1253, 1980.

93. Connor, J., Yatvin, M. B., and Huang, L., pH-sensitive liposomes: acid-induced liposome fusion, *Proc. Natl. Acad. Sci. U.S.A.*, 81, 1715, 1984.

94. Litzinger, D. C. and Huang, L., Phosphatidylethanolamine liposomes: drug delivery, gene transfer and immunodiagnostic applications, *Biochim. Biophys. Acta*, 1113, 201, 1992.

95. Düzgünes, N., Straubinger, R. M., Balwin, P. A., Friend, D. S., and Papahadjopoulos, D., Proton-induced fusion of oleic acid-phosphatidylethanolamine liposomes, *Biochemistry*, 24, 3091 1985.

96. Marsh, M. and Helenius, A., Virus entry into animal cells, *Adv. Virus Res.*, 36, 107, 1989.

97. Hoekstra, D., Membrane fusion of enveloped viruses: especially matter of proteins, *J. Bioenerg. Biomem.*, 22, 121, 1990.

98. White, J., Martin, K., and Helenius, A., Cell fusion by Semliki forest, influenza, and vesicular stomatitis virus, *J. Cell. Biol.*, 89, 674, 1981.

99. Lear, J. D. and De Grado, W. F., Membrane binding and conformational properties of peptides representing the NH2 terminus of influenza HA-2, *J. Biol. Chem.*, 262, 6500, 1987.

100. Clague, M. J., Knutson, J. R., Blumenthal, R., and Herrmann, A., Interaction of influenza hemagglutinin amino-terminal peptide with phospholipid vesicles: a fluorescence study, *Biochemistry*, 30, 5491, 1991.

101. Rafalski, M., Ortiz, A., Rockwell, A., Van Ginkel, L. C., Lear, J. D., De Grado, W. F., and Wilschut, J., Membrane fusion activity of the influenza virus hemagglutinin: interaction of HA2 N-terminal peptides with phospholipid vesicles, *Biochemistry*, 30, 10211, 1991.

102. Murata, M., Sugahara, Y., Takahashi, S., and Ohnishi, S. I., pH-dependent membrane fusion activity of a synthetic twenty amino acid peptide with the same sequence as that of the hydrophobic segment of influenza virus hemagglutinin, *J. Biochem.*, 102, 957, 1987.

103. Wagner, E., Plank, C., Zatloukal, K., Cotten, M., and Birnstiel, M. L., Influenza virus hemagglutinin HA-2 N-terminal fusogenic peptides augment gene transfer by transferrin-polylysine-DNA complexes: toward a synthetic virus-like gene-transfer vehicle, *Proc. Natl. Acad. Sci. U.S.A.*, 89, 7934, 1992.

104. Sechoy, O., Philippot, J. R., and Bienvenüe, A., Targeting of loaded fusogenic vesicles bearing specific antibodies towards leukemic T cells, *Exp. Cell Res.*, 185, 122, 1989.

105. Compagnon, B., Milhaud, P., Bienvenue, A., and Philippot, J. R., Targeting of poly(rI)-poly(rC) by fusogenic (Fprotein) immunoliposomes, *Exp. Cell Res.*, 200, 333, 1992.

106. Earl, R. T., Hunneyball, I. M., Bilelett, E. E., and Mayer, R. J., Evaluation of reconstituted Sendai virus envelopes as intra-articular drug vectors: effects on normal and experimentally arthritic rabbit knee joints, *J. Pharma. Pharmacol.*, 40, 166, 1988.

107. Düzgünes, N. and Shavnin, S. A., Membrane destabilization by N-terminal peptides of viral envelope proteins, *J. Membrane Biol.*, 128, 71, 1992.

108. Burger, K. N. J., Wharton, S. A., Demel, R. A. and Verkleij, A. J., Interaction of influenza virus hemagglutinin with a lipid monolayer. A comparison of the surface activities of intact virions, isolated hemagglutinins, and a synthetic fusion peptide, *Biochemistry*, 1991, 11173, 1991.

109. Martin, I., Defrisequertain, F., Mandieau, V., Nielsen, N. M., Saermark, T., Burny, A., Brasseur, R., Ruysschaert, J. M., and Vandenbranden, M., Fusogenic activity of SIV (Simian Immunodeficiency Virus) peptides located in the gp32 NH2 terminal domain, *Biochem. Biophys. Res. Commun.*, 175, 872, 1991.

110. Rafalski, M., Lear, J. D., and DeGrado, W. F., Phospholipid interactions of synthetic peptides representing the N-terminus of HIV gp41, *Biochemistry*, 29, 7917, 1990.

111. Garcia, L. A. M., Araujo, P. S., and Chaimovich, H., Fusion of small unilamellar vesicles induced by serum albumin fragment of molecular weight 9000, *Biochim. Biophys. Acta*, 772, 231, 1984.

112. Morgan, D. G., Williamson, H., Fuller, S., and Hudson, B., Melittin induces fusion of unilamellar phospholipid vesicles, *Biochim. Biophys. Acta*, 732, 668, 1983.

113. Eytan, G. D., Broza, R., and Shalitin, Y., Gramicidin S and dodecylamine induce leakage and fusion of membranes at micromolar concentrations, *Biochim. Biophys, Acta*, 937, 387, 1988.

114. Parente, R. A., Nir, S., and Szoka, F. C., Jr., pH-dependent fusion of phosphatidylcholine small vesicles, *J. Biol. Chem.*, 263, 4724, 1988.

115. Kono, K., Kimura, S., and Imanishi, Y., pH-dependent interaction of amphiphilic polypeptide poly(Lys-Aib-Leu-Aib) with lipid bilayer membrane, *Biochemistry*, 29, 3631, 1990.

116. Syenaga, M., Lee, S., Park, N.G., Aayagi, H., Kato, T., Umeda, A., and Amako, K., *Biochim. Biophys. Acta*, 981, 143, 1989.

117. Gad, A. E., Cationic polypeptides induced fusion of acidic liposomes, *Biochim. Biophys. Acta*, 728, 377, 1983.

118. Beigel, M., Keren-Zur, M., Laster, Y., and Loyter, A., Poly(aspartic acid)-dependent fusion of liposomes bearing the quaternary ammonium detergent (1,1,3,3-tetramethylbutyl)cresoxy ethoxyethyl-dimethylbenzylammonium hydroxide, *Biochemistry*, 27, 660, 1988.

119. Yoshimura, T., Goto, Y., and Aimoto, S., Fusion of phospholipid vesicles induced by an amphiphilic model peptide: close correlation between fusogenicity and hydrophobicity of the peptide in an a-helix, *Biochemistry*, 31, 6119, 1992.

120. Kato, T., Lee, S., Ono, S., Agawa, Y., Aoyagi, H., Ohno, M., and Nishino, N., Conformational studies of amphipathic a-helical peptides containing an amino acid with a long alkyl chain and their anchoring to lipid bilayer liposomes, *Biochim. Biophys. Acta*, 1063, 191, 1991.

121. Bondeson, J. and Sundler, R., Promotion of acid-induced membrane fusion by basic peptides. Amino acid and phospholipid specificities, *Biochim. Biophys. Acta*, 1026, 186, 1990.

122. Stegman, T., Delfino, J. M., Richards, F. M., and Helenius A., The HA2 subunit of influenza hemagglutinin inserts into target membrane prior to fusion, *J. Biol. Chem*, 266, 18404, 1991.

123. Farber, F. E., Melnick, J. L. and Janet, S. B., Optimal conditions for uptake of exogenous DNA by chinese hamster lung cells deficient in hypoxanthine-guanine phosphoribosyltransferase, *Biochim. Biophys. Acta*, 390, 298, 1975.

124. Ryser, H. J.-P., Wei-Chiang Shen, Conjugation of methrotrexate to poly(L-Lysine) increases drug transport and overcomes drug resistance in cultured cells. *Proc. Natl. Acad. Sci. U.S.A.*, 75, 3867, 1978.

125. Wei-Chiang Shen and Ryser H. J.-P., Poly(L-Lysine) has different membrane transport and drug-carrier properties when complexed with heparin, *Proc. Natl. Acad. Sci. U.S.A.*, 78, 7589, 1981.

126. Wei-Chiang Shen and Ryser, H. J.-P., Conjugation of poly-L-Lysine to albumin and horseradish peroxidase: a novel method of enhancing the cellular uptake of proteins, *Proc. Natl. Acad. Sci. U.S.A.*, 75, 1872, 1978.

127. Lemaitre, M., Bayard, B., and Lebleu, B., Specific antiviral activity of a poly(L-Lysine)-conjugated oligodeoxyribonucleotide sequence complementary to vesicular stomatitis virus N protein mRNA initiation site, *Proc. Natl. Acad. Sci. U.S.A.*, 84, 648, 1987.

128. Milhaud, P. G., Silhol, M. Salehzada,T., and Lebleu, B., Requirement for endocytosis of Poly(rI)-Poly(rC) to generate toxicity on interferon-treated LM cells, *J. Gen. Virol.*, 68, 1125, 1987.

129. Mulligan, R. C., The basic science of gene therapy, *Science*, 260, 926, 1993.

130. Felgner, P. L., Gadek, T. R., Holm, M., Roman, R., Chan, H. W., Wenz, M., Northrop, J. P., Ringold, G. M., and Danielsen, M., Lipofection: a highly efficient, lipid-mediated DNA-transfection procedure, *Proc. Natl. Acad. Sci. U.S.A.*, 84, 7413, 1987.

131. Felgner, P. L. and Holm, M., Cationic liposome-mediated transfection, *Focus*, 11, 21, 1989.

132. Rupert, L. A. M., Hoekstra, D., and Engberts, J. B. F. N., Fusogenic behavior of didodecyldimethylammonium bromide bilayer vesicles, *J. Am. Chem. Soc.*, 107, 2628, 1985.

133. Pinnaduwage, P., Schmitt, L., and Huang, L., Use of quaternary ammonium detergent in liposome mediated DNA transfection of mouse L-cells, *Biochim. Biophys. Acta*, 985, 33, 1989.

134. Gao, X. and Huang, L., A novel cationic liposome reagent for efficient transfection of mammalian cells, *Biochem. Biophys. Res. Commun.*, 179, 280, 1991.

135. Leventis, R. and Silvius, J.R., Interactions of mammalian cells with lipid dispersions containing novel metabolizable cationic amphiphiles, *Biochim. Biophys. Acta*, 1023, 124, 1990.

136. Behr, J.-P., Demeneix, B., Loeffler, J.-P. and Perez-Mutul, J., Efficient gene transfer into mammalian primary endocrine cells with lipopolyamine-coated DNA, *Proc. Natl., Acad. Sci., U.S.A.*, 86, 6982, 1989.

137. Zhu, N., Liggitt, D., Liu Y., and Debs, R., Systemic gene expression after intravenous DNA delivery into adult mice, *Science,* 261, 209, 1993.
138. Loke, S. L., Stein, C. A., Zhang, X. H., Mori, K., Nakanishi, M., Subasinghe, C., Cohen, J. S., and Neckers, L. M., Characterization of oligonucleotide transport into living cells, *Proc. Natl. Acad. Sci. U.S.A.,* 86, 3474, 1989.

Chapter 4

Liposomes at the Industrial Scale

Gérard Redziniak, Pierre Perrier, and Christian Marechal

CONTENTS

I. INTRODUCTION

Since its discovery in 1965 by Bangham and co-workers,[1] the liposome has not ceased to spark interest in many different fields of application. From pharmaceuticals to cosmetics, from diagnostics to food products, many industries have found or are now discovering that this "magic bullet" may improve the cosmetic or therapeutic properties of the encapsulated molecule and also increase the sensitivity or the rapidity of biochemical reactions.

Even if it is not always necessary to encapsulate a molecule "x" in a liposome to make it more "attractive", it is clear that the biocompatibility and bioavailability properties of this drug delivery system make it the carrier of choice for different routes of administration in humans and in animals: intravenous, oral, aerosol, topical (skin, eye) (for a review see Reference 2).

Industrial-scale production and the now reasonable cost of raw materials necessary for its manufacture, as well as the mastery of the industrial processes, have and will continue to open up many opportunities for this microvehicle.

Since 1986, when the first cosmetic products containing liposomes were launched, (Niosomes® from l'Oréal and Capture® from Christian Dior), the growth in the number of patents in different fields of application provides testimony to the economic boom that the liposome will create in years to come.

However, before discussing this economic boom, estimated at several millions of dollars by specialists, it should be pointed out that the pharmaceutical industry must manufacture liposomes on an industrial scale. To do so, the technician specialized in this type of formulation must rely on three key factors — (1) raw materials, (2) processing, and (3) controls — and at the same time, keep informed as to relevant patents in order to avoid "doing or repeating what has already been done."

II. RAW MATERIALS

Raw materials can be divided into 3 major categories:

Amphipathic:
- Phospholipids (sphingolipids...)
- Ionic or Nonionic detergents
- Glycolipids
- Drugs

Lipophilic:
- Solvents (chloroform, dichloromethane, methanol, ethanol...)
- Steroids (cholesterol, sitosterol...)
- Antioxidants (a-tocopherol...)
- Charged lipids (fatty acids, dicetylphosphate, fatty amines...)
- Hydrophobic drugs

Hydrophilic:
- Water
- Buffers (phosphate or citrate salts...)
- Antioxidants (ascorbic acid...)
- Drugs
- Others

Suppliers of these different ingredients are located all over the world, but it can be useful to list suppliers by specific molecules of the lipid bilayer. In particular these lipid derivatives are produced by:

United States:
- Avanti Polar Lipids, Inc.
- Lipitek
- Calbiochem
- Sigma Chemical Co.

Japan:
- Nikko Chemicals Co. Ltd.
- Nippon Fine Chemicals Co. Ltd.
- Showa Sangyo Co.
- Kao Kabushiki Kaisha
- Towa Chemical Industry Co. Ltd.

Europe:
- Lucas Meyer (Germany)
- Lipoid K.G. (Germany)
- Nattermann und Cie, GmbH (Germany)
- Enzymatix/Genzyme Ltd. (U.K.)
- Lipid Products (U.K.)

Cost vs. quality has not evolved much over the years and it is not unusual to find hemisynthetic phospholipids ($\geq$ 98% phosphatidylcholine; $\approx$ 15% palmitic acid, $\approx$ 85% stearic acid) at a cost of about $1000 per kilogram, while synthetic products cost over $10,000 and even $100,000!

Raw materials, as a whole, comply perfectly with the criteria requested by industries because the manufacturers of these raw materials work in close collaboration with the companies concerned and clearly answer to the specifications demanded. In the future, this joint cooperation should contribute to a considerable reduction in the cost of these materials, particularly products of synthetic origin, since despite their quality, they remain inaccessible for most finished consumer products manufactured in large quantities.

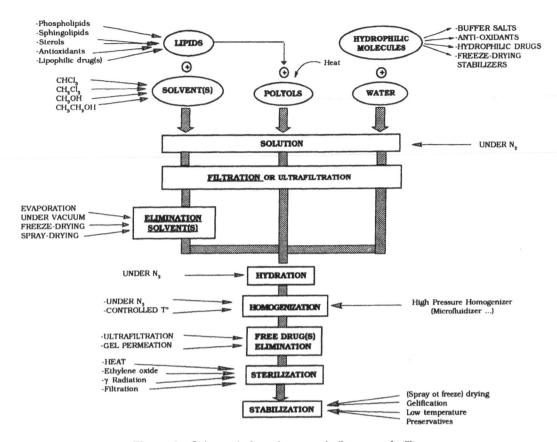

Figure 1 Schematic for a large-scale liposome facility.

III. PROCESSING

Large-scale manufacture of liposomes is discussed in several articles and reviews.[3-10] Whether the application be for the aging of cheese,[10] artificial blood,[7] or "high precision" liposomes in pharmaceuticals,[8] all the processes described, as well as the one developed by Christian Dior, are based on the same procedure (Figure 1):

1. Mixing
2. Separation
3. Hydration
4. Homogenization
5. Sterilization
6. Stabilization

A. MIXING

The mixing of hydrophobic molecules in one or more organic solvents (chloroform, dichloromethane, methanol, ethanol...) requires special equipment with impermeability qualities (especially with respect to gasket efficacy and wear) that must be inspected periodically, as well as flameproof rooms where all operations must be scrupulously controlled for vapor emissions (solvent detectors...).

The mixing of hydrophilic substances certainly does not require heavy investment in equipment, but at this initial stage of raw materials handling it is advisable to work under sterile conditions (sterile mixer, mixing of components under filtered inert gas...).

B. SEPARATION

Separation techniques are used at various stages of the industrial process. Following the initial "remixing" step, it is sometimes necessary to eliminate insoluble residues in the organic or aqueous solvents by

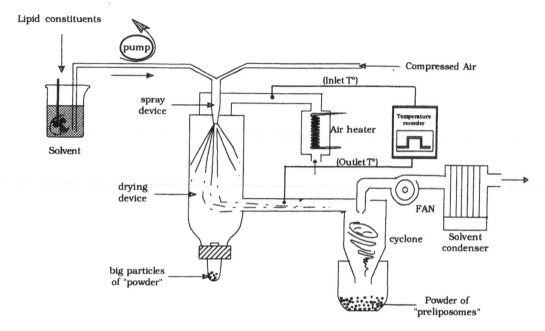

Figure 2 Spray-drying technique: the mixture of lipids (phospholipids, sterols, antioxidants, and amphiphilic or hydrophobic active molecules) to be embedded in the bilayer is dissolved in a suitable solvent (methylene chloride/methanol, etc.). After spray drying, a "preliposome" powder is obtained. The solvent(s) is removed in a activated charcoal condenser.

filtration. High-quality autoclavable filters that withstand organic solvents are commercially available.[11] In addition, ultrafiltration can be used to eliminate pyrogens such as LPS from hydrophobic and hydrophilic solutions.[8]

Separation techniques are especially relevant during the elimination phase of solvent(s), detergents (in the case of dialysis[5]), non-encapsulated drug(s), or active molecule(s) such as toxic substances.

1. Elimination of Solvents

Several techniques exist for the elimination of organic solvents. Solvent evaporation,[12, 13] the most classical, is limited to solvent volumes that produce only a small quantity of "lipid film". Besides, the evaporation time may be long and in association with elevated temperatures may cause denaturation of the products (oxidation ...).

A second technique, and one that requires lots of energy, is lyophilization. This technique has the advantage of preserving the lipidic substances during elimination of the solvent(s), but unfortunately it cannot be employed with mixtures containing antifreeze (methanol, ethanol...).

A final technique, which we were the first to develop, is the spray-drying method[14] (Figure 2). It is based on the atomization of a solvent mixture (dichloromethane-methanol), containing the lipidic and lipophilic constituents of the liposome bilayer, into air or a hot inert gas ($\approx 65°C$). The contact and evaporation time is a few hundredths of a second, and the lipid mixture is recovered at the bottom of the cyclone chamber in the form of a powder we call the "preliposome powder" (Figure 3).

Each particle of this powder may be likened to a "dry liposome" which hydrates almost instantaneously when placed in contact with an aqueous solution, whether or not this solution contains the hydrophilic molecule to be encapsulated.

2. Elimination of Detergents

Dialysis is the most widely used technique to eliminate detergents.[5] The Liposomat, an apparatus sold by Dianorm (Germany), is used to obtain large quantities of liposomes by this technique.

3. Elimination of Non-Encapsulated Drugs

Industrially, techniques such as gel filtration,[16] ultrafiltration,[17] and centrifugation[8] are used to rid liposomes of non-encapsulated molecules.

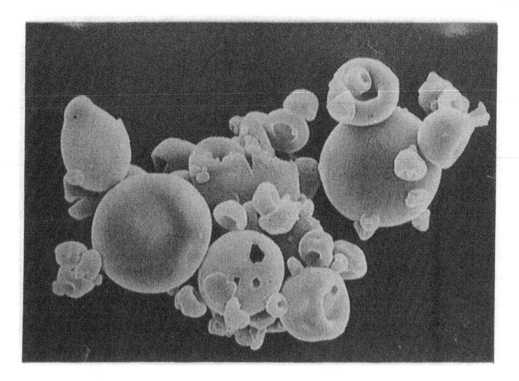

Figure 3 Scanning electron microphotograph of "preliposome" powder. (Magnification × 500.)

C. HYDRATION

The hydration step may seem trivial and yet, depending on whether the lipid mixture obtained is a film or a powder, the quantity of aqueous phase or buffer encapsulated per mole of lipid can vary by 50%.[15] With film, the hydration time should be long — at least 20 hours — and so special attention must be paid to the equipment used (sterilizable, thermostatted). With powders, this is not a problem. Hydration requires little time (1 to 2 hours) and the encapsulation rate of hydrophilic molecules can reach 30 to 40%, as in the case of our "preliposome powder" (unpublished result) prior to the homogenization step.

D. HOMOGENIZATION

Mostly in order to avoid problems with stability over time (such as sedimentation and/or de-encapsulation), the production of liposomes of uniform size is an essential parameter in the industrial manufacture of these microvesicles.

There are many types of machines to precalibrate or calibrate liposomes; the industrial equipment is all based on either agitation (Ultra-Turrax — Janke and Kunkel, Germany) or pressure with or without extrusion (Microfluidizer — Microfluidics Corp., U.S. and Avestin, Canada; Nanojet — Nanojet Engineering, Germany; Schleicher and Schuell extruder[18]).

We have discovered that the properties of high pressure extrusion reduce liposome size[19] (Figure 4).

E. STERILIZATION

Sterilization of liposomal suspensions represents an important and essential step for use in certain applications or treatments in humans. Moreover, sterilization makes it possible to dispense with the use of certain preservative agents incompatible with some therapeutic routes of administration.

The sterilization techniques should be adapted to the type of liposomes developed. The various sterilization techniques include treatments by:

- Heat[20]
- Ethylene oxide[21]
- Filtration on a 0.22 μm membrane
- High pressure homogenization[22]

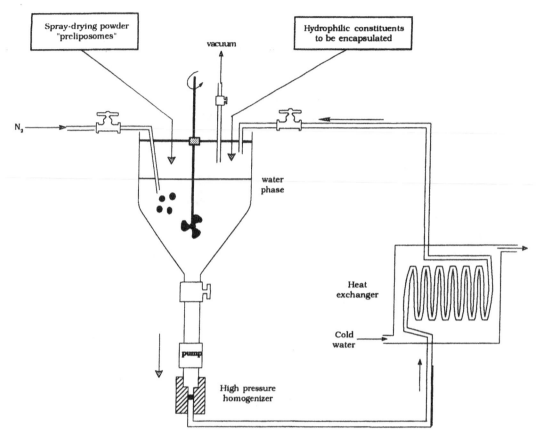

Figure 4 Liposome preparation: "preliposome" powder is hydrated together with the hydrophilic constituents to be encapsulated; the lamellar phase is then passed through a high-pressure homogenizer.[19]

The latter homogenization technique, in combination with Tyndallization [3 × (1 h at 60°C and 1 h at 4°C)] enabled us to obtain excellent results in terms of sterility and stability for liposomal suspensions manufactured from unsaturated phospholipids hydrated in nitrogen-saturated aqueous solutions (unpublished results).

F. STABILIZATION

In parallel with sterilization or addition of preservative agents to stabilize the liposomal suspensions against microbial contamination, the liposomes may also be stabilized physically (with respect to sedimentation, aggregation, melting...) and chemically (oxidation, hydrolysis) by adjusting the quality and quantity of raw materials that make up the bilayer (charge, sterols, antioxidants) or that are present in the aqueous phase to be encapsulated (antioxidants, buffers...).

In the latter phase it is possible to gelatinize with synthetic polymers, for example, carbomers or natural polymers (polysaccharides). Some authors have even proposed the encapsulation of liposomes in cross-linked proteins such as collagen.[23] But alongside these gelatification or embedding techniques, there exists a good stabilization method involving lyophilization in the presence of cryoprotectants.[24-29]

IV. CONTROLS

Once the liposome is manufactured, it must of course be controlled chemically, physically, optically, and biologically.

A. CHEMICAL

The quality and quantity of raw materials that make up the liposome bilayer can be checked by thin layer chromatography or HPLC.[30] We also check the oxidation state (TBARS, iodine value, peroxide value) and for the presence of lysophospholipids which may appear in unbuffered suspensions whose pH deviates from neutrality.[8]

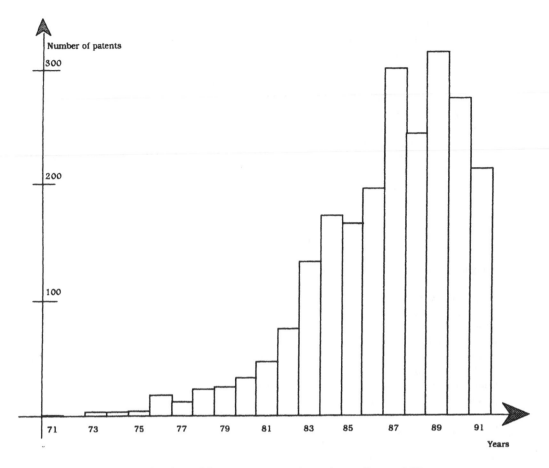

Figure 5 Number of liposome patents issued over the past 21 years.

B. PHYSICAL

Size measurements are performed by dynamic light scattering. Many different instruments exist to do this (Brookhaven Instruments Corp., U.S.A.; Leeds and Northrup Microtrac Lab. Serv., U.S.A.; Malvern Instruments, Ltd., U.K.; Nicomp Particle Sizing Systems, U.S.A.; Coultronics, U.S.A.; Hiac-Royco, U.S.A.).

The encapsulation rate of the molecule(s) is obtained after gel permeation or ultracentrifugation. The zeta potential, which gives an indication of the liposome charge as well as the fluidity of the bilayer, can also be determined and provides an idea as to the shelf life of these vesicles.

C. OPTICAL

Liposomes may be visualized by electron microscopy after negative staining for suspensions or after freeze fracture for complex media such as gels and emulsions. This provides an opportunity to quantify the size and number of lamellae and monitor the physical stability of the liposomes over time.

D. BIOLOGICAL

Microbiological controls for viruses, bacteria, and yeasts should be performed on raw materials entering into the manufacturing process as well as on the manufactured liposomes. For intravenous use, the limulus test[32] is used to detect pyrogens.

V. PATENTS AND CONCLUSIONS

The liposome literature today contains thousands of references. Twenty-one years of liposome patents (1971–1991) represents about 2000 patents filed and published, of which 88% is shared by 5 countries: U.S.A., Japan, Germany, France, and the U.K.

66

The yearly breakdown (Figure 5) shows, if worth recording, the extraordinary explosion of industrial interest in the "magic bullet". From the pharmaceutical industry to the food industry, from the chemical industry to the cosmetics industry, the number of patents filed is contining proof of the present and future commercial impact of the liposome.

In the face of some 180 cosmetics patents, held in large part by France, there are already hundreds of products on the market that claim the presence of this microvehicle.

By the dawn of the 21st century, some 40 pharmaceutical (antifungals, antimitotics) and diagnostic products should see the light. In the latter product category, Schering A.G. Laboratories is preparing to market liposomes charged with iodine contrast products.

Therefore, thanks to the technological contribution of many firms and public organizations, obstacles to the affordable use of liposomes that were encountered a dozen years ago have now disappeared. Moreover, some of these industrial techniques make it possible to envisage the design of more sophisticated liposomal formulations, which present numerous advantages for future developments.

REFERENCES

1. Bangham, A.D., Standish, M.N., and Watkins, J.C., *J. Mol. Biol.,* 13, 138, 1965.
2. Gregoriadis, G., in *Liposomes as Drug Carriers. Recent Trends and Progress,* Gregoriadis, G., Ed., John Wiley & Sons, New York, 1988.
3. Fidles, F.J.T., in *Liposomes: From Physical Sturcture to Therapeutic Applications,* Knight, C.G., Ed., Elsevier, Holland, 1981.
4. Rao, L.S., in *Liposomes Technology,* Vol. 1, Gregoriadis, G., Ed., CRC Press, Florida, 247, 1984.
5. Weder, H.G. and Zumbuehl, O., in *Liposomes Technology,* Vol. 1, Gregoriadis, G., Ed., CRC Press, Florida, 79, 1984.
6. Ostro, M.J., in *Liposomes as Drug Carriers. Recent Trends and Progress,* Gregoriadis, G., John Wiley & Sons, New York, 855, 1988.
7. Vidalnaquet, A., Gossage, J.L., Sullivan, C.P., Haynes, J.W., Gilruth, B.H., Bessinger, R.L., Sehgal, L.R., and Rosen, A.L., *Biomaterials,* 17, 531, 1989.
8. Martin, F.J., *Drug Pharm. Sci.,* 41, 267, 1990.
9. Weiner, A.P., *Biopharm.,* April 16, 1990.
10. Vullemard, J.C., *J. Microencapsulation,* 8, 547, 1991.
11. Klimchak, R.J. and Lenk, R.P., *Biopharm.,* February 18, 1988.
12. Deamer, D. and Bangham, A.D., *Biochim. Biophys. Acta,* 443, 629, 1976.
13. Schieren, H., Rudolph, S., Finkelstein, M., Coleman, P., and Weissmann, G., *Biochim. Biophys. Acta,* 443, 629, 1976.
14. Redziniak, G. and Meybeck, A., Production of Pulverulent Mixtures of Lipidic and Hydrophobic Constituents, U.S. Patent 4,508,703, priority FR 82.02620, 1982.
15. Szoka, F. and Papahadjopoulos, D., *Biochemistry,* 75, 4194, 1978.
16. Juang, C.H., *Biochemistry,* 8, 344, 1969.
17. Magin, R.L. and Chan, H.C., *Biotechnol. Tech.,* 1, 158, 1987.
18. Amselem, S., Gabizon, A., and Barenholz, Y., *J. Liposome Res.,* 1, 287, 1990.
19. Redziniak, G. and Meybeck, A., Method of Homogenizing Dispersions of Hydrated Lipid Lamellar Phases and Suspensions Obtained by the Said Method, US Patent 4,621,023, priority FR 82.17311.
20. Kikuchi, H., Carlsson, A., Yachi, K., and Hirota, S., *Chem. Pharma. Bull.* (Tokyo), 39, 1018, 1991.
21. Ratz, H., Freise, J., Magerstedt, P., Schaper, A., Preugschat, W., and Keyser, D., *J. Microencapsulation,* 6, 485, 1989.
22. Mentrup, E., Butz, P., Stricker, H., and Ludwig, H., *Pharma Industry,* 50, 363, 1988.
23. Pajean, M., Huc., A., and Herbage, D., *Int. J. Pharm.,* 77, 31, 1991.
24. Shulkin, P.M., Seltzer, S.E., Davis, M.A., and Adams, D.F., *J. Microencapsulation,* 1, 73, 1984.
25. Crowe, L.M., Crowe, J.H., Rudolph, A., Womersley, C., and Appel, L., *Arch. Biochem. Biophys.,* 242, 240, 1985.
26. Franzen, G.J., Salemink, P.J.M., and Crommelin, D.J.A., *Int. J. Pharm.,* 33, 27, 1986.
27. Crowe, J.H., Crowe, L.M., Carpenter, J.F., and Aurell Wistrom, C., *Biochem. J.,* 242, 1, 1987.
28. Harrigan, P.R., Madden, T.D., and Cullis, P.R., *Chem. Phys. Lipids,* 52, 139, 1986.
29. Lloyd, A.W., Rutt, K.J., and Olliff, C.J., *J. Pharm. Pharmacol.,* 43, 97, 1991.

30. Becart, J., Chevalier, C., and Biesse, J.P., *J. High Resolution Chromatography,* 13, 126, 1990.
31. Talsma, H. and Crommelin, D.J.A., *Pharm. Tech. Int.,* January 26, 1993.
32. Pearson, F.C., *Pyrogens Endotoxins, LAL Testing and Depyrogenation,* Marcel Dekker, New York, 119, 1985.

PART II
Liposomes as Cell
or
Membrane Models

Chapter 5

Liposomes as Tools for the Reconstitution of Biological Systems

Jean-Louis Rigaud and Bruno Pitard

CONTENTS

I. INTRODUCTION

The complexity of most biological membranes makes it difficult to study its individual components *in situ*. Phospholipid vesicles incorporating purified membrane proteins (proteoliposomes) are therefore a powerful tool in order to elucidate at a molecular level functional as well as structural aspects of membrane associated enzymes. This approach allows the experimental control over parameters that are otherwise inaccessible or highly variable in the natural membrane and has been applied with success to a diverse range of membrane proteins. Although it is difficult to list all information published since the pioneering work that Racker and co-workers[1-3] initiated 20 years ago, a number of reviews are available including specific classes of membrane proteins such as receptors and transport proteins[4-11]. However, despite the extensive use and diverse applications of proteoliposomes, it has to be stressed that the mechanism of their formation is in many ways surprisingly ill-defined. This is partly understandable as most of the researchers were mainly interested in the development of reconstitution methods that "work for their proteins" and not in the physico-chemical parameters involved in the preparation procedures. Consequently the most important lesson to be learned from past reconstitution studies is that no one reconstitution procedure is likely to serve equally well for the reconstitution of all membrane proteins and that the experimental approach has to be as broad as possible. Whatever, in the last few years, important progress in the knowledge of the mechanisms of liposome formation[12] as well as in the understanding of the physical behavior of detergent-lipid systems,[13] was allowed to build on a set of basic principles that limits the number of experimental variables and thus the empirical approach of reconstitution experiments. In this context, in order to analyze some aspects of energy-transducing membrane proteins through the use of well characterized proteoliposomes, we have recently developed in our laboratory a broad and systematic assessement of the reconstitution of different classes of membrane proteins including bacteriorhodopsin[14], Ca^{2+}-ATPase,[15] and H$^+$-ATPase.[16] To this end a new experimental strategy has been developed, providing more insight into the mechanisms that trigger protein insertion into liposomes during the most employed technique, namely detergent-mediated reconstitutions. It is the purpose of this article to review, after a brief introduction of proteoliposome preparation methods, our personal experience in the reconstitution field. The main aim of the present article is to discuss and analyze our contribution to the knowledge of membrane-reconstitution studies and has been written with the hope that it will promote the increase of an integrated approach to this field that has often seemed more like art than science.

II. STRATEGIES FOR MEMBRANE PROTEINS RECONSTITUTION INTO LIPOSOMES

From the thousand publications concerning reinsertion of membrane proteins into liposomes, four main technical strategies can be outlined, mainly derived from the strategies used to prepare pure phospholipid vesicles.[17]

A. ORGANIC SOLVENT-MEDIATED RECONSTITUTIONS

Organic solvents have been widely used to prepare liposomes with large internal aqueous space and high capture efficiency (ethanol injection,[18] ether infusion,[19] reverse-phase evaporation[20]). However the usefulness of such strategies for studies of membrane proteins has been limited because it requires exposure to solvents that are often denaturing to amphiphilic membrane proteins.

One maneuver to reassemble membrane proteins into large vesicles is to extract a protein-lipid complex into apolar solvents and after solvent removal to rehydrate the drying films. This technique has been applied to some membrane proteins,[21] including squid and bovine rhodopsins, cytochrome C-oxydase, reactions centers and acetylcholine receptors extracted into ether or hexane in the presence of soybean phospholipids. Giant vesicles (5–300 µm) containing the protein in an active form have been obtained. However the inhomogeneity in the size distribution, the large proportion of multilamellar structures, and the osmotic fragility of such proteoliposomes precluded detailed functional studies of the reconstituted proteins. The advantage of this procedure would have layed to enable the study of the electrical properties of ion transport proteins with microelectrodes or patch clamp techniques. However in this context other techniques avoiding harmful organic solvents are preferred to prepare giant proteoliposomes.[22-23]

The only suitable method reported in organic solvent-mediated reconstitution of membrane proteins is the reverse phase evaporation method. Rhodopsin[24] and bacteriorhodopsin[25] have been efficiently incorporated into large unilamellar vesicles by this strategy. Vesicles containing membrane proteins are formed by depletion of the organic phase from water-in-oil emulsion of lipid-water-protein in an organic solvent (ether or pentane). Bacteriorhodopsin proteoliposomes have been characterized in detail in our laboratory[25-30] and were shown to meet all criteria to be very useful as model systems (see Table 1):

1. Large internal aqueous volume: the heterogeneous size distribution of the crude preparation (0.15–2 µm) can be reduced by sequential extrusion through 0.1 and 0.2 µm nucleopore filters leading to a homogeneous preparation of large unilamellar proteoliposomes with a mean size of 0.2 µm.[25,30]
2. Homogeneous protein distribution among the liposomes for lipid-protein ratios ranging from 1 to 160 (w/w).[30]
3. Low H$^+$ and counterion permeabilities leading to the generation of light-induced pH gradients as large as 2.5 pH units across the membranes.[26,27]
4. Good unidirectional orientation of the protein in the membrane (80–85% inside-out).

Detailed studies[28,29] indicated that the two orientations were at least partially distributed in different proteoliposomes. These two functional subclasses of liposomes could be separated using Sepacryl S 1000 gel filtration chromatography, and large proteoliposomes with pure inwardly pumping activities were fractionated. It was suggested that the reverse phase evaporation method led to 100% unidirectional orientation of bacteriorhodopsin, the heterogeneity resulting from extrusion through nucleopore filters by opening and reclosing in opposite directions of some of the largest proteoliposomes (unpublished results). Unfortunately, the lack of general procedures to transfer other more hydrophilic membrane proteins in active form into apolar solvents has precluded the general use of this method.

Table 1   **Criteria for ideal reconstituted proteoliposomes**

Criteria	Structural Advantages	Functional Advantages	Techniques
Large size of proteoliposomes	Lipid packing Low lipid-to-protein ratios	Intraliposomal bulk water High solute or ion accumulation	Freeze-fracture electron microscopy Determination of internal volume Gel chromatography
Uniformly sized proeoliposomes		Kinetic studies of transport	Freeze-fracture electron microscopy
Uniform lipid-to-protein ratios	Lipid-protein interactions Protein-protein interactions	Kinetic studies of transport Estimation of efficient internal volume	Density sucrose gradient Freeze-fracture electron microscopy
Asymetric protein orientation	Protein-protein interactions Transmembrane protein orientation	Kinetic studies of transport Stoichiometries (substratstransported species)	Proteolytic digestion One-sided inhibitors Enzymatic activity
Low passive permeability		High solute accumulation Intravesicular pH Transmembrane electrical potentials Countertransports	pH jumps Ionic permeability

B. MECHANICAL MEANS

Putative mechanisms of vesicle formation by these methods have been reviewed elsewhere.[12,31] Multilamellar lipid vesicles form spontaneously when dry phospholipid films swell in excess buffer. Large unilamellar vesicles (LUV) and small unilamellar vesicles (SUV) have higher free energies, and some energy must be dissipated into the system in order to produce them. The most widely used technique to prepare SUV involves sonication of MLVs. They can be also produced by forcing a MLV suspension through a French press.[32] The transformation of SUV's into LUV's, MLV's, or hydrated phospholipid aggrerates upon freeze-thawing[31] or dehydratation-rehydratation[22] cycles can also be envisaged from this kind of energetic considerations.

Sonication of a mixed suspension of lipids and isolated proteins has been widely used in the earlier stages of membrane proteins reconstitution to demonstrate the function of purified proteins.[4] The major disadvantages of the sonication procedure are variability and inactivation of many proteins by long sonication. Another crucial drawback is the resulting small size of the proteoliposomes (20–40 nm), which limits the transport capacity for transport proteins and may also perturb the lipid-protein interactions or the structure of the protein due to the high curvature of such proteoliposomes.

A great improvement was the freeze-thawing procedure originally described by Kasahara and Hinkle[33] for a glucose transporting protein. Rapid freezing of sonicated suspension with the solubilized protein, followed by thawing, resulted in the formation of large proteoliposomes up to 7 μm diameter. A brief sonication broke up the larger permeant structures to make unilamellar functional proteoliposomes with 20 to 200 nm. It has been suggested[31] that water molecules crystallize during the rapid freezing on the charged phospholipid interface, forming two frozen planes separated by the hydrophobic core of the membrane. The bilayer structure is easily fractured, and the exposed hydrophobic cores fuse to form large liposomes during the slow thawing. In this context, high ionic force, Ca^{2+}, or sucrose inhibit the fusion process that, on other hand, requires negatively charged lipids. The freeze-thawing-sonication strategy has been widely used since it is rapid and can be used for proteins that are sensitive to sonication or to detergents. However, very few systematic studies on the orientation of the proteins after reconstitution have been performed, which makes difficult further interpretation and generalization of this procedure. Furthermore, disadvantages are the relatively wide size distribution of the resulting proteoliposomes and more importantly the possibility of unfavorable perturbation of membrane protein structure (e.g., aggregation) due to subtle changes produced by the process. Actually the freeze-thaw process combined with dehydratation-rehydratation cycles is mainly used to produce giant proteoliposomes adapted to electro-physiological techniques.[22,23]

C. DIRECT INCORPORATION INTO PREFORMED LIPOSOMES

In early variations of this approach, spontaneous incorporation of membrane proteins into preformed unilamellar vesicles was shown by Eytan and co-workers to be catalyzed by low cholate or lysolecithin concentrations.[5,34] Membrane proteins have also been incorporated in the absence of added detergents, but, in this case, the liposomes had to be of defined composition, including phosphatidylethanolamine and acidic phospholipids.[35] Importantly, protein incorporation occurred preferentially into liposomes of small diameters (20 nm). From these observations, direct incorporation was envisaged as fusion of the lipid envelope of adhering proteins with liposomes.

The mechanisms of direct incorporation were further analyzed in details by Jain and Zakim[9] using different integral membrane proteins. Such incorporations, inefficient with large liquid crystalline liposomes, required SUV doped with amphipatic contaminants such as cholesterol, short chain lecithins, detergents, lysoderivatives, or fatty acids. Two events appeared to be involved in forming the phospholipid-protein complexes. The first was a rapid insertion of all proteins into a small percentage of total vesicles. The second was slower but continued fusion of proteoliposomes with remaining SUV. The features that promoted direct protein incorporation were described as "defects" that could arise from transbilayer asymmetry, lateral immiscibility, thermal motion, or geometrical constraints.

The main advantage of this strategy is that in all cases where it has been checked, the protein was found unidirectionally oriented in the membrane of the proteoliposomes. Proteins are inserted through the hydrophobic domain of the membrane, always with their more hydrophobic moiety first. Important shortcomings of the method are that proteoliposomes have a wide size distribution and that the protein is heterogeneously distributed among the liposomes. Furthemore the presence of those "impurities" needed for efficient reconstitution could be a problem in the case of transport proteins by increasing the basic permeability of the resulting proteoliposomes.

D. DETERGENT-MEDIATED RECONSTITUTIONS

The most successful and employed strategy to prepare proteoliposomes is that involving the use of detergents since most of membrane proteins are isolated and purified in the presence of detergents. In the standard procedure, membrane proteins are first cosolubilized with phospholipids in that appropriate detergent to form an isotropic solution of lipid-protein-detergent and lipid-detergent mixed micelles. Next the detergent is removed resulting in the spontaneous formation of bilayer vesicles with incorporated protein.

The methods for detergent removal are diverse and based on the physico-chemical properties of detergents: critical micelle concentration (cmc), micellar size related to the aggregation number of detergents in a micelle, and hydrophilic-lipophilic balance (HLB).[36]

Detergents with high cmc, which is defined as the concentration at which the monomers are forming micellar aggregates, are easily removed by dialyzing.[37] These detergents include cholate, deoxycholate, octylglucoside, and CHAPS. Such detergents that form small micelles can also be removed by gel exclusion chromatography in which the proteoliposomes are formed as the mixed micelle solution passes the column and eluted in the void volume before the detergent. Depending upon the size of the micelles, one can use different gel-sized columns ranging from Sephadex G25 to Sephadex G200.[37]

Detergents with low cmc and that consequently form large micelles are not readily removed by gel chromatography and even less by dialysis. Efficient removal of such detergents can be achieved through adsorption on hydrophobic resins (Biobeads SM2[38] or Amberlite XAD[39]). Such detergents include Triton X100 and Octaethylene glycol mono-N-dodecyl ether (C12E8), but it has to be stressed that the technique has been shown to be efficient for all kinds of detergents including those with high cmc. (see Section IV. B)

Another procedure to form proteoliposomes from lipid-protein-detergent micellar solutions consists in diluting the reconstitution mixture.[40] By dilution the detergent concentration is lowered below its cmc and proteoliposomes form spontaneously. The usefulness of the procedure is obviously related to detergent with high cmc's to allow the assay to be performed after dilution. Generally the dilution is followed by centrifugation of the diluted proteoliposomes. Importantly the proteoliposomes are very leaky due to residual detergent that should in any cases be removed by previously described procedures.

From the abundant literature it appears that reconstitution from lipid-protein-detergent mixtures yield proteoliposomes of different sizes and compositions depending upon the nature of the detergent, the peculiar procedure to remove it as well as the nature of the protein and lipid composition. Other factors influencing the final product of reconstitution are the ionic conditions, the size of the liposomes, and the precise conditions of protein solubilization. Thus not surprisingly and as stated in the introduction, each membrane protein responds differently to the various reconstitution procedures, and the approach for a long time has been entirely empirical. This empirical approach appears even more striking when taking into account the ideal criteria imposed to fully optimize the potential use of reconstituted proteoliposomes in membrane protein research. Besides the conditions that preserve the integrity and the activity of the protein under study, the following important points should be considered: the morphology and the size of the proteoliposomes; the homogeneity in size and in protein distribution; the number of proteins units incorporated; the final orientation of the incorporated protein; and the permeability of the proteoliposomes.

All the criteria we imposed to our reconstituted systems are listed in Table 1 together with the techniques we used to control them. At this point, it is important to stress the necessity of such a thorough characterization of the reconstituted systems in addition to the functional criteria. This is the only way to carry out an optimal reconstitution and a key factor to obtain detailed information on membrane protein reconstitution. The lack of such systematic investigations with many of the previously reported reconstitutions has been a limit in our understanding of the mechanisms of lipid-protein association during detergent-mediated reconstitution.

III. LIPID-PROTEIN VESICLES FORMATION DURING DETERGENT REMOVAL

The molecular mechanisms for the formation of proteoliposomes upon detergent removal from a detergent-lipid-protein mixture are only partly known and relie on a model proposed by Lasic for the generation of pure phospholipid vesicles.[12] They comprise three distinct aspects: micellar equilibration, vesiculation, and postvesiculation size transformation. The basic concepts are that as detergent molecules are removed, a series of micelles-micelles interactions is initiated to minimize the unfavorable energy resulting from consequent exposure of lipids and proteins hydrophobic regions to the aqueous medium. This results in large, mixed, disk-like structures whose edges are coated by detergent. When they have grown past a critical radius, a subsequent bending of large micelles to form curved micelles occurs. At a critical micellar size, the amplitude of the bending is sufficient to cause bilayer closure and thus vesicle formation. Ultimately these initially formed vesicles still undergo size transformation process as long as the level of residual detergent remains high. This postvesicular size transformation has been suggested to occur through detergent-promoted intervesicular mass transfer.[41-42] Many results reported in the literature are consistent with this simplest proposed scheme of proteoliposome formation but are not absolutely unambiguous due to inherent difficulty in analyzing the unstable intermediate structures throughout the course of detergent removal. In this context slightly different models were proposed involving the formation of intermediates between micelles and closed bilayers such as large bilayered aggregates[43] or extended rod-like structures.[44]

A key step in reconstitution experiments is the insertion of the protein into the bilayer of the formed vesicles. With regard to protein-lipid association during detergent removal from mixed binary and ternary micelles two mechanisms were proposed by Eytan[5]: (1) the protein simply participates in the membrane formation process that corresponds to the micellar-lamellar phase transformation or (2) liposomes are first formed by partial detergent removal and only after further removal is the protein inserted into the preformed detergent-doped liposomes.

Since formation of lipid-protein complexes from a micellar solution depends on the changes in local composition within the particles formed upon detergent removal, the aqueous solubility of each constituent and the composition of the initial mixed micelles can be key factors in determining the products of reconstitution. On the other hand, Eytan[5], who compared cytochrome oxidase incorporation into proteoliposomes by different reconstitution procedures, suggested that the rate of detergent removal might be crucial in proteoliposome formation. He proposed that upon slow detergent removal, vesicle formation might precede incorporation of protein into the lipid bilayer and that upon rapid detergent removal, the proteins incorporated during the liposome formation which corresponded to the micellar-lamellar transition. In this framework the rate of detergent removal was shown to drastically influence the homogeneity of protein distribution[45] during octylglucoside-mediated reconstitution of rhodopsin: rapid detergent removal caused simultaneous coalescence of lipid-detergent and rhodopdsin-lipid-detergent micelles leading to a homogeneous end product while gradual detergent removal allowed protein-free and protein-containing micelles to assemble into vesicles at different stages, leading to a more heterogeneous population of proteoliposomes. In addition, Helenius and co-workers[46] demonstrated that a critical factor in determining the mechanism of protein reconstitution was the state of aggregation of the proteins when membranes began to form from the solubilized lipids. Other factors involved include vesicle size, lipid composition, and ionic conditions.

Importantly, as pointed out by Eytan, the mechanism by which protein associates with phospholipid seems to be a key factor in determining the final orientation of proteins in the reconstituted proteoliposomes. In this connection, it has to be stressed that in most cases where direct incorporation of protein into preformed liposomes has been described, the proteins were found inserted unidirectionally. On the opposite, a scramble protein orientation is expected when proteoliposomes are formed by a coalescence of mixed micelles containing lipid, protein, and detergent.

Thus from all these considerations, it is obvious that our comprehension of the mechanism of lipid-protein association during detergent-mediated reconstitutions is unclear and makes it difficult to rationalize all the reported information or to establish basic principles that would limit variables that must be controlled.

IV. MECHANISM OF LIPID-PROTEIN ASSOCIATION DURING DETERGENT-MEDIATED RECONSTITUTIONS: A STEP-BY-STEP RECONSTITUTION PROCEDURE

In a practical detergent-mediated reconstitution procedure, experimental monitoring of the mechanism by which proteins are inserted into liposomes is problematic. Indeed the mixed micelles are inherently unstable structures that make their size, shape, and composition analysis very difficult or ambiguous. Furthermore, lipid-protein-detergent dispersions are heterogeneous populations of binary and ternary micelles of different sizes and compositions that may coalesce at different stages of detergent removal. Finally, one cannot reduce the detergent concentration slowly enough to follow the equilibration process and organization of the vesicles and the micelles during detergent removal.

To allow some realistic experimental monitoring of the mechanisms by which proteins may associate to lipids, we developed in our laboratory a new strategy based on the idea that reassociation of lipids and proteins upon selective removal of detergent from the detergent-lipid-protein mixture is the mirror image of the solubilization process.[11,46] Indeed, addition of detergents to preformed liposomes destabilizes them, causing them to open and disintegrate into mixed micelles, following probably the same sequence of intermediate structures. Accordingly our strategy was to first add detergent to preformed liposomes through all the range of detergent concentrations that causes the transformation of lamellar structures into mixed micelles. The protein that has to be reconstituted was then added, and its incorporation could be suitably studied in each accurately adjusted step of the lamellar to micellar transition.

A. SOLUBILIZATION OF LIPOSOMES BY DETERGENTS

Solubilization by different detergents of large unilamellar liposomes prepared by reverse-phase evaporation was analyzed in details. The methods used included turbidity and centrifugation experiments, magnetic resonnance spectroscopy, freeze-facture electron microscopy, and permeability determinations.[14,47-48] The results of these studies were related to the three-stage model describing the interactions of detergents with lipidic bilayers. For a given concentration of liposomes, three stages in the solubilization process are apparent depending on the nature and the concentration of the detergent[13] (Figure 1).

In stage I, an increase in the total detergent concentration increases both the concentrations of monomeric detergent in the aqueous phase and the mole fraction of detergent in the bilayer. The incorporation of detergent into the bilayer phase can be descibed with a well-defined partition coefficient. Although our studies with C12E8[48] and Triton X-100 (unpublished results) have shown an ideal mixing of lipids and detergents throughout all this stage, data with other detergents indicate that a true equilibrium partition of detergents remains valid only at low mole fractions of bilayer-incorporated detergent. It seems to depend upon the nature of the phospholipids, the kind of vesicles used, the nature of the detergent, and probably the method used.[49-50] It is of interest to note that while bilayer solubilization does not occur in this region, the detergents induce structural perturbations as evidenced by a great increase in membrane permeability and in some cases changes in size and morphology of the vesicles. In particular, small sonicated vesicles exhibit a complex behavior in the presence of sub-solubilizing detergent concentrations:

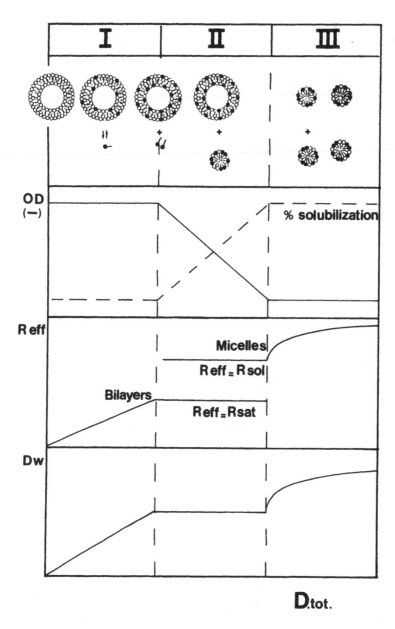

Figure 1 Solubilization of liposomes by increasing concentration of detergents. The regions designated by I, II, and III correspond, respectively, to ranges of detergent concentration in which liposomes only, liposomes and micelles, or micelles only are present. Panels from top to bottom: diagram of the solubilization process (first panel); changes in turbidity and percent of lipid incorporated into mixed micelles (second panel); mole fractions of detergent in bilayers and/or micelles (third panel); and concentration of monomeric detergent (fourth panel).

the detergents act as "wedges" disrupting the vesicle structure and thus releasing it from the lateral strain imposed by the high curvature of these liposomes. These very unstable open vesicles display a natural tendency to aggregate and/or fuse.[47,51]

During stage II solubilization of detergent-saturated liposomes occurs with the appearance of a coexisting population of mixed micelles. We have quantitatively determined[47-48] that the effective detergent to phospholipid ratios in the vesicles and in the mixed micelles are constant throughout most of stage II and correspond to the critical detergent to phospholipid ratios at which stage II begins (Rsat) and finishes (Rsol), respectively. During all the bilayer to micelle transition both of these amphiphilic

Table 2 **Parameters describing the solubilization process of liposomes by different detergents**

Detergent	R_{Sat} (mol/mol)	R_{Sol} (mol/mol)	D_{water} (mM)
Triton X-100	0.64	2.5	0.18
Sodium cholate	0.3	0.9	2.18
C12E8	0.6	2.1	0.2
Octylglucoside	1.3	2.6	17
Octylthioglucoside	2.8	5.3	7
Octylgalactoside	1.1	2.8	10.2

$D_{Total} = D_{Bilayer} + D_{Water}$
$D_{Total} = D_{Water} + R_{eff}.(Lipid)$
$R_{eff} = D_{Bilayer}/(Lipid)$
$R_{eff} = R_{Sat}$ (onset of solubilization)
$R_{eff} = R_{Sol}$ (total solubilization)

structures coexist, and only their relative proportion varies with increasing detergent concentrations. It should be noted that under certain conditions, in particular in octylglucoside-mediated solubilization, a dramatic increase in turbidity occurred in this solubilizing range attributed to a micellar aggregation.[42,47]

Finally in region III, the phospholipids are completely solubilized into mixed micelles. As the total detergent concentration increases, the mole fraction of detergent in the micelles increases with concomitant decrease in the size of these micelles.[47,52]

The results of systematic studies allowed to define quantitatively and accurately the precise composition of a phospholipid-detergent mixture depending upon the initial phospholipid concentration and the nature and the concentration of the detergent. The process of solubilization was described by the equation:

$$(Dt) = Dw + Reff(L)$$

where (L) and (Dt) are the total lipid and detergent concentrations, Dw the monomeric detergent concentration and Reff the effective detergent to phospholipid ratios at which the phase transformations occur (Rsat at the onset of solubilization, Rsol at the total solubilization). Table 2 summarizes the parameters which characterize the solubilization process of liposomes by different detergents.

We have also analyzed in detail the process of liposome reconstitution upon detergent removal from mixed detergent-phospholipid micelles using SM2 Biobeads as the detergent removing agent.[48] This protocol (see Section IV. B) is well suited for investigating the time course of vesiculation and for following the composition and relative proportions of the different aggregates present in lipid-detergent samples. The most interesting finding is that liposome formation during detergent removal takes place in three distinct stages that are the symmetrical opposites of those observed during the solubilization process. Furthermore, the striking similarities of the critical detergent to phospholipid ratios at which phase transformations occur as well as the composition of the lamellar and micellar structures present in solution during the reconstitution and the solubilization process demonstrate that the two processes are quantitatively symmetrical. However, at this point it is important to stress that the present results regarding the quantitative description of the intermediates stages during liposome formation do not preclude any information concerning the structure (size and shape) of the different intermediates formed, which can be very different in solubilization and reconstitution experiments.

B. MECHANISMS OF PROTEIN INCORPORATION INTO LIPOSOMES

The standard reconstitution procedure for studying the incorporation of membrane protein is carried out in three distinct steps:

1. **Stepwise solubilization of preformed liposomes** — liposomes prepared by reverse evaporation were resuspended at the desired concentration and treated with different amounts of detergent through all the range of lamellar to micellar transition as described above.

2. **Protein addition** — after 5–10 minutes incubation, sufficient for detergent equilibration, a solution of detergent-solubilized monomeric protein was added to give the desired final lipid to protein ratio. The detergent-protein-phospholipid mixtures were kept at room temperature for 1 minute to 2 hours under gentle stirring.

3. **Detergent removal** — this was generally performed by successive addition of SM2 Biobeads. After detergent removal the resulting proteoliposomes were analyzed with respect to protein incorporation and orientation by freeze-fracture electron microscopy, sucrose density gradients, and activity measurements.

Such systematic studies were performed using three prototypic energy transducing membrane proteins: bacteriorhodopsin (MW: 27 000; 7 transmembrane α-helices; mainly hydrophobic; for a review see Reference 53), Ca^{2+}ATPase (MW: 110 000; 8–10 transmembrane α-helices and a large hydrophilic part; for a review see Reference 54), H^+ATPase from chloroplasts (MW: 550 000; composed of a F_0 hydrophobic part and a very large F_1 hydrophilic part; both F_0 and F_1 are composed of many subunits; for a review see Reference 55).

The results from our reconstitution studies allowed us to identify, depending upon the nature of the detergent, three mechanisms by which proteins can associate with phospholipid to give functional proteoliposomes. (see Figure 2)

The results from the reconstitutions with sodium cholate demonstrated that proteoliposome formation only arose from detergent depletion of protein-lipid-detergent micelles. No protein incorporation into preformed liposomes, even destabilized by saturating levels of cholate, could be detected. Proteoliposome formation was linearly related to the percentage of lipid solubilization in the starting lipid-detergent suspension. Maximal activities and homogeneous protein distribution were measured in samples reconstituted from isotropic micellar solutions.

In the case of Triton X-100-mediated reconstitutions, although no protein was found associated with phospholipids until the starting material contained mixed micelles, the efficiency of the reconstitution was not related to the amount of mixed micelles initially present in the incubation medium. Optimal reconstitutions were detected in samples reconstituted from Triton X-100-phospholipid-protein suspensions where about 60 to 70% of the phospholipid was still present as Triton X-100 saturated liposomes. A time dependent incorporation (about 1 hour) of the proteins was observed suggesting a transfer of the protein molecules initially present in the micelles to detergent-saturated liposomes still present in the incubation medium. In the case of bacteriorhodopsin and/or H^+ATPase, the transfer was total leading to the formation of an homogeneous proteoliposome preparation with a final lipid to protein ratio similar to the initial ratio.[14,16]

The results from octylglucoside-mediated reconstitutions indicated that reconstitution was optimal when starting from a suspension at a detergent to phospholipid ratio around the critical ratio for the onset of liposome solubilization. Thus proteoliposomes could be formed by direct incorporation of the protein into preformed liposomes provided they were destabilized by saturating levels of octylglucoside. It was observed that such an incorporation occurred very rapidly since after 2 minutes of incubation of solubilized protein with octylglucoside-saturated liposomes the resulting activities were already maximal. Furthermore when two reconstitutions were performed, one such that solubilized protein was added to octylglucoside-treated liposomes and the other such as the appropriate amount of detergent was first added to the protein and next to preformed pure liposomes, identical results were obtained after 2 minutes incubation. Therefore there must have been rapid equilibration, i.e., movement of protein and/or phospholipid between micelles and untreated liposomes on one hand and micelles and detergent-saturated liposomes on the other hand. Finally, density gradient centrifugation and electron-microscopy revealed also that protein incorporation was complete and relatively homogeneous among the liposomes.

Table 3 summarizes the results concerning the mode of optimal lipid-protein association depending upon the nature of the detergent for the different proteins studied. The first clear implication of these results is that the mode of protein-lipid association relies more on the nature of the detergent than on the structure of the protein. However, further experiments to clarify the "inconsistent" observation that the Triton X-100-mediated mechanism reported for bacteriorhodopsin and/or H^+ATPase was inefficient for the Ca^{2+}ATPase, demonstrated that another key factor in determining the final proteoliposome composition was the state of aggregation of the protein in the incubation medium.[15] More precisely it was shown that the different mechanisms of reconstitution of the Ca^{2+}ATPase were predominantly driven by the tendency for self-aggregation of this protein upon slight changes in detergent concentration. Centrifugation experiments and freeze fracture electron microscopy indicated that upon slow detergent removal from Triton X-100 solubilized samples Ca^{2+}ATPase molecules aggregated and were finally found

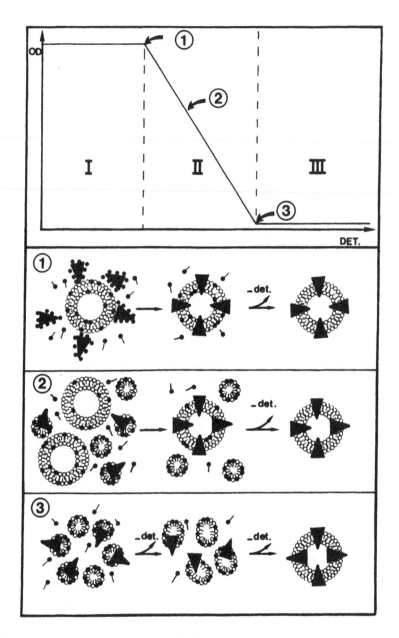

Figure 2 Schematic representation of the different mechanisms of protein-lipid association during step-by-step detergent-mediated reconstitutions. Top panel: steps in the lamellar-to-micellar transitions where optimal reconstitution of membrane proteins are observed depending upon the nature of the detergent. Panel 1: direct incorporation into detergent-saturated liposomes; panel 2: protein transfer from mixed micelles to detergent-saturated liposomes; panel 3: proteoliposome formation from ternary mixed micelles. Note that final protein orientation is unidirectional in mechanism 1 and random in mechanism 3.

segregated in very few liposomes while upon rapid detergent removal compositionally homogeneous proteoliposomes were obtained with high transport activities. From all these results a model was proposed describing the sequence of events that led to incorporation of Ca^{2+}ATPase[15] and that could be generalized to the reconstitution of other membrane proteins.

This model is schematically depicted in Figure 3. Initially, mixed lipid-detergent and lipid-protein-detergent micelles are present. Depending upon the rate of detergent removal two main processes may occur.

Table 3   **Steps in the lamellar-to-micellar transitions where optimal reconstitutions of different membrane proteins are observed[14-15], (unpublished results)**

Protein	Detergent	Step of Liposome Solubilization		
		Onset	Intermediate	Total
Bacteriorhodopsin	Octylglucoside	+		
(Halobacterium Halobium)	Sodium cholate			+
	Triton X-100		+	
H+-ATPase	Octylglucoside	+		
(Spinach Chloroplast)	Sodium cholate			+
	Triton X-100		+	
Ca2+-ATPase	Octylglucoside	+		
(Sarcoplasmic Reticulum)	Sodium cholate			+
	Triton X-100			+
Bacteriorhodopsin				
+	Octylglucoside	+		
H+-ATPase	Sodium cholate			+
(Termophilic Bacterium PS3)	Triton X-100		+	

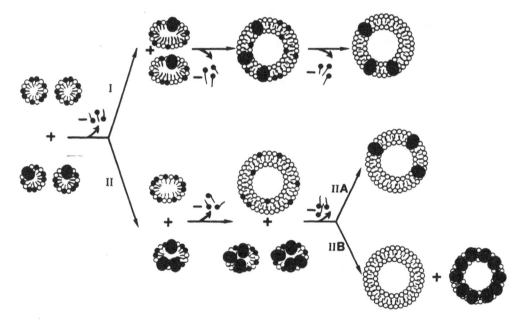

Figure 3   Schematic representation of the mechanisms by which proteins can associate with phospholipids upon detergent removal from initial micellar detergent-lipid-protein solutions (see text for details). (From Levy, D., Gulik, A., Bluzat, A., and Rigaud, J.L., *Biochim. Biophys. Acta,* 1107, 283, 1992. With permission.)

Upon slow detergent removal (process II in Figure 3) it is proposed that as the detergent is initially removed protein-rich structures are first formed. Then as the detergent concentration is further lowered, micelles containing lipids are disrupted leading to detergent-saturated liposomes. At this stage it is proposed that the dispersal of the protein among the liposomes will depend upon the ability of the detergent to incorporate the "aggregated" protein into the preformed liposomes. The results indicate that such a transfer is negligible, partial, or total in the presence of C12E8 (process II B), Triton X-100, or octylglucoside (process II A), respectively.

Upon rapid detergent removal (process I) the simultaneous disruption of both types of micelles results in a mixing of their components. Under these conditions whatever the nature of the detergent, the resulting proteoliposomes are more nearly a reflection of the overall composition of the starting solutions and thus more homogeneous. Finally, in the case of cholate-mediated reconstitution, it is proposed that proteoliposomes only arise from micelle coalescence whatever the rate of detergent removal.

The model we propose for reconstitution from micellar solution is in line with that previously reported by Eytan[5] (see also References 45, 46). However, importantly, our experimental results demonstrate that slow detergent removal leads to the formation of protein-rich structures prior to liposome formation and not to a sequence of events where liposomes could be first formed. Exactly what happens at the beginning of detergent removal is not clear, but could be simply related to one of the following factors: either lipid-protein-detergent micelles are less stable than lipid-detergent micelles so that the detergent is preferentially removed from the ternary micelles resulting in the initial formation of protein-rich structures; or detergent is removed from both types of micelles, but protein molecules tend to aggregate upon destabilization of the ternary micelles. Although there is no definite proof it is tempting to generalize about the formation of such protein-rich structures since most ternary micelles are unstable and most of the membrane proteins tend to aggregate upon small variations in detergent, protein, and/or phospholipid contents.

In regard to the ability of the detergent to incorporate the protein into the preformed liposomes, extension of the model proposed to other membrane proteins is also supported by our findings obtained under similar experimental conditions with bacteriorhodopsin and H$^+$ATPase. As shown in Table 2, these two proteins could be incorporated into octylglucoside-saturated preformed liposomes. It was also demonstrated that spontaneous insertion of crystalline arrays of bacteriorhodopsin[14] or string-like structures of H$^+$ATPase (unpublished results) occurred in octylglucoside-saturated liposomes. Since in many instances octylglucoside proved to be useful in facilitating the direct incorporation of membrane proteins into bilayer membranes, even in the aggregated state,[46,56,57] it is tempting to extend this mechanism to the reconstitution of most membrane proteins that tend to form aggregates or oligomers in the presence of low amounts of detergents. The results from Triton X-100-mediated reconstitutions indicated that membrane proteins could also be incorporated into preformed detergent-doped liposomes but by a mechanism that required the presence of some micellar structures. It could be proposed that the amount of detergent above that was needed for liposome saturation is necessary to allow the dissociation of protein oligomers and/or their further incorporation into liposomes. In the case of Ca^{2+}ATPase, due to its high propensity for self-association, the size and/or the stability of these protein complexes could be so important that they could not be dissociated by Triton X-100 and very few proteins could be directly incorporated. Finally, the results of reconstitution studies with sodium cholate demonstrated that proteoliposome formation only arose from mixed micelles coalescence, probably in relation with the very small size of the micelles in the presence of cholate and the relative high number of micelles needed to form a closed vesicle.

V. A NEW EFFICIENT RECONSTITUTION PROCEDURE

Besides providing information about the way in which proteins may associate to phospholipids during detergent-mediated reconstitutions, we believe that an important benefit of our study is the finding that the reconstitution method described is a method of choice for protein reconstitution and is more suitable than the usual methods using detergents. The almost standard procedure for detergent-mediated reconstitution consists in cosolubilization of membrane proteins and lipids with detergent to form a suspension of mixed micelles, followed by detergent removal. Since phospholipid and protein solubilities are generally small, it is therefore expected that starting from lipid and protein in the same micelle, detergent removal will lead to the formation of proteoliposomes with a lipid to protein ratio similar to that in the initial ternary micelle. The consequence is that, even if a protein can be inserted into a preformed liposome, this step can be missed during detergent removal from micellar solutions since proteoliposomes can be readily and preferentially formed from the ternary phospholipid-detergent-protein micelles initially present. This is clearly what is observed in octyglucoside and Triton X-100-mediated reconstitution (Table 4).[14,15,16] The advantage of the method described in this paper, i.e., incubating a protein in detergent-treated liposome suspensions at each step of the solubilization process, is to allow a rapid and easy determination of the experimental conditions for optimal detergent-mediated reconstitution of any protein. The relative ease and reproducibility of this method should make it a useful assay tool in the optimal reconstitution of other membrane proteins. We already know that the method is useful for reconstitution of bacteriorhodopsin, Ca^{2+}-ATPase and H^+-ATPase. Fairly high activities were observed for the three different proteins, thus proving the efficiency and general validity of the procedure in creating highly functional proteoliposomes (see Figure 4). Interestingly the activities of the reconstituted proteoliposomes are among the best reported up to now for each of these proteins and are relevant to physiological conditions.

A. PROTEIN ORIENTATION

Another important aspect of our study was that the mechanism by which proteins associated with phospholipid to give proteoliposomes was shown to critically affect the final orientation of the protein into the bilayer (Table 4; References 14 to 16). Better asymmetric orientations and consequently higher activities were always observed for the samples reconstituted by incorporation of the protein into preformed liposomes. Optimal unidirectional orientations (85 to 95%) were obtained after direct incorporation of proteins into octylglucoside-saturated liposomes. Although less efficient, good asymmetric orientations (70 to 80%) were also observed in Triton X-100-mediated reconstitutions where the starting detergent-lipid mixtures contained a large amount of preformed detergent-doped liposomes.

Noteworthy these results support the idea that the insertion of a protein into preformed liposomes leads to proteoliposomes with better asymmetric protein insertion than when proteoliposomes are formed by detergent removal from ternary phospholipid-detergent-protein micelles. A possible mechanism explaining unidirectional orientation of a membrane protein when it is incorporated into preformed liposomes is that protein inserts through the hydrophobic domain of the membrane always with its more hydrophobic moiety first. This seems obvious for asymmetric membrane proteins such as Ca^{2+}-ATPase or H^+-ATPase. In the case of bacteriorhodopsin, its carboxylic tail is the most hydrophilic containing at least five COOH groups, while the NH_2-terminal region is more hydrophobic. Thus the latter will be first to insert the membrane leading to almost inside-out orientation of bacteriorhodopsin into the resulting proteoliposomes. Other parameters, such as the respective charges of the proteins and the lipids, or steric factors should be taken into account when dealing with the mechanisms of direct incorporation.

B. DETERGENT REMOVAL

Besides the advantage of determining rapidly the optimal reconstitution conditions in regard to the nature of the detergent and the phospholipid-to-detergent ratio during protein insertion, an added advantage of the procedure described relies on the method of detergent removal. The efficient removal of detergent from the reconstituted proteoliposomes is an absolute necessity: residual detergent may either inhibit enzyme activity and/or increase drastically the passive permeability of the liposomes.[27,47,58] The method used to remove detergent was based on the procedure originally described by Holloway,[38] namely detergent adsorption onto hydrophobic Biobeads SM2. Using these polystyrene beads, it was shown that Triton X-100, Octylglucoside, cholate, and C12E8 could be removed efficiently.[47,48,58,59] Systematic studies demonstrated the importance of the initial detergent concentration, the amount of beads, the nature of detergent, the temperature, and the presence of phospholipids in determining the rates of detergent adsorption onto Biobeads. One of the main findings of our studies was that Biobeads allowed the almost complete removal of detergents whatever the initial experimental conditions. This is particularly important for detergents with low critical micelle concentration such as Triton X-100 and C12E8. Furthermore, we have demonstrated that all the inherent drawbacks of the method reported in the literature (high residual detergent, large loss of lipids, formation of multilamellar structures) could be minimized or overcome.[58]

Another important benefit of our systematic studies was the elaboration of a reconstitution method, which, starting from a micellar solution, provided a suitable and reproducible way of obtaining large, unilamellar, and fairly homogeneous liposomes. The principles of the procedure are as follows: (a) enough detergent must be added for complete lipid solubilization; (b) initial Biobead concentration must be adjusted to promote micellar-to-lamellar transition in not less than about 3 hours; (c) after such transition, Biobead concentration can be increased to accelerate detergent removal; (d) final Biobead concentration must never exceed about 50 times the lipid concentration (w/v).

Among the parameters exerting a predominant influence on the results of the reconstitution (temperature, Biobeads to detergent ratio), we believe that the most important is the low bead-to-detergent ratio used in the first step of reconstitution: this is essential, both to avoid phospholipid losses and to allow slow detergent removal during the micellar-to-lamellar transition, thus giving homogeneous and relatively

Table 4 Comparison of the H⁺ pumping efficiencies of bacteriorhodopsin proteoliposomes reconstituted from Triton X-100, octylglucoside or sodium cholate[14]

Detergent	ΔpH	Total H⁺ (nequiv H⁺)	Initial Rate (nequiv H⁺/min)	Inside Out (Right Side Out) Orientation %		Size (nm)
Triton X-100 (partial solubilization)	2.2	520	430	80–85	(20–15)	160
Triton X-100 (total solubilization)	1.65	345	220	65–70	(35–30)	200
Octylglucoside (onset of solubilization)	2.45	400	530	95	(5)	100
Octylglucoside (total solubilization)	1.9	300	370	70–75	(30–25)	200
Sodium cholate (total solubilization)	2.2	150	360	70–75	(30–25)	70

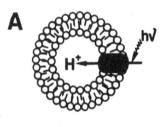

Bacteriorhodopsin

Initial rate: 5300neq H + /min/mg prot
Total H + extent: 4000neq H+/mg prot
△pH: 2.45

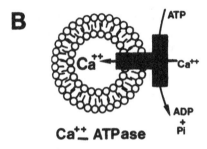

Ca⁺⁺ ATPase

Initial rate: 1.5-2μmol Ca + +/min/mg prot
Total Ca + + extent:1.5-2μmol/mg prot
(Ca + +)int: 15-20 mM

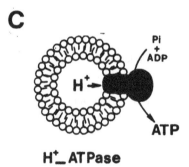

H⁺_ATPase

ATP turnover: 30-50 ATP/F1 F0/sec

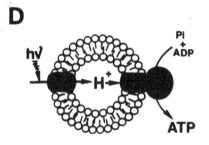

BR + H⁺_ ATPase

Light-induced ATP synthesis
0.5 μmol ATP/mg F1 F0/min

Figure 4 Efficiency of optimal detergent-mediated reconstitutions of different membrane proteins. **(A)** Light-induced H⁺ movements in bacteriorhodopsin proteoliposomes were assayed by the pH meter and ¹⁴C-methylamine distribution.[14] **(B)** Ca²⁺ uptake by Ca²⁺-ATPase proteoliposomes were measured in the absence of oxalate using murexide to monitor external Ca²⁺ concentration.[15] **(C)** ATP synthesis by chloroplast H⁺-ATPase were measured using the pH jump technique.[16] **(D)** Light-induced ATPsynthesis by coreconstituted proteoliposomes containing bacteriorhodopsin and H⁺-ATPase from thermophilic bacterium PS3 were measured under continuous light at 40°C.

large liposome preparations. In this connection, freeze-fracture electron microscopy studies demonstrated that the rate of detergent removal critically affected the final size distribution of the reconstituted liposomes. Small unilamellar liposomes are formed by rapid detergent removal, while larger liposomes are formed by slow removal.[48,58]

Finally, a more technical feature emerges from this study and is specifically related to membrane protein reconstitution into liposomes: it is, that as pointed out in a recent workshop on membrane protein reconstitution,[60] the rate of detergent removal is a key factor for final protein orientation, lipid-protein ratio homogeneity and/or liposomal morphology. Obviously no one strategy is likely to serve equally well for all membrane proteins.[5,15,45] Thus, useful detergent-mediated reconstitution procedures must be capable of varying and controlling the rate of detergent removal: on the basis of the results, this is clearly the case for the batch procedure using Biobeads SM2 as detergent removing agent and this is the case whatever the nature of the detergent.

C. LOW PASSIVE PERMEABILITY

Implication of the almost complete detergent removal is the relatively low permeability of the proteoliposomes reconstituted whatever the detergent used. In particular, ion passive permeability of proteoliposomes containing transport proteins such as bacteriorhodopsin, Ca^{2+}-ATPase, or H^+-ATPase is an important parameter in active transport processes since it partially determines the range of attainable $\Delta\bar{\mu}_{H^+}$ or $\Delta\bar{\mu}_{Ca^{2+}}$ values as well as the rates of proton gradient-driven ADP phosphorylation by H^+-ATPases. Thus in all the reconstituted systems analyzed in the laboratory, passive ion permeability was investigated in details. Proton and counterion fluxes generated by external acid pulses were monitored using the fluorescence of the pH-sensitive probe pyranine trapped inside reconstituted liposomes.[26,27,58,61] Permeabilities for anions and for cations were systematically analyzed as a function of the ionic composition of the medium, temperature, presence of ionophores (valinomycin, protonophores), and lipid-to-protein ratios. The most striking features of our studies is that proteoliposomes obtained by the strategy described in this paper are relatively tight: impermeant to SO_4^{2-}, PO_4^{2-}, Ca^{2+} and slightly permeant to H^+, K^+, Na^+. Observed H^+ permeabilities were in the range of 10^{-5} cm sec^{-1} larger than those for liposomes prepared by reverse-phase evaporation but still much smaller than those of other liposomal preparations.[62]

Such low permeabilities are clearly an improvement in the reconstitution procedures. Among the benefits of such non-leaky reconstituted systems one can cite: (1) accumulation of massive amounts of H^+ by bacteriorhodopsin liposomes[14] or Ca^{2+} by Ca^{2+}-ATPases liposomes[15] or high ATP synthesis rates by ATPase-H^+ liposomes[16]; (2) discrimination between the effects of electrical potentials or proton gradients on the rate of ATP synthesis by H^+-ATPases or on the functioning of bacteriorhodopsin[27]; (3) rates of proton transport-coupled ATP synthesis constant for long periods;[16] and (4) evidence for an obligatory H^+ counter transport during the uptake of Ca^{2+} ions by the sarcoplasmic reticulum Ca^{2+}-ATPases,[61] electrogenicity of the Ca^{++}-H^+ counter transport.[63]

VI. CONCLUSION

Reconstitutions of membrane proteins into liposomes have played and should stay a potentially powerful tool that can be used to identify the mechanism of action of membrane proteins. As a prerequisite, a sound charaterization of the reconstitution is required to lead to optimal reconstituted proteoliposomes. As appears for this short review, it is obvious that an optimistic perspective can be foreseen provided the experimental analysis is systematically performed. In light of the results, these conclusions are worth mentioning:

1. Three mechanisms of association between lipids and proteins were evidenced and mainly related to the nature of the detergent used for reconstitution. Proteins can be inserted into detergent-saturated liposomes (octylglucoside-mediated reconstitutions) or simply participate in the proteoliposome formation during the micellar-to-lamellar transition (cholate-mediated reconstitutions) or can be transferred from mixed micelles to detergent-saturated liposomes (Triton X-100-mediated reconstitutions).

2. Although the products of reconstitution are clearly mainly dependent upon the particular detergent used, the tendency for self-association and/or aggregation of the protein can be a key factor in determining the composition of the final proteoliposomes.

3. The final orientation of the protein in the membrane was critically dependent upon the mode of protein-lipid association. Proteins were found more asymmetrically oriented in the reconstituted liposomes when integrated into preformed liposomes than when integrated during proteoliposome formation by detergent depletion of ternary mixed micelles.

4. The strategy employed to study the mechanism of protein incorporation was revealed to be a powerful reconstitution procedure more suitable than the usual methods using detergents. It allowed a "snapshot" on each step of the lamellar-to-micellar transition leading to rapid and easy determination of the optimal reconstitution conditions of any membrane protein. Furthermore an added advantage of the reconstitution procedure relies on the batch procedure using SM2 Biobeads as detergent removing agent: it provided a reproducible and easy way to achieve unilamellar, relatively large and impermeable vesicles. Therefore proteoliposomes which satisfy most of the criteria for efficient reconstitution could be produced and sustained the highest transport activities reported to date and comparable to those measured in the native membranes.

However, it has to be stressed from the results that the underlying processes that lead to the formation of specific reconstituted form is not well understood. During the last decade, information on the processes of liposome solubilization as well as on the mechanisms of vesicle formation upon detergent-depletion appeared in the literature. It is hoped that the gradual accumulation of such information together with future detailed studies of the kind described in this review will likely result in the formulation of a general set of principles that would serve as a guide in the formulation of a reconstitution experiment.

ACKNOWLEDGMENTS

Part of this research was supported by Grant #BIO2CT-930073 from the EEC (Biotechnology).

REFERENCES

1. Kagawa, Y. and Racker, E., Partial resolution of the enzymes catalyzing oxidative phosphorylation, *J. Biol. Chem.*, 246, 5477, 1971.
2. Racker, E., Reconstitution of calcium pump with phospholipids and a purified Ca++-Adenosine triphosphatase from sarcoplasmic reticulum, *J. Biol. Chem.*, 247, 8198, 1972.
3. Hinkle, P. C., Kim, J.J., and Racker, E., Ion transport and respiratory control in vesicles formed from cytochrome oxidase and phospholipids, *J. Biol. Chem.*, 247, 1338, 1972.
4. Racker, E., Reconstitution of membrane processes, *Methods Enzymol.*, 55, 699, 1979.
5. Eytan, G. D., Use of liposomes for reconstitution of biological functions, *Biochim. Biophys. Acta*, 694, 185, 1982.
6. Casey, R. P., Membrane reconstitution of the energy-conserving enzymes of oxidative phosphorylation, *Biochim. Biophys. Acta*, 768, 319, 1984.
7. Levitzki, A., Reconstitution of membrane receptor systems, *Biochim. Biophys. Acta*, 822, 127, 1985.
8. McNamee, M. G., Jones, O. T., and Fong, T. M., Function of acetylcholine receptors in reconstituted liposomes, in *Channels, Carriers, and Pumps: an introduction to membrane transport*, Stein, W. D., Ed., Academic Press, 1990, chapter 10.
9. Jain, M. K. and Zakim, D., The spontaneous incorporation of proteins into preformed bilayers, *Biochim. Biophys. Acta*, 906, 33, 1987.
10. Cornelius, F., Functional reconstitution of the sodium pump. Kinetics of exchange reactions performed by reconstituted Na/K-ATPase, *Biochim. Biophys. Acta*, 1071, 19, 1991.
11. Silvius, J. R., Solubilization and functional reconstitution of biomembrane components, *Annu. Rev. Biophys. Biomol. Struct.*, 21, 323, 1992.
12. Lasic, D. D., The mechanism of vesicle formation, *Biochem. J.*, 256, 1, 1988.
13. Lichtenberg, D., Robson, R. J., and Dennis, E. A., Solubilization of phospholipids by detergents: structural and kinetic aspects, *Biochim. Biophys. Acta*, 737, 285, 1983.
14. Rigaud, J. L., Paternostre, M. T., and Bluzat, A., Mechanisms of membrane protein insertion into liposomes during reconstitution procedures involving the use of detergents. 2. Incorporation of the light-driven proton pump bacteriorhodopsin, *Biochemistry*, 27, 2677, 1988.
15. Levy, D., Gulik, A., Bluzat, A., and Rigaud, J.L., Reconstitution of the sarcoplasmic reticulum Ca^{2+}-ATPase: mechanisms of membrane protein insertion into liposomes during reconstitution procedures involving the use of detergents, *Biochim. Biophys. Acta*, 1107, 283, 1992.
16. Richard, P., Rigaud, J.L., and Gräber, P., Reconstitution of CF0F1 into liposomes using a new reconstitution procedure, *Eur. J. Biochem.*, 193, 921, 1990.
17. Szoka, F. and Papahadjopoulos, D., Comparative properties and methods of preparation of lipid vesicles (liposomes), *Ann. Rev. Biophys. Bioeng.*, 9, 467, 1980.

18. Batzri, S. and Korn, E. D., Single bilayer liposomes prepared without sonication, *Biochim. Biophys. Acta*, 298, 1015, 1973.

19. Deamer, D. and Bangham, A. D., Large volume liposomes by ether vaporization method, *Biochim. Biophys. Acta*, 443, 629, 1976.

20. Szoka, F. and Papahadjopoulos, D., Procedure for preparation of liposomes with large internal aqueous space and high capture by reverse-phase evaporation, *Proc. Natl. Acad. Sci. U.S.A.*, 75, 4194, 1978.

21. Darszon, A., Vandenberg, C. A., Schönfeld, M., Ellisman, M.H., Spitzer, N. C., and Montal, M., Reassembly of protein-lipid complexes into large bilayer vesicles: perspectives for membrane reconstitution, *Proc. Natl. Acad. Sci.*, 77, 239, 1980.

22. Riquelme, G., Lopez, E., Garcia-Segura, L. M., Ferragut, J.A., and Gonzalez-Ros, J.M., Giant liposomes: a model system in which to obtain patch-clamp recordings of ionic channels, *Biochemistry*, 29, 11215, 1990;

23. Oku, N. and MacDonald, R. C., Differential effects of alkali metal chlorides on formation of giant liposomes by freezing and thawing and dialysis, *Biochemistry*, 22, 855, 1983.

24. Darszon, A., Vandenberg, C. A., Ellisman, M.H., and Montal, M., Incorporation of membrane proteins into large single bilayer vesicles, *J. Cell. Biol.*, 81, 446, 1979.

25. Rigaud, J. L., Bluzat, A., and Büschlen, S., Incorporation of bacteriorhodopsin into large unilamellar liposomes by reverse phase evaporation, *Biochem. Biophys. Res. Commun.*, 111, 373, 1983.

26. Seigneuret, M. and Rigaud, J.L., Analysis of passive and light-driven ion movements in large bacteriorhodopsin liposomes reconstituted by reverse-phase evaporation. 1. factors governing the passive proton permeability of the membrane, *Biochemistry*, 25, 6716, 1986.

27. Seigneuret, M. and Rigaud, J.L., Analysis of passive and light-driven ion movements in large bacteriorhodopsin liposomes reconstituted by reverse-phase evaporation. 2. influence of passive permeability and back-pressure effects upon light-induced proton uptake, *Biochemistry*, 25, 6723, 1986.

28. Seigneuret, M. and Rigaud, J.L., Use of the fluorescent pH probe pyranine to detect heterogeneous directions of proton movement in bacteriorhodopsin reconstituted large liposomes, *FEBS Lett.*, 188, 101, 1985.

29. Seigneuret, M. and Rigaud, J.L., Partial separation of inwardly pumping and outwardly pumping bacteriorhodopsin reconstituted liposomes by gel filtration, *FEBS Lett.*, 228, 79, 1988.

30. Gulik-Krzywicki, T., Seigneuret, M., and Rigaud, J.L., Monomer-oligomer equilibrium of bacteriorhodopsin in reconstituted proteoliposomes, *J. Biol. Chem.*, 262, 15580, 1987.

31. Pick, U., Liposomes with a large trapping capacity prepared by freezing and thawing of sonicated phospholipid mixtures, *Arch. Biochem. Biophys.*, 212, 186, 1981.

32. Barenholzt, Y., Amselem, S., and Lichtenberg, D., A new method for preparation of phospholipid vesicles (liposomes)-French press, *FEBS Lett.*, 99, 210, 1979.

33. Kasahara, M. and Hinkle, P. C., Reconstitution and purification of the D-Glucose transporter from human erythrocytes, *J. Biol. Chem.*, 252, 7384, 1977.

34. Eytan, G., Matheson, M. J., and Racker, E., Incorporation of biologically active proteins into liposomes, *FEBS Lett.*, 57, 121, 1975.

35. Eytan, G., Matheson, M. J., and Racker, E., Incorporation of mitochondrial membrane proteins into liposomes containing acidic phospholipids, *J. Biol. Chem.*, 251, 6831, 1976.

36. Helenius, A. and Simons, K., Solubilization of membranes by detergents, *Biochim. Biophys. Acta*, 415, 29, 1975.

37. Allen, T. M., Romans, A. Y., Kercret, H., and Segrest, J. P., Detergent removal during membrane reconstitution, *Biochim. Biophys. Acta*, 601, 328, 1980.

38. Holloway, P.W., A simple procedure for removal of triton X-100 from protein samples, *Anal. Biochem.*, 53, 304, 1973.

39. Ueno, M., Tanford, C., and Reynolds, J. A., Phospholipid vesicle formation using nonionic detergents with low monomer solubility. Kinetic factors determine vesicle size and permeability, *Biochemistry*, 23, 3070, 1984.

40. Racker, E., Violand, B., O'Neal, S., Alfonzo, M., and Telford, J., Reconstitution, a way of biochemical research; some new approaches to membrane-bound enzymes, *Arch. Biochem. Biophys.*, 198, 470, 1979.

41. Almog, S., Kushnir, T., Nir, S., and Lichtenberg, D., Kinetic and structural aspects of reconstitution of phosphatidylcholine vesicles by dilution of phosphatidylcholine-sodium cholate mixed micelles, *Biochemistry*, 25, 2597, 1986.

42. Almog, S., Litman, B. J., Wimley, W., Cohen, J., Wachtel, E.J., Barenholz, Y., Ben-shaul, A., and Lichtenberg, D., States of aggregation and phase transformations in mixtures of phosphatidylcholine and octyl glucoside, *Biochemistry*, 29, 4582, 1990.

43. Wrigglesworth, J. M., Wooster, M. S., Elsden, J., and Danneel, H. J., Dynamics of proteoliposome formation, *Biochem. J.*, 246, 737, 1987.

44. Vinson, P. K., Talmon, Y., and Walter, A. Vesicle-micelle transition of phosphatidylcholine and octyl glucoside elucidated by cryo-transmission electron microscopy, *Biophys. J.*, 56, 669, 1989.

45. Jackson, M. L. and Litman, B. J., Rhodopsin-phospholipid reconstitution by dialysis removal of octyl glucoside, *Biochemistry*, 21, 5601, 1982.

46. Helenius, A., Sarvas, M., and Simons, K., Asymmetric and symmetric membrane reconstitution by detergent elimination, *Eur. J. Biochem.*, 116, 27, 1981.

47. Paternostre, M. T., Roux, M., and Rigaud, J. L., Mechanisms of membrane protein insertion into liposomes during reconstitution procedures involving the use of detergents. 1. Solubilization of large unilamellar liposomes (prepared by reverse-phase evaporation) by triton X-100, octyl glucoside, and sodium cholate, *Biochemistry*, 27, 2668, 1988.

48. Levy, D., Gulik, A., Seigneuret, M., and Rigaud, J.L., Phospholipid vesicle solubilization and reconstitution by detergent. Symmetrical analysis of the two processes using octaethylene glycol mono-N-dodecyl ether (C12E8), *Biochemistry*, 29, 9480, 1990.

49. Ueno, M., Partition behavior of a nonionic detergent, octyl glucoside, between membrane and water phases, and its effect on membrane permeability, *Biochemistry*, 28, 5631, 1989.

50. Bayerl, T. M., Werner, G. D., and Sackmann, E., Solubilization of DMPC and DPPC vesicles by detergent below their critical micellization concentration: high-sensitivity differential scanning calorimetry, Fourier transform infrared spectroscopy and freeze-fracture electron microscopy reveal two interaction sites of detergents in vesicles, *Biochim. Biophys. Acta*, 984, 214, 1989.

51. Alonso, A., Villena, A., and Goni, F.M., Lysis and reassembly of sonicated lecithin vesicles in the presence of triton X-100, *FEBS Lett.*, 123, 200, 1981.

52. Lichtenberg, D., Robson, R. J., and Dennis, E. A., Solubilization of phospholipids by detergents. Structural and kinetic aspects, *Biochim. Biophys. Acta*, 737, 285, 1983.

53. Oesterhelt, D., Tittor, J., and Bamberg, E., A unifying concept for ion translocation by retinal proteins, *J. Bioenerg. Biomembranes*, 24, 181, 1992.

54. Bigelow, D.J. and Inesi, G., Contributions of chemical derivatization and spectroscopic studies to the characterization of the Ca^{2+} transport ATPase of sarcoplasmic reticulum, *Biochim. Biophys. Acta*, 1113, 323, 1992.

55. Gräber, P., Kinetics of proton-transport coupled ATP synthesis in chloroplasts. *Bioelectrochemistry III*, Milazzo, G. and Blank, M., Eds., Plenum Press, New York, 1990, 277.

56. Bullock, J. O. and Cohen, F. S., Octylglucoside promotes incorporation of channels into neutral planar phospholipid bilayers. Studies with colicin Ia, *Biochim. Biophys. Acta*, 856, 101, 1986.

57. Curman, B., Klareskog, L., and Peterson, P. A., On the mode of incorporation of human transplantation antigens into lipid vesicles, *J. Biol. Chem.*, 255, 7820, 1980.

58. Levy, D., Bluzat, A., Seigneuret, M., and Rigaud, J.L., A systematic study of liposome and proteoliposome reconstitution involving Bio-Bead-mediated triton X-100 removal, *Biochim. Biophys. Acta*, 1025, 179, 1990.

59. Philippot, J., Mutaftschiev, S., and Liautard, J.P., A very mild method allowing the encapsulation of very high amounts of macromolecules into very large (1000nm) unilamellar liposomes, *Biochim. Biophys. Acta*, 734, 137, 1983.

60. Silvius, J. R. and Allen, T. M., Reconstitution of membrane proteins: a selected bibliography from Biophysical Society workshop on membrane protein reconstitution, *Biophys. J.*, 55, 207, 1989.

61. Levy, D., Seigneuret, M., Bluzat, A., and Rigaud, J.L., Evidence for proton countertransport by the sarcoplasmic reticulum Ca^{2+}-ATPase during calcium transport in reconstituted proteoliposomes with "low" ionic permeability, *J. Biol. Chem.*, 265, 19524, 1990.

62. Deamer, D. W., Proton permeation of lipid bilayers, *J. Bioenerg. Biomembranes*, 19, 457, 1987.

63. Yu, X., Carroll, S., Rigaud, J. L., and Inesi, G., H^+ Countertransport and electrogenicity of the sarcoplasmic reticulum Ca^{++} pump in reconstituted proteoliposomes, *Biophys. J.*, 64, 1232, 1993.

Chapter 6

Liposomes as Tools in Immunological Studies

Peter Walden

CONTENTS

I. INTRODUCTION

Aquired immunity is a function of specific lymphocytes which are capable of recognizing antigen and discriminating between self and foreign, and of executing effector functions which lead to elimination of the antigen. These lymphocytes bear receptors, immunoglobulins or T cell receptors which are generated in a huge number of varieties by rearrangement of gene segments from different pools.[1,2] Each cell clone expresses a single specificity. During the ontogeny of the different types of lymphocytes from this randomly created repertoire, specificities are eliminated or suppressed that are directed against the body's own structures, or they are expanded in the course of immune responses to establish self-restricted repertoires and fight infections. Lymphocytes contribute in different ways to maintain the integrity of an individuum and establish immunity. B cells as producers of antibodies and cytotoxic T lymphocytes exert effector functions, but they are themselves dependent upon the activity of regulatory or helper T cells. Effective immune responses are the result of a complicated network of interactions between these cell types and other accessory or antigen bearing cells.

Collaboration of these cells is mediated by cell surface molecules with various specialized functions.[3] Surface immunoglobulins and T cell receptors are responsible for specific recognition of antigen, the coreceptors CD4 and CD8 are involved in the induction of cellular responses,[4] adhesion molecules like the intercellular adhesion molecule (ICAM) and the lymphocyte function-associated antigen (LFA) support cell-cell binding,[5] and other receptors mediate specific cell- or organo-tropisms or act as modulators on signals which are induced through antigen specific receptors. Cytokines as soluble mediators modify the cellular functions which are induced through such membrane proteins.[6]

The result of immune responses is the elimination of the antigen, and in case that antigen is cell-associated the destruction of the cell. There are two effector systems which are employed for this purpose: the complement system which is based on immunoglobulins as the detector molecule[7] and the cytolytic machinery of T cells which use the T cell receptor to identify the antigen.[8] In both cases, upon specific recognition pore-forming molecules are activated that perforate the cell membrane and thereby initiate destruction of that cell.

Investigations of events that takes place at cell membranes and at the interphase between cells are the key to understanding immunity, and they are equally exciting and challenging. One approach which has been applied successfully to this field is to isolate from one cell a component which is believed to be responsible for a particular aspect of the collaboration of two cells, and test its effect on the other cell. Since the components in question are membrane proteins or proteins that act on cell membranes, subcellular fragments and artificial membrane structures like liposomes have emerged as indispensable tools in this field.

A. MHC RESTRICTED ANTIGEN RECOGNITION BY T LYMPHOCYTES

T cells recognize protein antigens as short peptides complexed with cell surface molecules that are encoded in the major histocompatibility complex (MHC).[9,10] This is referred to as MHC restricted antigen recognition. The MHC spans the sets of genes for two different types of highly polymorphic cell surface molecules: MHC class I molecules which are found on almost every cell of the body of vertebrates and MHC class II molecules which are restricted to some few cells of the immune system like B cell, macrophages, dendritic cells, and certain endothelial cells.[11] These molecules control antigen recognition by T lymphocytes. Recently, the elucidation of the crystal structure of MHC class I molecules,[12-15] and the identification of peptides which were isolated from[16-18] and analysis of the binding of peptides to these molecules,[19-24] have cleared our view for the molecular basis of antigen presentation by MHC proteins. In the case of MHC class I restriction of antigen recognition peptides of 8 or 9 amino acids, which are generated inside the presenting cell from protein precursors, are incorporated into a peptide binding groove at the top of the MHC structure during the biosynthesis of this protein. In the case of MHC class II restricted antigen presentation, the peptides are less well defined and originate from internal as well as external proteins; they are processed and picked up by MHC class II molecules in endolysosomes. The peptides are not just freely exchangeable ligands but appear to be an essential part of the structure of the MHC molecule.[25-30] Since T cells recognize a complex of the MHC molecule and antigenic peptide, two binding steps determine the specificity of T cells for antigen: peptide binding to MHC molecules and recognition of the adduct by the T cell receptor.

The coreceptors, CD4 or CD8, which bind to MHC molecules simultaneously with the T cell receptor, are involved in signal induction.[4] MHC class II molecules, CD4, and the T cell receptor for antigen form a signal induction complex for helper cells; class I molecules and CD8 together with the T cell receptor form the corresponding structure for cytolytic T lymphocytes.

B. ADHESION AND ACTIVATION MOLECULES

Like all cells of the blood system, lymphocytes originate from omnipotent stem cells in the bone marrow and go through stages of differentiation before they become immune competent cells. Precursors of T lymphocytes migrate first to the thymus where important steps in their development take place, and they leave that organ as a mature T cell with a functional receptor for antigen. These cells proliferate when they meet antigen, collaborate with other lymphocytes like B or cytolytic T cells, and thereby induce effective immune responses. Many of these events require special lymphatic organs, lymph nodes or the spleen, in order to be efficient. Moreover, activated T cells pass through the wall of blood vessels to execute their function in immune surveillance, or they gather in specific lymphatic organs or areas of the body that are affected by infections. These organo-tropic behaviors and collaboration with other cell types are mediated by cell surface receptors known as adhesion or activation molecules.[3,31,32] Such proteins act as adhesion molecules in as much as they are involved in the induction of signal transduction pathways and in the modulation of cellular responses. MEL-14 is an example for a homing receptor that guides T cells to the high-endothelial venules.[32] B7,[33] CD21,[34] CD40,[35] and CD72[36] are adhesion molecules which are expressed on B cells and mediate B-T cell collaboration by binding to the corresponding counter receptors CD28,[33] CD23,[34] CD40L,[35] and CD5 (Lyt-1)[36] on T cells.

C. EFFECTOR MECHANISMS OF THE IMMUNE SYSTEM

The final goal of the immune responses is the elimination of the antigen. To this end, structures tagged with antibodies are taken up by macrophages and digested. In the case of cell associated antigens like viral or tumor antigens, not only the actually accessible proteins are destroyed but also the cells that brought them to expression. This way, the reproduction of the antigen is stopped. All cellular antigens which are exposed to the immune system and can cause humoral responses will be labeled with specific antibodies and thereby trigger the complement system. This effector system is composed of a set of proteins of which many are serine proteases that activate one another and finally C9, which forms pores in the cell membrane. The cell is destroyed, and its content released and cleared by macrophages.[7]

Antigens which are not expressed at the surface of cells still can give rise to peptides which are displayed as part of the MHC class I structures. Such antigens are the target for cytolytic T lymphocytes which, upon recognition of their cognate antigen MHC class I molecule complex, release the content of cytolytic granules into the contact area between T and target cells. These granules contain, among other reagents, perforin, which in the presence of calcium forms holes in the membrane and initiates cytolysis.[8] Phenotypically, cell death caused by antibody and complement or by cytolytic T lymphocytes are different and are described as necrosis in the first instance and apoptosis in the second case. The reasons for these differences are still a matter of controversy.[37, 38]

II. LIPOSOMES IN STUDIES ON ANTIGEN RECOGNITION BY LYMPHOCYTES

A. ISOLATION AND RECONSTITUTION OF MHC MOLECULES

MHC molecules are heterodimeric type I membrane glycoproteins. The heavy chain of class I molecules has three extracellular domains: the outer two form a peptide binding pocket, the third bears the site for CD8 binding. A single transmembrane portion and a short cytoplasmatic tail anchor the protein to the membrane. This polymorphic α-subunit is noncovalently associated with the non-polymorphic β_2-microglobulin which is not encoded in the MHC. The molecular weight of the two chains is about 45,000 and 12,000 Da, respectively.[11] Crystal structure data were published for the extracellular fragments of four different MHC class I molecules.[12,14,39-41] The two subunits of MHC class II molecules are around 33,000 and 28,000 Da and similar in size. The domain organization of the two polypeptides shows two extracellular, one transmembrane, and a small cytoplasmatic section for each of the subunits. The N-terminal domains of the two chains fold into one binding pocket for peptide that appears to be similar to the corresponding structure in class I molecules.[42] Crystal structure data have recently been published.

Three different approaches were developed to isolate MHC molecules. For use in studies with liposomes MHC molecules are isolated by detergent solubilization from tumor cells, spleens, mitogen induced T or lymphocyte blasts, or membrane preparations from one of the above cell sources.[43-47] Alternative methods to obtain these proteins in a form that is suitable for biochemical studies are either to cleave them off the membrane with proteases[44] or to induce the expression of genes that code for products which lack the transmembrane portion and therefore are secreted.[30,48] We will concern ourselves here only with the first approach. Detergents used for purification of MHC molecules should be mild and nonionic in order to avoid denaturation of the protein.[49] Triton® X-100,* β-octylglycoside, and desoxycholate are the most frequently used detergents. All three are very efficient for extracting MHC molecules from cell membranes. Because of its low critical micelle concentration (CMC), however, Triton® X-100 cannot be removed easily, and since it contains an ether group, peroxides can form which are harmful to all kinds of biological material. Only batches which were tested for biochemical applications should be used. β-octylglycoside, a detergent with excellent biochemical properties, has a high CMC and can be removed easily, but for the same reason high concentrations of that detergent are required. However, β-octylglycoside is one of the best choices for the production of liposomes. Desoxycholate has been used in many cases to avoid the disadvantages of the other two detergents. For a detailed discussion of the properties of different detergents and their use for solubilizing membrane proteins see the reviews by Helenius and Simons[50] and by Hjelmeland and Chrambach.[51]

MHC molecules are purified from the mixtures of solubilized membrane proteins by immune affinity chromatography using allele specific monoclonal antibodies.[52] The conditions required to elute the protein from the immune absorbent are unspecific and quite drastic. Usually, buffers are used with pH values between 10.5 to 12.[53,54] Such conditions can affect the polypeptide structure of the proteins, for instance, cause deamidization of glutamine or asparagine. Proteins should be exposed to such conditions as briefly as possible. A preceding enrichment of the glycoprotein fraction that contains MHC molecules with immobilized lentil lectin and α-methylmannoside as eluant results in a higher purity of the MHC preparation. Still, contamination can be found in these preparations.[44] Actin, which is often copurified with MHC molecules, is particularly confusing in MHC class I isolates because of its similar molecular weight. Pober and colleagues describe possibilities for avoiding such contamination.[55]

Purified MHC molecules are reconstituted into liposomes by mixing them with detergent solubilized lipids and removing the detergent from the mixed micelle preparation by dialysis[56,57] or gelfiltration.[58-63] The low CMC of Triton® X-100 requires the addition of lipophilic raisins to the outer compartment during dialysis in order to aid removal of this detergent.[64,65] Desoxycholate and β-octylglycoside dialyze out easily. The properties of the liposomes, like size and fluidity of the membrane, will be influenced by several factors. The most important parameters are lipid composition, detergent, ratio of lipid and

---

* Triton is a registered trademark of Union Carbide Chemicals and Plastics, Co., Inc.

detergent, temperature at which vesicles are formed, rapidity of removing the detergent, the characteristics of the protein, and ratio of lipid to protein.[56] Obviously, not all of these variables are independent, but they can be controlled to some extent in contrast to contaminants carried over from the protein isolation that can exert strong influence on the result of reconstitution experiments. The size of the liposomes and the mobility of the incorporated proteins are strongly influenced by the choice of lipids. Among the most frequently used lipids, dipalmitoyl-phosphatidylethanolamine will yield rigid membranes, dipalmitoyl-phosphatidylcholine results in phase transition temperatures which are around the incubation temperatures for many bioassays, high cholesterol contents result in intermediate, gel like behaviors of lipid bilayers, and the addition of unsaturated lipids drastically reduces the phase transition temperatures. These rough rules for the relationship of lipid composition and membrane properties are often applied when liposomes are defined as rigid or fluid. Special care should be taken when cholesterol or unsaturated lipids are used. Oxidized cholesterol shows suppressive activity on cell mediated immune reactions.[66-69] Also, oxidation of unsaturated lipids, which can occur easily, has undesired effects on the properties of membranes and proteins. These problems can be controlled by adding trace amounts of an antioxidant like butylhydroxytoluene to the buffers. Further details on the general characteristics of liposomes, their production, and the reconstitution of membrane proteins into lipid bilayers are dicussed in several other chapters of this volume.

MHC molecules incorporated in such membranes can have different orientations.[70] The proportion of outward oriented MHC molecules will increase when the liposomes become smaller. The transmembrane portion can have the same conformation as in natural membranes or form a hairpin. In most published cases, however, such features were not investigated and biological activity was taken as proof for successful reconstitution. For biophysical studies on MHC molecules, planar membranes offer advantages over liposomes in that the proteins are distributed on a planar surface with defined dimensions.[71] For production of such membranes MHC molecules are first incorporated into liposomes which are incubated with extensively cleaned glass cover slips. The glass surface becomes coated with a lipid bilayer as described by Brian and McConnell.[72]

Several techniques are available for coupling proteins or peptides to the surface of liposomes, like crosslinking with glutaraldehyde or ethyldimethylaminopropylcarbodiimide or after thiolation of the protein by introducing a disulfide bridge.[73,74] The latter approach is preferred by many investigators because of the controllability of every step: ethanolamine headgroups of phospholipids are modified with succinimidylpyridyldithiopropionate (SPDP) prior to their incorporation into liposomes. Proteins which are thiolated with Traut's reagent or with SPDP, followed by reduction with dithiothreitol, readily couple to these lipids. The release of thiopyridine, which is an excellent leaving group, can be followed photometrically. These procedures are discussed in detail in Chapter 2 of this volume.

Liposomes have been used successfully as vehicles to introduce membrane components which are derived from one cell into the plasma membrane of a different cell.[75] Membrane fusion was achieved in these experiments with polyethyleneglycol[76-78] or with the help of Sendai virus fusion protein.[79-84] Successful fusion was controlled by using fluorescent membrane components or by encapsulating fluorescent dyes in the liposomes which are released into and can be detected in the cytosol.[75] Among the first applications of this technique in immunology were experiments that were carried out to analyze the mechanism of T cell dependent cytolysis. For this purpose, target cells were created that carried foreign MHC molecules, or viral or tumor antigens incorporated into their membranes, and tested for their capacity to stimulate cytotoxic T lymphocytes.[85] Chapter 7 and Chapter 8 summarize the current knowledge on membrane fusion and give an overview of the application of liposomes in this field.

B. INDUCTION OF MHC CLASS I RESTRICTED T CELL RESPONSES BY LIPOSOMES

Early attempts to investigate the molecular basis of antigen recognition by cytolytic T lymphocytes employed membrane vesicles[86] or reconstituted membrane components derived from potential stimulator or target cells for the induction of responses to allogenic,[87-90] xenogenic,[91] tumor[92,93] or viral antigens[94-103] and also used such vesicles to introduce antigens into the plasma membrane of other cells by membrane fusion. Among other points, probably the most interesting findings in these early studies were that target antigens for cytolytic T cells in fact are associated with the membranes of the stimulator or target cells. The crucial role of MHC class I molecules could be explored in detail and their association demonstrated with particular antigens like the crossreactive, strain independent influenza virus antigen.[98] In subsequent studies purified MHC class I molecules were incorporated into liposomes and tested for their capacity to induce cytotoxic T cell responses.[66,104-116] Induction of cytolytic functions turned out to be dependent upon accessory and helper T cells. Highly purified CD8 cell populations that contained cytolytic precursor cells proliferated to MHC molecules in liposomes, especially when interleukin-1 (IL-1) was provided, but they did not develop cytolytic functions. Such activities required the addition of interleukin-2 (IL-2) to the induction cultures. Experiments of this kind established the requirement for T cell help during the induction of cytolytic immune responses and led to the identification of lymphokines that are essential in this process.[117] Antigen and MHC molecules had to be present on the same vesicle. However, attempts failed to prove physical association of the two proteins in fluorescence bleaching recovery studies.[118,119] Since the ligand for the T cell receptor is a complex between the MHC protein and antigen peptide, we now understand why intact protein antigens cannot be found in association with the MHC molecule.

The current dogma for antigen presentation to cytolytic T lymphocytes is that the antigen has to be produced inside the stimulator or target cell.[120] Based on this concept pH-sensitive liposomes were developed for cytosolic delivery of protein antigens.[121,122] Further details on this topic are discussed in Chapter 13 of this volume. Recently however, also other types of liposomes have been employed successfully in vivo to induce MHC class I restricted responses to protein antigens that were encapsulated in these liposomes.[123,124] This capacity of liposomes to deliver antigen for MHC class I restricted presentation is both of great practical significance and an illustration of theoretical considerations that try to explain collaboration of helper and cytolytic T cells. Several observations point to a three cell cluster as the structural basis for T-T collaboration. Accessory cells that present MHC class II restricted as well as MHC class I restricted antigens bring together the two T-cell types and thereby link regulatory and effector stages of cytolytic T cell responses.[125] Since these presenting cells are only in exceptional cases the source of the antigens, they must be able to pick up extracellular antigen and process it appropriately for MHC class I restricted presentation.

C. LIPOSOMES AS MODEL SYSTEM FOR MHC CLASS II RESTRICTED ANTIGEN PRESENTATION

Similar to the results found for MHC class I, MHC class II molecules inserted into artificial membranes also elicit reactions in T cell.[126-128] Antigen specific, MHC restricted T cells were stimulated by liposomes that carried the corresponding MHC molecule when peptides that represented the antigenic epitope were added to the cultures[129-134] or when intact protein antigens were coupled covalently to lipid components of the liposome.[135-137] The response in the latter situation was MHC dependent at low antigen densities, but it was possible to override this dependency by increasing the antigen density. That these results really represent MHC independent recognition of antigen by T cells could be shown by adding monoclonal antibodies to the cultures. Only the MHC class II dependent T cell response at low antigen density could be inhibited with antibodies with specificity for the MHC molecules. The MHC independent response was not affected by these reagents but could be blocked with antibodies directed against the antigen.[139] Also, noncovalent adsorption of antigen to liposomes to which monoclonal antibodies with specificity for the same antigen had been coupled were found to be effective.[140] Thus, gross structural alterations of the antigen protein due to the coupling chemistry did not determine the results. In contrast to the experiments done with peptides, intact proteins had be to bound covalently to the same membrane that carried the MHC molecules. Free protein was as inefficient as a mixture of MHC and antigen bearing liposomes.[135-137] These findings excluded a possible processing and presentation of antigen and MHC by the test cells added to the cultures. Coupling peptides to liposomes rendered them incapable of stimulating T cells, either because antigenic determinants are easily destroyed in small molecules or because of the vicinity of the membrane that could prevent binding by the T cell receptor.[132,136,141] It was concluded from these experiments that T cells are capable of recognizing unprocessed antigen. This conclusion implies that, at least for the examples tested, aminoacids which are exposed on the surface of the parental protein are also exposed on the surface of the MHC peptide complex, and that the T cell receptor has sufficient affinity for antigen alone without the MHC molecule to cause induction of T cell responses. The possibility of overriding the MHC restriction requirement with a high density of antigen on liposomes might be useful for the development of synthetic vaccines.[142,143]

The lipids in the membranes of the liposomes strongly influenced the efficiency of antigen presentation.[136-144] Liposomes composed of phosphatidylethanolamine as the only type of lipid were the most efficient stimulators for T cell. Liposomes with high cholesterol contents were intermediate and vesicles that contained unsaturated phospholipid were the least efficient stimulators. Since these liposomes differed in size as well as in the phase transition temperature of their membranes, and thereby in the mobility of incorporated proteins, it could not be decided which of the two parameters is more influential. Phospholipids also influence MHC restricted antigen presentation by cells. Phospholipase treatment of antigen presenting cells, for instance, strongly interferes with T cell stimulation.[145] In all those experiments a typical sharp optimum of the T cell response was observed. Why at a high density of antigen the response is strongly reduced is not yet known. Steric hindrance of antigen recognition or even induction of negative signals in the cells could be explanations.

IL-1 can be demonstrated as soluble or as membrane associated activity.[146] It is an extremely pleiotropic cytokine which is important for the induction of helper T cell responses. Membrane IL-1 activity can be isolated and transferred to liposomes.[147] Such IL-1 bearing liposomes show improved capacity to stimulate antigen specific T cells.[141] They even elicit T cell responses to liposomes that carry covalently attached peptides. These observations are in agreement with the demonstration of IL-1 activity on chemically fixed antigen presenting cells, and are another example of the complex interrelationships between receptor affinity and ligand density in multivalent interactions at the interphases between two cells.

In modification of the liposome approach, planar membranes have been used by the groups of McConnell and Watts and associates to study various biophysical aspects of antigen presentation by MHC class II molecules. They combined liposome technology with the methodology of evanescent wavefields as the excitation light source for fluorescence measurements.[148-151] Equilibrium and kinetic constants for peptide binding to MHC molecules were determined using this system.[152] A complicated two-step kinetics of peptide binding was found that indicates peptide induced conformational changes in the MHC molecules.[149] Using fluorescein modified peptides bound to MHC class II molecules, Watts contructed planar membranes with different densities of MHC-molecule peptide adducts and tested their capacity to stimulate T cells.[151] An average distance of 20 nm between two specific complexes was found to be optimal. This can be interpreted as a demonstration of receptor crosslinking being the basis for induction of T cell receptor dependent responses.

D. LIPOSOMES IN STUDIES ON SPECIFIC B CELL RESPONSES

Liposomes have been used to study lipid antigens, and many publications report the use of liposomes in studies on the binding of antibodies to multivalent antigen on liposomes as well as activation of complement by such vesicles. The current knowledge on these topics was reviewed recently by Alving.[153] In contrast to the work done with T cells, very recently liposomes were employed to investigate cellular responses of B cells.[154,155] In experiments reported by Ohyama and colleagues,[155] trinitrophenyl, which is a popular hapten in immunological studies, was coupled to liposomes and calcium changes in B cells that specifically responded to that hapten were analyzed. Varying the length of the spacer that connected the hapten and membrane and changing the fluidity of that membrane, the authors found that the degree of cellular responses increased with the decrease of flexibility of the anchor for the hapten. This is the opposite of that found for antibody binding and complement activation, but very much resembles the results published for T cell activation by liposomes.[156-158] Flexible anchors for antigen allow high affinity binding of antibodies because the ligand can be forced into an optimal orientation. The interaction between cells or cells and liposomes, however, is based on a multivalent binding situation which can make up for reduced affinity at an individual binding site so that even rigid membranes are bound efficiently. Rigid structures are better for receptor crosslinking and therefore good inducers of cellular responses. The relationship between density and distribution of ligands for cellular receptors on membrane surfaces as the basis for multivalent interactions and cellular responses poses many unresolved problems. Liposomes and other artificial membrane structures are valuable instruments in this field.

III. ANALYSIS OF THE FUNCTION OF ADHESION AND ACTIVATION MOLECULES

Questions about the structure and functional size of receptors, identity of the counter receptor, and mechanisms of signal transduction are recurring problems in membrane receptor research and are often approached with the help of vesicles or liposomes. Membrane vesicles were employed by Simister and Rees[159] to determine the functional size of a receptor for the constant portion of immunoglobulins (Fc receptor) that very much resembles MHC class I molecules. This receptor transports antibodies from the lumen of the intestines of neonates into the blood vessels and thereby provides newborn animals with maternal antibodies until their own production is sufficient. The membrane vesicles were prepared from brush border cells and exposed to high doses of radiation. The degree of reduction of the antibody binding capacity is a function of the irradiation dose and size of the target molecule. The authors established the functional size of the Fc receptor with this technique and concluded that it is a tetradimer. Since doses in the megarad range are needed for this kind of experiment cells can not be used, but any kind of subcellular structure like vesicles or liposomes is usable if it carries the protein of interest. This approach is particularly interesting for the investigation of membrane proteins which are active only as part of a lipid bilayer.

Membrane vesicles and liposomes are very well-suited tools for identifying receptor counter-receptor pairs. Activated T helper cells express on their surface a protein dubbed CD40L (the ligand for CD40) which interacts with CD40 on B cells and induces the activation and differentiation of those cells.[35] The first indication for the existence of such receptors came from studies in which vesicles were prepared from activated T cells and were successfully used to stimulate B cells. Further investigations led to the identification of CD40 as the ligand for this activation molecule and finally of CD40L itself. Another recent example for the application of membrane vesicles in this field was published by Bonnefoy's group.[34] CD23, which is a receptor for immunoglobulin E (IgE), was incorporated into fluorescent liposomes and tested for binding to other cells. This way, the existence of a cell surface receptor as an additional ligand for CD23 was demonstrated.[160] Subsequent studies identified CD21 as this counter receptor molecule. In these studies CD23 was isolated by extraction with Triton® X-100 from the insect cell Sf9 that was infected with baculo virus which carried the cDNA for CD23, and the membrane protein was purified by immune affinity chromatography. The detergent was changed to β-octylglycoside during this step. The purified protein was mixed with lipids to which a fluorescent probe was added, and detergent was removed by dialysis. Binding of these liposomes to cells was detected in a fluorescence flow cytometer. Tunicamycin treatment of cells expressing CD23 significantly reduces their capacity to stimulate the partner cell. This effect is also reflected in reduced binding of CD23 bearing fluorescent liposomes to that cell, indicating a significant role of protein glycosylation in this process. CD21 can induce different signals in a cell that expresses this molecule. It seems that CD23/CD21 mediated cell contacts are involved in IgE responses and allergic reactions. A soluble form of CD23 efficiently inhibits IgE secretion and might be applicable to the treatment of allergies. This example of liposome usage for the identification of receptor counter-receptor pairs marks a fruitful direction for future applications of liposomes in immunological research.[161]

IV. LIPOSOMES AS TARGETS FOR IMMUNOLOGICAL EFFECTOR SYSTEMS

A. COMPLEMENT MEDIATED MEMBRANE ATTACK

In addition to recognition of membrane associated antigens or receptors, immunological effector machineries bring into action a second aspect of membrane directed activities of the immune system: the insertion of pore forming proteins in the lipid bilayer, whereby the main function of biomembranes as barriers between intra- and extracellular compartments is cancelled out with fatal effect for the cell.[7,8,38] Formation of defined pores as the cause for cell death was proven in experiments with liposomes. Various molecules, including big proteins like β-galactosidase that were encapsulated in vesicles, were released after complement attack. Early studies have concentrated on the effects of membrane constitution and type and density of antigen at the suface of the lipid bilayer, as well as the class of the antibody and the kind of complement used to trigger lysis. For a discussion of these topics see the review by Tyrrell and colleagues.[158]

Recently, interest has turned to the antibody independent alternative complement pathway.[162] This route is initiated by certain surface components of bacteria and parasites and results in destruction of the microorganisms before antibody responses can become effective. One important intermediate in the complement activation cascade is C3b, which is a potent opsonin and promotes phagocytosis by macrophages. In studies which employed liposomes composed of lipids with phosphatidylethanolamine headgroups, the alternative pathway was activated. However, no effect was seen with phosphatidylcholine or phosphatidylserine.[163] Gangliosides were found to be potent inhibitors of complement activation.[164] These investigations are particularly important in light of the key role complement plays in the defense against pathogenic microorganisms. Complement evasion has evolved as an important immune escape strategy for many of these pathogens.[165] Consequently, understanding the factors that stimulate or inhibit this immunological effector system is crucial for designing strategies against infections by such microorganisms.

B. CELL MEDIATED LYMPHOLYSIS

The pore-forming molecule of the cytolytic machinery of T cells is perforin, which shows extensive sequence similarity with C9, the corresponding component of the complement system.[38] Perforin is concealed in compact cytolytic granules and is released into the interphase between T and target cells after specific engagement of the T cell receptor by MHC-antigen complexes. Polymerization of perforin into the membrane of the target cell is calcium dependent. Liposomes composed of lipids with phosphatidyl-choline headgroups efficiently absorb perforin when calcium is present.[166] Phosphatidylethanolamine or phosphatidylserine containing liposomes have no effect. The pore forming, N-terminal peptides of perforin and C9 insert into liposome membranes without preference for any lipid and independent from calcium.[167] Perforin and C9 form pores of similar size, but nevertheless the sequence of events that lead to cell death is different for the two effector systems, indicating that other factors which are delivered by the T cells are involved.

REFERENCES

1. Adams, J. M., The organisation and expression of immunolglobulin genes, *Immunol. Today*, 1, 10, 1980.
2. Davis, M. M. and Bjorkman, P. J., T-cell antigen receptor genes and T-cell recognition, *Nature*, 334, 385, 1988.
3. Springer, T. A., Adhesion receptors of the immune system, *Nature*, 346, 425, 1990.
4. Janeway, C. A., The T cell receptor as a multicomponent signalling machine: CD4/CD8 coreceptors and CD45 in T cell activation, *Ann. Rev. Immunol.*, 10, 645, 1992.
5. Dustin, M. L. and Springer, T. A., Role of lymphocyte adhesion receptors in transient interactions and cell locomotion, *Ann. Rev. of Immunol.*, 9, 27, 1991.
6. Arai, K., Cytokines: coordinators of immune and inflammatory responses, *Ann. Rev. Biochem.*, 59, 783, 1990.
7. Müller-Eberhard, H.-J., The membrane attack complex of complement, *Ann. Rev. Immunol.*, 4, 503, 1986.
8. Podack, E. R., Hengartner, H., and Lichtenheld, M. G., A central role of perforin in cytolysis?, *Ann. Rev. Immunol.*, 9, 129, 1991.
9. Bjorkman, P. J. and Parham, P., Structure, function and diversity of class I major histocompatibility complex molecules, *Ann. Rev. Biochem.*, 59, 253, 1990.
10. Brodsky, F. M. and Guagliardi, L. E., The cell biology of antigen processing and presentation, *Ann. Rev. Immunol.*, 9, 443, 1991.
11. Klein, J., *Natural History of the Major Histocompatibility Complex*, John Wiley & Sons, New York, 1986.
12. Bjorkman, P. J., Saper, M. A., Samraoui, B., Bennet, W.S., Strominger, J. L., and Wiley, D. C., Structure of the human class I histocompatibility antigen, HLA-A2, *Nature*, 329, 506, 1987.
13. Bjorkman, P. J., Saper, M. A., Samraoui, B., Bennet, W. S., Strominger, J. L., and Wiley, D. C., The foreign antigen binding site and T cell recognition region of class I histocompatibility antigens, *Nature*, 329, 512, 1987.
14. Fremont, D. H., Matsumura, M., Stura, E. A., Peterson, P. A., and Wilson, I. A., Crystal structures of two viral peptides in complex with murine MHC class I H-2Kᵇ, *Science*, 257, 919, 1992.
15. Matsumura, M., Fremont, D. H., Peterson, P. A., and Wilson, I. A., Emerging principles for the recognition of peptide antigens by MHC class I molecules, *Science*, 257, 927, 1992.
16. Falk, K., Rötzschke, O., Stevanovic´, S., Jung, G., and Rammensee, H.-G., Allele-specific motifs revealed by sequencing of self-peptides eluted from MHC molecules, *Nature*, 351, 290, 1991.
17. Van Bleek, G. M. and Nathenson, S. G., Isolation of an endogenously processed immunodominant viral peptide from the class I H-2Kᵇ molecule, *Nature*, 348, 213, 1990.
18. Rudensky, A. Y., Preston-Hurlburt, P., Al-Ramadi, B. K., Rothbard, J., and Janeway, C. A., Jr., Truncation variants of peptides isolated from MHC class II molecules suggest sequence motifs, *Nature*, 356, 429, 1992.
19. Babbitt, B. P., Allen, P. M., Matsueda, G., Haber, E., and Unanue, E. R., Binding of immunogenic peptides to Ia histocompatibility molecules, *Nature*, 317, 359, 1985.
20. Sadegh-Nasseri, S. and McConnell, H. M., A kinetic intermediate in the reaction of an antigenic peptide and I-Eᵏ, *Nature*, 337, 274, 1989.

21. Bouillot, M., Choppin, J., Cornille, F., Martinon, F., Papo, T., Gomard, E., Fournie-Zaluski, M.-C., and Levy, J.-P., Physical association between MHC class I molecules and immunogenic peptides, *Nature*, 339, 473, 1989.

22. Roche, P. A. and Cresswell, P., High-affinity binding of an influenza hemagglutinin-derived peptide to purified HLA-DR, *J. Immunol.*, 144, 1849, 1990.

23. Luescher, I. F., Romero, P., Cerottini, J.-C., and Maryanski, J. L., Specific binding of antigenic peptides to cell-associated MHC class I molecules, *Nature*, 351, 72, 1991.

24. Cerundolo, V., Elliott, T., Elvin, J., Bastin, J., Rammensee, H.-G., and Townsend, A., The binding affinity and dissociation rates of peptide for class I major histocompatibility complex molecules, *Eur. J. Immunol.*, 21, 2069, 1991.

25. Townsend, A., Öhlen, C., Bastin, J., Ljunggren, H.-G., Foster, L., and Kärre, K., Association of class I major histocompatibility heavy and light chains induced by viral peptides, *Nature*, 340, 443, 1989.

26. Lie, W.-R., Myers, N. B., Gorka, J., Rubocki, R. J., Connolly, J. M., and Hansen, T. H., Peptide ligand-induced conformation and surface expression of the L^d class I molecule, *Nature*, 344, 439, 1990.

27. Silver, M. L., Parker, K. C., and Wiley, D. C., Reconstitution by MHC-restricted peptides of HLA-A2 heavy chain with β2-microglobulin, *in vitro*, *Nature*, 350, 619, 1991.

28. Elliott, T., Cerundolo, V., Elvin, J., and Townsend, A., Peptide-induced conformational change of the class I heavy chain, *Nature*, 351, 402, 1991.

29. Sadegh-Nasseri, S. and Germain, R. N., A role for peptide in determining MHC class II structure, *Nature*, 353, 167, 1991.

30. Fahnestock, M. L., Tamir, I., Narhi, L., and Bjorkman, P. J., Thermal stability comparison of purified empty and peptide-filled forms of a class I MHC molecule, *Science*, 258, 1658, 1992.

31. Makgoba, M. W., Bernard, A., and Sanders, M. E., Cell adhesion/signalling: biology and clinical application, *Eur. J. Clin. Invest.*, 22, 443, 1992.

32. Gellatin, W. M., St. John, T. P., Siegelman, M., Reichert, R., Butcher, E. C., and Weissman, I. L., Lymphocyte homing receptors, *Cell*, 44, 673, 1986.

33. Linsley, P. S., Brady, W., Grosmaire, L., Aruffo, A., Damle, N. K., and Ledbetter, J. A., Binding of the B cell activation antigen B7 to CD28 costimulates T cell proliferation and interleukin-2 mRNA accumulation, *J. Exp. Med.*, 173, 721, 1991.

34. Aubry, J.-P., Pochon, S., Graber, P., Jansen, K. U., and Bonnefoy, J.-Y., CD21 is a ligand for CD23 and regulates IgE production, *Nature*, 358, 505, 1992.

35. Noelle, R. J., Roy, M., Shepherd, D. M., Stamenkovic, I., Ledbetter, J. A., and Aruffo, A., A 39-kDa protein on activated helper T cells binds CD40 and transduces the signal for cognate activation of B cells, *Proc. Natl. Acad. Sci. U.S.A.*, 89, 6550, 1992.

36. Van de Velde, H., von Hoegen, I., Luo, W., Parnes, J. R., and Thielemans, K., The B-cell surface protein CD72/Lyb-2 is the ligand for CD5, *Nature*, 351, 661, 1991.

37. Berke, G., Lymphocyte-triggered internal target disintegration, *Immunol. Today*, 12, 396, 1991.

38. Krähenbühl, O. and Tschopp, J., Perforin-induced pore formation, *Immunol. Today*, 12, 399, 1991.

39. Garrett, T. P. J., Saper, M. A., Bjorkman, P. J., Strominger, J. L., and Wiley, D. C., Specific pockets for the side chains of peptide antigens in HLA-Aw68, *Nature*, 342, 692, 1989.

40. Saper, M. A., Bjorkman, P. J., and Wiley, D. C., Refined structure of the human histocompatibility antigen HLA-A2 at 2.6 A resolution, *J. Mol. Biol.*, 219, 277, 1991.

41. Madden, D. R., Gorga, J. C., Strominger, J. L., and Wiley D. C., The structure of HLA-B27 reveals nonamer self-peptides bound in an extended conformation, *Nature*, 353, 321, 1991.

42. Brown, J. H., Jardetzky, T., Saper, M. A., Samraoui, B., Bjorkman, P. J., and Wiley, D. C., A hypothetical model of the foreign antigen binding site of class II histocompatibility molecules, *Nature*, 332, 845, 1988.

43. Crumpton, M. J. and Snary, D., Isolation and structure of human histocompatibility (HLA) antigens, *Contemp. Topics Mol. Immunol.*, 6, 53, 1977.

44. Lopez de Castro, J. A., Purification of human HLA-A and HLA-B class I histocompatibility antigens, *Meth. Enzymol.*, 108, 582, 1984.

45. Bridgen, J., Snary, D., Crumpton, M. J., Barnstable, C., Goodfellow, P., and Bodmer, W. F., Isolation and N-terminal amino acid sequence of membrane-bound human HLA-A and HLA-B antigens, *Nature*, 261, 200, 1976.

46. Freed, J. H., Sears, D. W., Brown, J. L., and Nathenson, S. G., Biochemical purification of detergent solubilized H-2 alloantigens, *Mol. Immunol.*, 16, 9, 1979.

47. Rogers, M. J., Robinson, E. A., and Appella, E., The purification of murine histocompatibility antigens (H-2b) from RBL-5 tumor cells using detergent, *J. Biol. Chem.*, 254, 11126, 1979.

48. Masazumi, M., Saito, Y., Jackson, M. R., Song, E. S., and Peterson, P. A., *In vitro* peptide binding to soluble empty class I major histocompatibility complex molecules isolated from transfected *Drosophila melanogaster* cells, *J. Biol. Chem.*, 267, 23589, 1992.

49. Pincus, J. H., Solubilization of biologically active cell surface antigens, *Contemp. Topics Mol. Immunol.*, 5, 87, 1976.

50. Helenius, A. and Simons, K., Solubilization of membrane by detergents, *Biochim. Biophys. Acta*, 415, 29, 1975.

51. Hjelmeland, L. M. and Chrambach, A., Solubilization of functional membrane proteins, *Meth. Enzymol.*, 104, 305, 1984.

52. Stallcup, K. C., Springer, T. A., and Mescher, M. F., Characterization of an anti-H-2 monoclonal antibody and its use in large-scale antigen purification, *J. Immunol.*, 127, 923, 1981.

53. Turkewitz, A. P., Sullivan, C. P., and Mescher, M. F., Large scale purification of murine I-Ak and I-Ek antigens and characterisation of the purified proteins, *Mol. Immunol.*, 20, 1139, 1983.

54. Herrmann, S. H. and Mescher, M. F., Purification of the H-2Kk molecule of the murine major histocompatibility complex, *J. Biol. Chem.*, 254, 8713, 1979.

55. Pober, J. S., Guild, B. C., Strominger, J. L., and Veatch, W. R., Purification of HLA-A2 antigen, fluorescent labeling of its intracellular region, and demonstration of an interaction between fluorescently labeled HLA-A2 antigen and lymphoblastoid cell cytoskeleton proteins in vitro, *Biochemistry*, 20, 5625, 1981.

56. Szoka, F. and Papahadjopoulos, D., Comparative properties and methods of preparation of lipid vesicles (liposomes), *Ann. Rev. Biophys. Bioeng.*, 9, 467, 1980.

57. Hauser, H., Methods of preparation of lipid vesicles: assessment of their suitability for drug encapsulation, *Trends Pharmacol. Sci.*, 3, 274, 1982.

58. Zumbuehl, O. and Weder, H.-G., Liposomes of controllable size in the range of 40 to 180 nm by defined dialysis of lipid/detergent mixed micelles, *Biochim. Biophys. Acta*, 640, 252, 1980.

59. Mimms, L. T., Zampighi, G., Nozaki, Y., Tanford, C., and Reynolds, J. A., Phospholipid vesicle formation and transmembrane protein incorporation using octyl glucoside, *Biochemistry*, 20, 833, 1981.

60. Engelhard, V. H., Guild, B. C., Helenius, A., Terhorst, C., and Strominger, J. L., Reconstitution of purified detergent-soluble HLA-A and HLA-B antigens into phospholipid vesicles, *Proc. Natl. Acad. Sci. U.S.A.*, 75, 3230, 1978.

61. Klareskog, L., Banck, G., Forsgren, A., and Peterson, P. A., Binding of HLA antigen-containing liposomes to bacteria, *Proc. Natl. Acad. Sci. U.S.A.*, 75, 6197, 1978.

62. Albert, F., Boyer, C., Leserman, L. D., and Schmitt-Verhulst, A.-M., Immunopurification and insertion into liposomes of native and mutant H-2Kb: Quantification by solid phase radioimmunoassay, *Mol. Immunol.*, 20, 655, 1983.

63. Curman, B., Klareskog, L., and Peterson, P. A., On the mode of incorporation of human transplantation antigens into lipid vesicles, *J. Biol. Chem.*, 255, 7820, 1980.

64. Volsky, D. J. and Loyter, A., An efficient method for reassembly of fusogenic Sendai virus envelopes after solubilization of intact virions with Triton X-100, *FEBS Lett.*, 92, 190, 1978.

65. Eriksson, H., Spontaneous association of purified major histocompatibility class I antigens with recipient cells after removal of detergent, *J. Immunol. Meth.*, 115, 133, 1988.

66. Humphries, G. M. K., Evidence for direct control of an *in vitro* plaque-forming cell response by quantitative properties of intact, fluid, haptenated liposomes. A potential model system for antigen presentation by macrophages, *J. Immunol.*, 126, 688, 1981.

67. Yasuda, T., Tadakuma, T., Pierce, C. W., and Kinski, S. C., Primary *in vitro* immunogenicity of liposomal model membranes in mouse spleen cell cultures, *J. Immunol.*, 123, 1535, 1979.

68. Tadakuma, T., Yasuda, T., Kinski, S. C., and Pierce, C. W., The effect of epitope density on the *in vitro* immunogenicity of hapten-sensitized liposomal model membranes, *J. Immunol.*, 124, 2175, 1980.

69. Humphries, G. M. K. and McConnell, H. M., Potent immunosuppression by oxidized cholesterol, *J. Immunol.*, 122, 121, 1979.

70. Cardoza, J. D., Kleinfeld, A. M., Stallcup, K. C., and Mescher, M. F., Hairpin configuration of H-2Kk in liposomes formed by detergent dialysis, *Biochemistry*, 23, 4401, 1984.

71. Ishiguro, T. and Nakanishi, M., Preparation of supported planar membranes containing transmembrane proteins, *J. Biochem.*, 95, 581, 581.

72. Brian, H. H. and McConnell, H. M., Allogeneic stimulation of cytotoxic T cells by supported planar membranes, *Proc. Natl. Acad. Sci. U.S.A.*, 81, 6159, 1984.
73. Carlsson, J., Drevin, H., and Axen, R., Protein thiolation and reversible protein-protein conjugation, *Biochem. J.*, 173, 723, 1978.
74. Barbet, J., Machy, P., and Leserman, L. D., Monoclonal antibody covalently coupled to liposomes: specific targeting to cells, *J. Supramol. Struct. Cell. Biochem.*, 16, 243, 1981.
75. Blumenthal, R., Weinstein, J. N., Shorrow, S. O., and Henkart, P., Liposome-lymphocyte interaction: saturable sites for transfer and intracellular release of liposome content, *Proc. Natl. Acad. Sci. U.S.A.*, 74, 5603, 1977.
76. Reed. M. L. and Herrmann, S. H., Generation of targets for alloreactive CTL using purified H-2K^k in liposomes and polyethylene glycol, *Mol. Immunol.*, 23, 1339, 1986.
77. Szoka, F., Magnusson, K.-E., Wojcieszyn, J., Hou, Y., Derzko, Z., and Jacobson, K., Use of lectins and polyethylene glycol for fusion of glycolipid-containing liposomes with eukaryotic cells, *Proc. Natl. Acad. Sci. U.S.A.*, 78, 1685, 1981.
78. Engelhard, V. H., Powers, G. A., Moore, L. C., Holterman, M. J., and Correa-Freire, M. C., Cytotoxic T lymphocyte recognition of HLA-A/B antigens introduced into EL-4 cells by cell-liposome fusion, *J. Immunol.*, 132, 76, 1984.
79. Prujansky-Jakobovits, A., Volsky, D. J., Loyter, A., and Sharon, N., Alteration of lymphocyte surface properties by insertion of foreign functional components of plasma membrane, *Proc. Natl. Acad. Sci. U.S.A.*, 77, 7247, 1980.
80. Jacobovits, A., Fremkel, A., Sharon, N., and Cohen, I. R., Inserted H-2 gene membrane products mediate immune response phenotype of antigen-presenting cell, *Nature*, 291, 666, 1981.
81. Jacobovits. A., Sharon, N., and Zan-Bar, I., Acquisition of mitogenic responsiveness by nonresponding lymphocytes upon insertion of appropriate membrane components, *J. Exp. Med.*, 156, 1274, 1982.
82. Gething, M.-J., Koszinowski, U., and Waterfield, M., Fusion of Sendai virus with the target cell membrane is required for T cell cytotoxicity, *Nature*, 274, 689, 1978.
83. Morein, B., Barz, D., Koszinowski, U., and Schirrmacher, V., Integration of a virus membrane protein into the lipid bilayer of target cells as a prerequisite for immune cytolysis, *J. Exp. Med.*, 150, 1383, 1979.
84. Volsky, D., Shapiro, I. M., and Klein, G., Transfer of Epstein-Barr virus receptors to receptor-negative cells permits virus penetration and antigen expression, *Proc. Natl. Acad. Sci. U.S.A.*, 77, 5453, 1980.
85. Correa-Freire, M. C., Powers, G. A., and Engelhard, V. H., Introduction of HLA-A/B antigens into lymphoid cell membranes by cell-liposome fusion, *J. Immunol.*, 132, 69, 1984.
86. Lemonnier, F., Mescher, M., Sherman, L., and Burakoff, S., The induction of cytolytic T lymphocytes with purified plasma membrane, *J. Immunol.*, 120, 1114, 1978.
87. Mescher, M., Sherman, L., Lemonnier, F., and Burakoff, S., The induction of secondary cytolytic T lymphocytes by solubilized membrane proteins, *J. Exp. Med.*, 147, 946, 1978.
88. Fast, L. D. and Fan, D. P., Dissociated and reconstituted subcellular alloantigen capable of stimulating mouse cytotoxic T lymphocytes *in vitro*, *J. Immunol.*, 120, 1092, 1978.
89. Sherman, L., Burakoff, S. J., and Mescher, M. F., Induction of allogeneic cytolytic T lymphocytes by partially purified membrane glycoproteins, *Cell. Immunol.*, 51, 141, 1981.
90. Hale, A. H., Evans, D. L., and McGee, M. P., H-2 antigens incorporated into phospholipid vesicles interact specifically with allogenetic cytotoxic T lymphocytes, *Cell. Immunol.*, 63, 42, 1981.
91. Raphael, L. and Tom, B. H., *In vitro* induction of primary and secondary xenoimmune responses by liposomes containing human colon tumor antigens, *Cell. Immunol.*, 71, 224, 1982.
92. Alaba, O. and Law, L. W., Secondary induction of cytotoxic T lymphocytes with solubilized syngeneic tumor cell plasma membranes, *J. Exp. Med.*, 148, 1435, 1978.
93. Acuto, O., Pugliese, O., Müller, M., and Tosi, R., Preparation of liposomes incorporating membrane components from human lymphoid cells, *Tissue Antigens*, 14, 385, 1979.
94. Loh, D., Ross, A. H., Hale, A. H., Baltimore, D., and Eisen, H. N., Synthetic phospholipid vesicles containing a purified viral antigen and cell membrane proteins stimulate the development of cytotoxic T lymphocytes, *J. Exp. Med.*, 150, 1067, 1979.
95. Finberg, R., Mescher, M., and Burakoff, S. J., The induction of virus-specific cytotoxic T lymphocytes with solubilized viral and membrane proteins, *J. Exp. Med.*, 148, 1620, 1978.
96. Hale, A. H., Lyles, D. S., and Fan, D. P., Elicitation of anti-Sendai virus cytotoxic T lymphocytes by viral and H-2 antigens incorporated into the same lipid bilayer by membrane fusion and by reconstitution into liposomes, *J. Immunol.*, 124, 724, 1980.

97. Koszinowski, U. H. and Gething, M.-J., Generation of virus-specific cytotoxic T cells *in vitro*. II. Induction requirements with functionally inactivated virus preparations, *Eur. J. Immunol.*, 10, 30, 1980.

98. Koszinowski, U. H., Allen, H., Gething, M.-J., Waterfield, M. D., and Klenk, H.-D., Recognition of viral glycoproteins by influenza A-specific cross-reactive cytotoxic T lymphocytes, *J. Exp. Med.*, 151, 945, 1980.

99. Lawman, M. J. P., Naylor, P. T., Huang, L., Courtney, R. J., and Rouse, B. T., Cell-mediated immunity to Herpes simplex virus: induction of cytotoxic T lymphocyte responses by viral antigens incorporated into liposomes, *J. Immunol.*, 126, 304, 1981.

100. Hackett, C. J., Taylor, P. M., and Askonas, B. A., Stimulation of cytotoxic T cells by liposomes containing influenza virus or its components, *Immunology*, 49, 255, 1983.

101. Duprez, V., Mescher, M. F., and Burakoff, S. J., Stimulation of secondary anti-MSV cytolytic T lymphocytes with MBL-2 reconstituted membranes, *J. Immunol.*, 130, 493, 1983.

102. Mercier, D. O., Perrin, P., Joffret, M. L., Sureau, P., and Thibodeau, L., The association of the Rabies glycoprotein with liposomes (Immunosomes) induces an *in vitro* specific release of interleukin-2, *Cell. Immunol.*, 108, 220, 1987.

103. Noguchi, Y., Noguchi, T., Sato, T., Yokoo, Y., Itoh, S., Yoshida, M., Yoshiki, T., Akiyoshi, K., Sunamoto, J., Nakayama, E., and Shiku, H., Priming for in vitro and in vivo anti-human T lymphotropic virus type 1 cellular immunity by virus-related protein reconstituted into liposomes, *J. Immunol.*, 146, 3599, 1991.

104. Engelhard, V. H., Strominger, J. L., Mescher, M., and Burakoff, S., Induction of secondary cytotoxic T lymphocytes by purified HLA-A and HLA-B antigens reconstituted into phospholipid vesicles, *Proc. Natl. Acad. Sci. U.S.A.*, 75, 5688, 1978.

105. Siliciano, R. F., Collelo, R. M., Keegan, A. D., Dintzis, R. Z., Dintzis, H. M., and Shin, H. S., Antigen valency determines the binding of nominal antigen to cytolytic T cell clones, *J. Exp. Med.*, 162, 768, 1985.

106. Curman, B., Östberg, L., and Peterson, P. A., Incorporation of murine MHC antigens into liposomes and their effect in the secondary mixed lymphocyte reaction, *Nature*, 272, 545, 1978.

107. Engelhard, V. H., Kaufman, J. F., Strominger, J. L., and Burakoff, S. J., Specificity of mouse cytotoxic T lymphocytes stimulated with either HLA-A and -B or HLA-DR antigens reconstituted into phospholipid vesicles, *J. Exp. Med.*, 152, 54s, 1980.

108. Weinberger, O., Herrman, S. H., Mescher, M. F., Benacerraf, B., and Burakoff, S. J., Cellular interactions in the generation of cytolytic T lymphocyte responses: role of Ia-positive splenic adherent cells in the presentation of H-2 antigen, *Proc. Natl. Acad. Sci. U.S.A.*, 77, 6091, 1980.

109. Weinberger, O., Herrman, S., Mescher, M. F., Benacerraf, B., and Burakoff, S. J., Antigen-presenting cell function in induction of helper T cells for cytotoxic T-lymphocyte responses: evidence for antigen processing, *Proc. Natl. Acad. Sci. U.S.A.*, 78, 1796, 1981.

110. Herrman, S. H. and Mescher, M. F., Secondary cytolytic T lymphocyte stimulation by purified H-2K^k in liposomes, *Proc. Natl. Acad. Sci. U.S.A.*, 78, 2488, 1981.

111. Goldstein, S. A. N. and Mescher, M. F., Carbohydrate moieties of major histocompatibility complex class I alloantigens are not required for their recognition by T lymphocytes, *J. Exp. Med.*, 162, 1381, 1985.

112. Herrman, S. H., Weinberger, O., Burakoff, S. J., and Mescher, M. F., Analysis of the two-signal requirement for precursor cytolytic T lymphocyte activation using H-2K^k in liposomes, *J. Immunol.*, 128, 1968, 1982.

113. Herrman, S. H. and Mescher, M. F., Lymphocyte recognition of H-2 antigen in liposomes, *J. Supramol. Cell. Biochem.*, 16, 121, 1981.

114. Nakanishi, M., Brian, A. A., and McConnell, H. M., Binding of cytotoxic T-lymphocytes to supported lipid monolayers containing trypsinized H-2K^k, *Mol. Immunol.*, 20, 1227, 1983.

115. Herrman, S. H. and Mescher, M. F., The requirement for antigen multivalency in class I antigen recognition and triggering of primed precursor cytolytic T lymphocytes, *J. Immunol.*, 136, 2816, 1986.

116. Goldstein, S. A. N. and Mescher, M. F., Cell-sized supported artificial membranes (pseudocytes): response of precursor cytotoxic T lymphocytes to class I MHC proteins, *J. Immunol.*, 137, 3383, 1986.

117. Burakoff, S. J., Weinberger, O., Krensky, A. M., and Reiss, C. S., A molecular analysis of the cytolytic T lymphocyte response, *Adv. Immunol.*, 36, 45, 1984.

118. Cartwright, G. S., Smith, L. M., Heinzelmann, E. W., Ruebush, M. J., Parce, J. W., and McConnell, H. M., H-2K^k and vasicular stomatitis virus G proteins are not extensivly associated in reconstituted membranes recognized by T cells, *Proc. Natl. Acad. Sci. U.S.A.*, 79, 1506, 1982.

119. Geiger, B., Rosenthal, K. L., Klein, J., Zinkernagel, R. M., and Singer, S. J., Selective and unidirectional membrane redistribution of an H-2 antigen with an antibody-clustered viral antigen: relationship to mechanisms of cytotoxic T-cell interactions, *Proc. Natl. Acad. Sci. U.S.A.*, 76, 4603, 1979.

120. Moore, M. W., Carbone, F. R., and Bevan, M. J., Introduction of soluble protein into the class I pathway of antigen processing and presentation, *Cell*, 54, 777, 1988.

121. Reddy, R., Zhou, F., Huang, L., Carbone, F., Bevan, M., and Rouse, B. T., pH sensitive liposomes provide an efficient means of sensitizing target cells to class I restricted CTL recognition of a soluble protein, *J. Immunol. Meth.*, 141, 157, 1991.

122. Reddy, R., Zhou, F., Nair, S., Huang, L., and Rouse, B., In vivo cytotoxic T lymphocyte induction with soluble proteins administered in liposomes, *J. Immunol.*, 148, 1585, 1992.

123. Collins, D. S., Findlay, K., and Harding, C. V., Processing of exogenous liposome-encapsulated antigens in vivo generates class I MHC-restricted T cell responses, *J. Immunol.*, 148, 3336, 1992.

124. Lopes, L. M. and Chain, B. M., Liposome-mediated delivery stimulates a class I-restricted cytotoxic T cell response to soluble antigen, *Eur. J. Immunol.*, 22, 287, 1992.

125. Fischer, A. G., Goff, L. K., Lightstone, L., Marvel, J., Mitchison, N. A., Poirier, G., Stauss, H., and Zamoyska, R., Problems in the physiology of class I and class II MHC molecules and of CD45, *CSH Symp. Quant. Biol.*, 54, 667, 1989.

126. Gorga, J. C., Foran, J., Burakoff, S. J., and Strominger, J. L., Use of the HLA-DR antigens incorporated into liposomes to generate HLA-DR specific cytotoxic T lymphocytes, *Meth. Enzymol.*, 108, 607, 1984.

127. Gorga, J. C., Knudsen, P. J., Foran, J. A., Strominger, J. L., and Burakoff, S. J., Immunochemically purified DR antigens in liposomes stimulate xenogeneic cytolytic T cells in secondary *in vitro* cultures, *Cell. Immunol.*, 103, 160, 1986.

128. Coeshott, C. M., Chesnut, R. W., Kubo, R. T., Grammer, S. F., Jenis, D. M., and Grey, H. M., Ia-specific mixed leukocyte reactive T cell hybridomas: analysis of their specificity by using purified class II MHC molecules in a synthetic membrane system, *J. Immunol.*, 136, 2832, 1986.

129. Watts, T. H., Brian, A. A., Kappler, J. W., Marrack, P., and McConnell, H. M., Antigen presentation by supported planar membranes containing affinity-purified I-A^d, *Proc. Natl. Acad. Sci. U.S.A.*, 81, 7564, 1984.

130. Babbitt, B. P., Matsueda, G., Haber, E., Unanue, E. R., and Allen, P. M., Antigenic competition at the level of peptide-Ia binding, *Proc. Natl. Acad. Sci. U.S.A.*, 83, 4509, 1986.

131. Fox, B. S., Quill, H., Carlson, L., and Schwartz, R. H., Quantitative analysis of the T cell response to antigen and planar membranes containing purified Ia molecules, *J. Immunol.*, 138, 3367, 1987.

132. Bakouche, O. and Lachman L. B., Synthetic macrophages: liposomes bearing antigen, class II MHC and membrane IL-1, *Lymphokine Res.*, 9, 259, 1990.

133. Watts, T. H., Gariépy, J., Schoolnik, G. K., and McConnell, H. M., T-cell activation by peptide antigen: Effect of peptide sequence and method of antigen presentation, *Proc. Natl. Acad. Sci. U.S.A.*, 82, 5480, 1985.

134. Gay, D., Coeshott, C., Golde, W., Kappler, J., and Marrack, P., The major histocompatibility complex-restricted antigen receptor on T cells, *J. Immunol.*, 136, 2026, 1986.

135. Walden, P., Nagy, Z. A., and Klein, J., Induction of regulatory T-lymphocyte responses by liposomes carrying major histocompatibility complex molecules and foreign antigen, *Nature*, 315, 327, 1985.

136. Dal Monte, P. R. and Szoka, F. C., Antigen presentation by B cells and mocrophages of cytochrome c and its antigenic fragment when conjugated to the surface of liposomes, *Vaccine*, 7, 401, 1989.

137. Walden, P., Nagy, Z. A., and Klein, J., Major histocompatibility complex-restricted and unrestricted activation of helper T cell lines by liposome bound antigen, *J. Mol. Cell. Immunol.*, 2, 191, 1985.

138. Walden, P., Nagy, Z. A., and Klein, J., Antigen presentation by liposomes, *Haematol. Blood Transfus.*, 29, 481, 1985.

139. Walden, P., Nagy, Z. A., and Klein, J., Antigen presentation by liposomes: inhibition with antibodies, *Eur. J. Immunol.*, 16, 717, 1986.

140. Walden, P., Antigen presentation by liposomes as model system for T-B cell interaction, *Eur. J. Immunol.*, 18, 1851, 1988.

141. Bakouche, O. and Lachman, L. B., Antigen presentation by liposome bearing class II MHC and membrane IL-1, *Yale J. Biol. Med.*, 63, 95, 1990.

142. Watari, E., Dietzschold, B., Szokan, G., and Heber-Katz, E., A synthetic peptide induces long-term protection from lethal infection with herpes simplex virus 2, *J. Exp. Med.*, 165, 459, 1987.

143. Krowka, J., Stites, D., Debs, R., Larsen, C., Fedor, J., Brunette, E., and Düzgünes, N., Lymphocyte proliferative responses to soluble and liposome-conjugated envelope peptide of HIV-1, *J. Immunol.*, 144, 2535, 1990.

144. Roof, R. W., Luescher, I. F., and Unanue, E. R., Phospholipids enhance the binding of peptides to class II major histocompatibility molecules, *Proc. Natl. Acad. Sci. U.S.A.*, 87, 1735, 1990.

145. Mecheri, S., Dannecker, G., Denning, D., and Hoffmann, M.K., Immunogenic peptides require an undisturbed phospholipid cell membrane environment and must be amphipathic to immobilize Ia on B cells, *J. Immunol.*, 144, 1369, 1990.

146. Kurt-Jones, E. A., Beller, D. I., Mizel, S. B., and Unanue, E. R., Identification of a membrane-associated interleukin 1 in macrophages, *Proc. Natl. Acad. Sci. U.S.A.*, 82, 1204, 1985.

147. Bakouche, O., Brown, D. C., and Lachman, L. B., Liposomes expressing IL-1 biological activity, *J. Immunol.*, 138, 4256, 1987.

148. Watts, T. H., Gaub, H. E., and McConnell, H. M., T-cell mediated association of peptide antigen and major histocompatibility complex protein detected by energy transfer in an evanescent wave-field, *Nature*, 320, 179, 1986.

149. Sadegh-Nasseri, S. and McConnell, H. M., A kinetic intermediate in the reaction of an antigenic peptide and I-E^k, *Nature*, 337, 274, 1989.

150. Watts, T. H., T cell activation by preformed, long-lived Ia-peptide complexes, *J. Immunol.*, 141, 3708, 1988.

151. Lee, J. M. and Watts, T. H., On the dissociation and reassociation of MHC class II-foreign peptide complexes, *J. Immunol.*, 144, 1829, 1990.

152. Watts, T. H. and McConnell, H. M., Biophysical aspects of antigen recognition by T cells, *Ann. Rev. Immunol.*, 5, 461, 1987.

153. Alving, C. R., Immunologic aspects of liposomes: presentation and processing of liposomal protein and phospholipid antigens, *Biochim. Biophys. Acta*, 1113, 307, 1992.

154. Furukawa, K. and Sahasrabuddhe, C. G., A novel method of human B cell activation by liposomes coated with Fab' fragments of anti-IgM, *J. Immunol. Meth.*, 131, 105, 1990.

155. Ohyama, N., Hamano, T., Hamakawa, N., Inagaki, K., and Nakanishi, M., Membrane fluidity and lipid hapten structure of liposomes affect calcium signals in antigen-specific B cells, *Biochemistry*, 30, 11154, 1991.

156. Brulet, P. and McConnell, H. M., Structural and dynamical aspects of membrane immunochemistry using model membranes, *Biochemistry*, 16, 1209, 1977.

157. Kimura, K., Arata, Y., Yasuda, T., Kinosita, K., and Nakanishi, M., Location of membrane-bound antigen with different length spacers, *Immunology*, 69, 323, 1990.

158. Tyrrell, D. A., Heath, T. D., Colley, C. M., and Ryman, B. E., New aspects of liposomes, *Biochim. Biophys. Acta*, 457, 259, 1976.

159. Simister, N. E. and Rees, A. R., Isolation and characterisation of an Fc receptor from neonatal rat small intestine, *Eur. J. Immunol.*, 15, 733, 1985.

160. Pochon, S., Graber, P., Yaeger, M., Jansen, K., Bernard, A. R., Aubry, J.-P., and Bonnefoy, J.-Y., Demonstration of a second ligand for the low affinity receptor for immunoglobulin E (CD 23) using recombinant CD23 reconstituted into fluorescent liposome, *J. Exp. Med.*, 176, 389, 1992.

161. Takai, Y., Reed, M. L., Burakoff, S. J., and Herrman, S. H., Direct evidence for a receptor-ligand interaction between the T-cell surface antigen CD2 and lymphocyte-function associated antigen 3, *Proc. Natl. Acad. Sci. U.S.A.*, 84, 6864, 1987.

162. Kinski, S. C. and Nicolotti, R. A., Immunological properties of model membranes, *Ann. Rev. Biochem.*, 46, 49, 1977.

163. Mold, C., Effect of membrane phospholipids on activation of the alternative complement pathway, *J. Immunol.*, 143, 1663, 1989.

164. Michalek, M. T., Bremer, E. G., and Mold, C., Effect of gangliosides on activation of the alternative pathway of human complement, *J. Immunol.*, 140, 1581, 1988.

165. Joiner, K. A., Complement evasion by bacteria and parasites, *Ann. Rev. Microbiol.*, 42, 201, 1988.

166. Tschopp, J., Schäfer, S., Masson, D., Peitsch, M. C., and Heusser, C., Phosphorylcholine acts as a Ca^{2+}-dependent receptor molecule for lymphocyte perforin, *Nature*, 337, 272, 1989.

167. Ojcius, D. M., Persechini, P. M., Zheng, L. M., Notaroberto, P. C., Adeodato, S. C., and Young, J. D.-E., Cytolytic and ion channel-forming properties of the N terminus of lymphocyte perforin, *Proc. Natl. Acad. Sci. U.S.A.*, 88, 4621, 1991.

168. Sauer, H., Pratsch, L., Tschopp, J., Bhakdi, S., and Peters, R., Functional size of complement and perforin pores compared by confocal laser scanning microscopy and fluorescence microphotolysis, *Biochim. Biophys. Acta*, 1063, 137, 1991.

Chapter 7

Liposomes as Tools for Elucidating the Mechanisms of Membrane Fusion

Nejat Düzgüneş and Shlomo Nir

CONTENTS

I. MANY BIOLOGICAL PROCESSES INVOLVE MEMBRANE FUSION

Membrane fusion is a central biophysical and biochemical reaction in numerous biological processes. Exocytosis involves the fusion of the secretory vesicle membrane with the plasma membrane in diverse biological systems, including neurotransmitter release at the neuromuscular junction, histamine release from mast cells, chromaffin granule extrusion from adrenal medullary cells, trichocyst discharge in *Paramecium*, endotoxin-induced degranulation in *Limulus* amebocytes, and the cortical reaction in sea urchin eggs.[1-4] The first stage in the formation of an endocytotic vesicle is the fusion of apposed regions of the invaginated plasma membrane. Later in the endocytotic pathway, endosomes and phagosomes fuse with lysosomes. Receptor recycling back to the plasma membrane proceeds through pinching off of receptor-containing vesicles from the compartment of uncoupling of receptor and ligand. Transport of newly synthesized membrane or secretory proteins from the endoplasmic reticulum to the Golgi apparatus,

0-8493-4569-3/95/$0.00+$.50
© 1995 by CRC Press, Inc.

within the *cis*, *medial*, and *trans* regions of the Golgi is thought to be mediated by transport vesicles that bud off from one compartment and fuse with another.[5-7]

Membrane fusion reactions also take place between cell membranes. During development, myoblasts fuse with each other to form myotubes.[8] Fertilization begins with the fusion of the sperm and egg cell membranes.[9] Multinucleated giant cells can form during the inflammatory process. Several lipid enveloped viruses, such as Sendai virus and human immunodeficiency virus, can cause syncytia formation via fusion of cell membranes.

Lipid-enveloped viruses infect their host cells by fusing either with the plasma membrane, or the endosome membrane following endocytosis of the virions. Because of their relatively simple membrane composition, viruses have proved to be a useful experimental system to study the mechanisms of membrane fusion.[2,10-13]

II. LIPOSOMES CONSTITUTE A SIMPLE MODEL SYSTEM FOR STUDIES ON THE MECHANISMS OF MEMBRANE FUSION

Early studies by Papahadjopoulos and co-workers showed that liposomes containing anionic phospholipids could undergo membrane fusion in the presence of calcium ions.[14-17] These observations were particularly relevant to the fusion of biological membranes, since Ca^{2+} is involved in the control of exocytosis, myoblast fusion, and other biological membrane fusion phenomena. Fusion of certain types of liposomes also displayed a specificity of Ca^{2+} over Mg^{2+}.[17,18] Molecules such as phosphate and polyamines were found to enhance Ca^{2+}-induced fusion of phospholipid vesicles.[19-21] Certain cytoplasmic proteins, such as synexin and clathrin, were shown to modulate or induce the fusion of liposomes.[22-29] These observations have emphasized the relevance of the liposome fusion model for biological membrane fusion. Clearly, liposome fusion is a simple model that enables the investigator to focus on individual components of biological membranes to learn about their behavior under conditions that induce membrane fusion. Some of the properties of these membrane components relevant to fusion include cation binding to phospholipid headgroups and membrane phase state.

Liposomes can fuse with intracellular membranes, such as secretory granules and Golgi membranes, under certain conditions.[30-32] Liposomes also serve as target membranes for viruses, and certain types of liposomes have been demonstrated to sustain fusion with viruses under conditions identical to those necessary for fusion with biological membranes.[13,33-36]

Viral envelope proteins that mediate membrane fusion can be studied in isolation, by reconstitution of the individual proteins in liposomes, and monitoring their fusion with target membranes. For example, the fusion activities of the Sendai virus F and HN proteins,[37-41] vesicular stomatitis virus G protein,[42,43] influenza hemagglutinin (HA),[44-47] and the combination of Sendai and influenza envelope proteins[48] have been characterized by functional reconstitution in liposomes.

III. MEMBRANE FUSION CAN BE MONITORED BY FLUORESCENCE ASSAYS

A variety of methods have been used to monitor the fusion of liposomes. These include:

1. Electron microscopy to visualize the fusion products and morphological changes during fusion[15,49-51]
2. Electron spin resonance to monitor the dilution of spin-labeled lipids from one membrane into another[52]
3. Dynamic light scattering to evaluate the size distribution of the fusion products[53-56]
4. Differential scanning calorimetry to monitor the intermixing of lipids with differing phase transition temperatures[57]
5. Fluorescence resonance energy transfer to monitor the intermixing or dilution of phospholipids labeled with fluorescent probes[58-66]
6. Fluorescence assays to monitor the intermixing of aqueous contents of liposomes[62,66-73]

Fluorescence assays for lipid mixing and for the intermixing of aqueous contents are the most widely used assays in studies of liposome fusion, particularly due to the sensitivity, reliability, and convenience of these methods. The criteria for the reliability of fusion assays have been discussed previously.[66,74] For example, in aqueous contents mixing assays, the fluorescence reactions must be faster than the fusion reaction, and the molecular components of the assay should not bind to the membranes or interact outside the liposomes. In lipid mixing assays, the fluorescent lipid molecules should not transfer molecularly

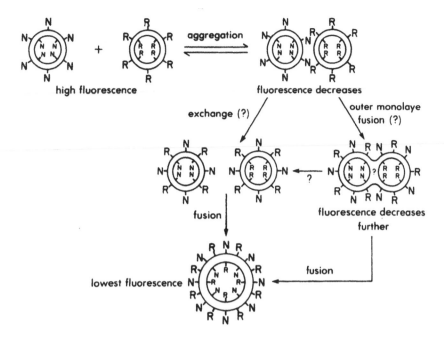

Figure 1 The "probe dilution" assay for the intermixing of lipids between fusing membranes (From Düzgüneş, N., Allen, T.M., Fedor, J., and Papahadjopoulos, D., *Biochemistry*, 26, 8435, 1987. With permission.)

between membranes when the liposomes are dispersed or even when aggregated (but not fused), and their fluorescent properties should not be altered as a result of the interaction of fusion-inducing molecules with the liposome membrane.

A. LIPID MIXING ASSAYS

One of the most widely used assays for lipid mixing is one employing fluorescence resonance energy transfer between N-(nitrobenzoxadiazol)-phosphatidylethanolamine (NBD-PE) and N-(lissamine rhodamine B sulfonyl)-phosphatidylethanolamine (Rh-PE) when they are in close proximity.[60,61,65] The assay can be utilized in two different configurations. In the "probe dilution" method,[62] both probes are initially incorporated in one population of vesicles. Fusion of these vesicles with unlabeled vesicles results in the dilution of the probes and the decrease in the efficiency of resonance energy transfer between NBD-PE and Rh-PE (Figure 1). In the "probe mixing" configuration,[62] the probes are placed in two separate liposome populations. Fusion of the membranes results in resonance energy transfer between the probes due to the close proximity of the fluorophores (Figure 2). One of the shortcomings of this method is that the aggregation of liposomes in the absence of fusion (as monitored by independent techniques) also results in resonance energy transfer between the fluorophores in apposed membranes.[62,74]

B. AQUEOUS CONTENTS MIXING ASSAYS

The intermixing of aqueous contents between fusing liposomes constitutes the most stringent measurement of membrane fusion, since lipid mixing assays are also sensitive to the phenomenon of semi-fusion,[2] where the outer monolayers of two liposomes intermix, without concomitant communication between the internal aqueous spaces.[61,70] Several assays have been developed for monitoring contents mixing during membrane fusion.[66] The most widely used assays for the intermixing of aqueous contents are the terbium/dipocolinic acid (Tb/DPA)[66,68,69,73] and the aminonaphthalene trisulfonic acid/p-xylylene bis(pyridinium) bromide (ANTS/DPX)[66,70,71,73] assays. In the Tb/DPA assay, Tb citrate is encapsulated in one population of liposomes, and DPA in another. Membrane fusion results in an increase of Tb fluorescence (Figure 3). The presence of divalent cations and EDTA in the medium prevents the interaction of Tb and DPA outside the fusing liposomes, thus avoiding the registry of events that lead to contents leakage in the absence of membrane fusion. In the ANTS/DPX assay, ANTS and DPX are encapsulated in different populations of vesicles, and membrane fusion results in the collisional quenching of ANTS fluorescence by DPX (Figure 4). Any

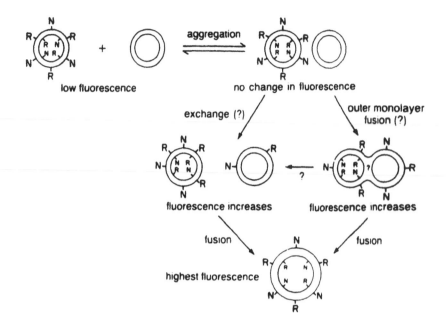

Figure 2 The "probe mixing" assay for the intermixing of lipids between fusing membranes (From Düzgüneş, N., Allen, T.M., Fedor, J., and Papahadjopoulos, D., *Biochemistry*, 26, 8435, 1987. With permission.)

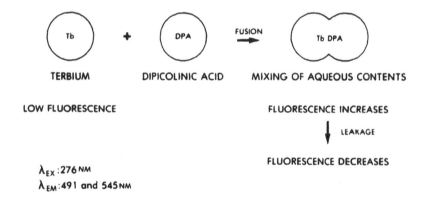

Figure 3 Elements of the Tb/DPA assay for intermixing of aqueous contents (From Düzgüneş, N., Hong, K., Baldwin, P.A., Bentz, J., Nir, S., and Papahadjopoulos, D., *Cell Fusion*, Sowers A.E., Ed., Plenum Press, New York, 1987, 241. With permission.)

contents that are released into the medium do not cause a decrease in fluorescence, since the quenching effect of DPX is concentration-dependent and the diluted DPX is unable to quench ANTS fluorescence.

The kinetics of the release of liposome contents can be monitored independently by

1. The relief of self-quenching of encapsulated carboxyfluorescein[19,69,75,76]
2. The release of Tb from liposomes, which then interacts with DPA in the medium[77]
3. The dissociation of encapsulated Tb/DPA complex as it is released into the medium[78-80]
4. The increase in fluorescence as ANTS and DPX co-encapsulated inside liposomes are diluted into the medium[70,81]

The relative kinetics of intermixing of aqueous contents and of the release of contents depends on the liposome composition and type, nature of the fusogen, and environmental condition, such as ionic strength and temperature. Figure 5 shows that for large unilamellar phosphatidyl serine (PS) liposomes

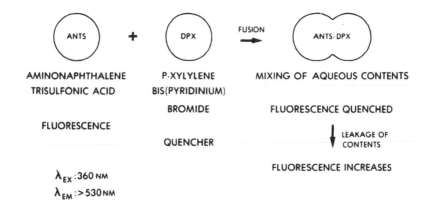

Figure 4 The ANTS/DPX assay for intermixing of aqueous contents (From Düzgüneş, N., Hong, K., Baldwin, P.A., Bentz, J., Nir, S., and Papahadjopoulos, D., *Cell Fusion*, Sowers A.E., Ed., Plenum Press, New York, 1987, 241. With permission.)

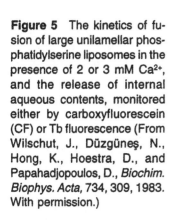

Figure 5 The kinetics of fusion of large unilamellar phosphatidylserine liposomes in the presence of 2 or 3 mM Ca^{2+}, and the release of internal aqueous contents, monitored either by carboxyfluorescein (CF) or Tb fluorescence (From Wilschut, J., Düzgüneş, N., Hong, K., Hoestra, D., and Papahadjopoulos, D., *Biochim. Biophys. Acta*, 734, 309, 1983. With permission.)

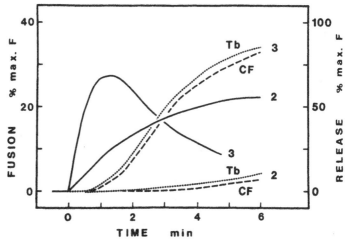

in the presence of Ca^{2+}, the kinetics of leakage of aqueous contents is much slower than that of contents mixing.[77] In fact, at the initial stages, fusion is essentially non-leaky. However, in some liposome systems, the leakage of contents may be too rapid to enable the accurate determination of the rate of fusion by contents mixing assays.[82,83]

C. COMPARISON OF ASSAYS

When the lipid mixing and contents mixing assays described above are compared in the same liposome system, PS liposomes fusing in the presence of Ca^{2+}, the lipid mixing assays reveal a faster process than the contents mixing assays[62,84] (Figure 6). This observation is most likely due to liposome-liposome interactions leading to the intermixing of outer monolayers, without concomitant intermixing of aqueous contents. Lipid "probe mixing" is also more rapid than lipid "probe dilution," indicating that the initial aggregation of the liposomes results in resonance energy transfer between the probes on two aggregated liposomes but does not necessarily lead to lipid mixing. The difference between the Tb/DPA and ANTS/DPX assays has been discussed in detail previously.[62,74] In contrast, in another liposome fusion system, cardiolipin (CL)/dioleoylphosphatidylcholine (DOPC) (1:1) in the presence of Ca^{2+}, the Tb/DPA assay reveals a faster fusion process than the ANTS/DPX assay.[74] It was speculated that Tb citrate or DPX interact to a limited extent with the lipid bilayer in these two systems, but in a different manner with the two negatively charged phospholipids, thus affecting their mobility or response to the fusogenic agent, and ultimately influencing the outcome of the fusion reaction.[74] Recently Walter and Siegel[85] have reported that DPX interacts with PS membranes, causing a lowering of the phase transition temperature,

Figure 6 Fusion of large (approx. 0.1 μm diameter) unilamellar phosphatidylserine liposomes in the presence of 2 mM Ca^{2+}, monitored by four different assays. The fluorescence scale for the solid lines is on the left and that for the dashed lines is on the right hand ordinate. Intermixing of aqueous contents was measured by either the Tb/DPA or the ANTS/DPX assay. Lipid mixing was monitored by the NBD-PE/Rh-PE assay, either in the probe dilution or the probe mixing mode (From Düzgüneş, N., Allen, T.M., Fedor, J., and Papahadjopoulos, D., *Biochemistry*, 26, 8435, 1987. With permission.)

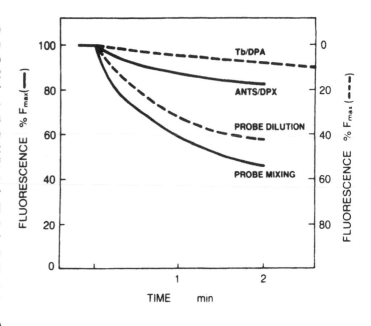

and that the presence of DPX in the vesicle interior results in faster lipid mixing rates. These authors have suggested that perturbation of the inner monolayer affects the reactivity of the liposome with respect to fusion.[85] That the monolayers in PS bilayers can affect each other (i.e., monolayer coupling) has indeed been demonstrated.[86]

D. FLUORESCENCE ASSAYS TO MONITOR THE FUSION OF VIRUSES WITH LIPOSOMES OR CELL MEMBRANES

The incorporation of octadecylrhodamine (R-18) into the membranes of lipid enveloped viruses at concentrations of a few mol% of the membrane lipids results in the self-quenching of rhodamine fluorescence.[87,88] Fusion of the virus membrane with liposome or cell membranes results in the dilution of the probe within the plane of the membrane and the dequenching of the fluorescence. This technique has been used in a number of studies on the mechanisms of fusion of influenza,[89-92] Sendai,[36,93-96] Epstein-Barr,[97] Rous sarcoma,[98] respiratory syncytial,[99] and simian immunodeficiency[100-102] and human immunodeficiency viruses[103-106] with liposomes or cell membranes. Some of these studies are described in detail by de Lima and Hoekstra,[13] and in Sections VIII and IX of this chapter.

Changes in the excimer/monomer ratio of pyrene that depend on the concentration of labeled phospholipids within the plane of the membrane can be utilized to monitor lipid mixing between viral and target (liposomal or cellular) membranes. Pyrene labeled lipids can not only be included in liposome membranes and their dilution into viral membranes followed by the increase in monomer/excimer ratio,[107] but they can also be incorporated metabolically into cell plasma membranes from which viruses bud off, thus incorporating the labeled phospholipid into their membrane.[108,109]

IV. MEMBRANE FUSION DEPENDS ON THE PHOSPHOLIPID COMPOSITION OF LIPOSOMES, THE IONIC ENVIRONMENT AND THE MEMBRANE PHASE STATE

One of the advantages of liposomes for studies on membrane fusion is that they can be used to investigate the role of individual phospholipids in the fusion reaction induced by a variety of agents such as Ca^{2+}, synthetic peptides and proteins.[2,110-117]

A. FUSION OF ACIDIC PHOSPHOLIPID LIPOSOMES

In the case of Ca^{2+}-induced fusion, the threshold concentration of the cation required to induce membrane fusion depends on the anionic phospholipid composition of liposomes and increases in the order phosphatidic acid (PA) > phosphatidylserine (PS) > phosphatidylglycerol (PG) >> phosphatidylinositol

Table 1 **The effect of liposome membrane composition on the threshold concentration of divalent cations required for membrane fusion. Liposomes of mixed composition are an equimolar mixture of the two phospholipids.**

Phospholipid Composition	Threshold Concentration	
	$[Ca^{2+}]_t$ (mM)	$[Mg^{2+}]_t$ (mM)
Phosphatidic acid	0.2	0.4
Phosphatidylserine	2	—
Phosphatidylglycerol	15	—
Phosphatidylinositol	—	—
Phosphatidic acid/Phosphatidylcholine	3	—
Phosphatidylserine/Phosphatidylcholine	—	—
Phosphatidylserine/Phosphatidylethanolamine	3	5
Phosphatidylinositol/Phosphatidylethanolamine	10	50

Adapted from References 111 and 114.

(PI)[111,114] (Table 1). In physiological concentrations of NaCl or other salt solutions, the surfaces of liposomes composed of anionic phospholipids do not approach each other to a close enough distance to undergo fusion because of electrostatic repulsion between the surfaces. Liposomes will form stable dimers only when there is a local minimum in the free energy of interaction between the particles, as defined by the sum of the attractive van der Waals and repulsive electrostatic forces.[112,114,118,119] Dimerization, or aggregation, of liposomes can occur in either the primary minimum, at intermembrane separations of <10 Å, or in the secondary minimum, which may be located 20–100 Å from the membrane surface.[112,120] Clearly, for membrane fusion to occur, the liposomes must approach each other within the primary minimum. Ca^{2+} binds anionic phospholipid membranes, both screening the surface charge and binding to negatively charged groups on the phospholipids thereby neutralizing the surface charge. It is intriguing, however, that Ca^{2+} can induce the fusion of PS membranes at concentrations much lower (1–2 mM)[17,69] than that required for complete charge neutralization (~ 80 mM).[121] The interaction of Ca^{2+} with the membrane is thought to alter the hydrophobicity of the surface,[122-124] both by replacing the water of hydration[75] and by exposing the hydrocarbon interior of the membrane at point defects in the membrane structure.[17,111,124] Collisions of liposomes resulting in transient intermembrane contact may lead to further defects in phospholipid packing (termed "contact-induced defects",[111] leading subsequently to membrane fusion between the two liposomes.

The specificity of the different anionic phospholipids in terms of their response to Ca^{2+}, probably arises from the affinity of the phospholipid head-group for the divalent cation, the proximity of the acidic moieties to the water-hydrocarbon interface, as well as the presence of other hydrated bulky groups. For example, the phosphate group on PA is directly exposed to the aqueous medium and is close to the glycerol backbone of the lipid; thus, the interaction of Ca^{2+} with the phosphate and the resulting local dehydration and charge neutralization may facilitate fusion at relatively low Ca^{2+} concentrations.[16,82] In contrast, the phosphate group on PI is hidden by the bulky and hydrated inositol group, which renders PI liposomes completely resistant to Ca^{2+}-induced fusion,[82] while in PG the phosphate is protected only by a glycerol group, allowing some interbilayer contact and fusion.[16,61]

B. CATION SPECIFICITY IN MEMBRANE FUSION

One of the most dramatic illustrations of cation specificity in membrane fusion is found in the differential effects of Ca^{2+} and Mg^{2+} on the fusion of PS large unilamellar vesicles (LUV).[18] While Ca^{2+} induces extensive fusion of such liposomes, Mg^{2+} merely causes aggregation and no fusion. PS small unilamellar vesicles (SUV), however, fuse up to a limiting size in the presence of Mg^{2+}, and then cease fusion activity. This differential effect may be attributed to the differences between the level of dehydration or hydrophobicity of the membrane surface in the presence of the different cations, as discussed in detail elsewhere.[2,111,114] The inclusion of 30 mol% cholesterol in PS LUV, however, enables Mg^{2+} to induce fusion, as long as the temperature is at least 30°C, suggesting that the Mg/PS coordination complex may be favored under these conditions, where cholesterol alters the spacing between PS molecules.[125] Since

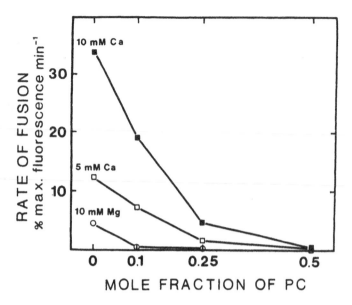

Figure 7 The effect of the phosphatidylcholine and phosphatidylethanolamine content on the initial rate of fusion of liposomes containing 50 mol% phosphatidylserine (From Düzgüneş, N., Wilschut, J., Fraley, R., and Papahadjopoulos, D., *Biochim. Biophys. Acta*, 642, 182, 1981. With permission.)

the coordination of Mg^{2+} has strict requirements for the coordination bonds, this particular spacing may be crucial in the case of Mg^{2+}-induced fusion. The ability of Mg^{2+} to induce rapid aggregation in the absence of fusion can be utilized to determine the effect of other divalent cations on the fusion reaction *per se*, independent of their effect on the rate of aggregation.[126] The sequence of effectiveness of divalent cations in inducing fusion of PS SUV under ionic conditions where both aggregation and destabilization affect the overall fusion process is $Ba^{2+} > Ca^{2+} > Sr^{2+} > Mg^{2+}$, while under conditions where fusion is rate-limiting, Ca^{2+} and Ba^{2+} appear to be equally effective.[78] When the amount of bound divalent cation is compared at the fusion threshold, based on their binding constants before aggregation of the vesicles, the sequence changes to $Ca^{2+} > Ba^{2+} > Sr^{2+} > Mg^{2+}$.

The trivalent cation La^{3+} also induces extensive and relatively non-leaky fusion of PS liposomes at concentrations between 10 and 100 μM, considerably lower than that required for Ca^{2+}.[127,128] An interesting aspect of La^{3+}-induced fusion is that addition of large concentrations of La^{3+} (e.g., 1 mM) can arrest fusion.[128,129] Furthermore, disruption of the fusion process by EDTA causes the extensive leakage of the aqueous contents of the fusion products, suggesting that the intermembrane contact site in the presence of La^{3+} is unstable.[128]

C. FUSION OF LIPOSOMES COMPOSED OF ACIDIC AND NEUTRAL PHOSPHOLIPIDS

When liposomes contain phospholipids with an overall neutral charge in addition to the acidic phospholipids, their ability to undergo fusion is altered, and in some cases this alteration can be drastic. Table 1 shows the change in threshold concentrations of Ca^{2+} and Mg^{2+} with the inclusion of phosphatidylcholine (PC) or phosphatidylethanolamine (PE) together with acidic phospholipids. The threshold Ca^{2+} concentration of PA/PC liposomes is an order of magnitude higher than that of pure PA liposomes, while large (0.1 μm diameter) PS/PC liposomes become completely resistant to fusion.[76,130] Similarly, the fusion activity of SUV in the presence of Ca^{2+} is diminished when some of the PS in the membrane is replaced by PC.[131] The sequence of effectiveness of divalent cations to induce the fusion of PA/PC SUV was found to be $Ca^{2+} > Mg^{2+} > Ba^{2+}$, suggesting that the structural intermediate in the fusion of these liposomes is a divalent cation-phospholipid complex with an optimal geometry to accommodate ions of intermediate ionic size.[132]

In contrast to PC, the presence of PE in acidic phospholipid membranes is not inhibitory to Ca^{2+}-induced fusion,[76] and even renders PI-containing liposomes susceptible to fusion, while pure PI liposomes are resistant to fusion under similar conditions.[130] In 3-component liposomes composed of 50 mol% PS and a mixture of PC and PE, as the mole fraction of PC is increased, the liposomes lose their ability to fuse in the presence of Ca^{2+} or Mg^{2+} (Figure 7). It is also interesting to note that although Mg^{2+} does not induce fusion of large PS liposomes, it becomes fusogenic when the membrane contains PE in addition to PS. This fusogenic activity is abolished, however, when 10 mol% PC is present in the membrane.[76]

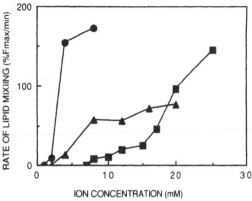

ION CONCENTRATION (mM)

Figure 8 Fusion of N-[1-(2,3-dioleyloxy)propyl]-N,N,N-trimethylammonium/phosphatidylethanolamine (1:1) liposomes in the presence of anions. The initial rate of fusion is given as a function of the anion concentration. (●): citrate; (▲): EDTA; (■): phosphate (From Düzgüneş, N., Goldstein, J.A., Friend, D.S., and Felgner, P.L., *Biochemistry*, 28, 9179, 1989. With permission.)

The difference in the ability of PC and PE to affect fusion may be attributed to a number of factors.[76,114] These include the lower affinity of PE for water,[133] the electrostatic and hydrogen bonding interactions of the phosphate oxygens and ammonium nitrogens of PE and the absence of these interactions in PC,[134] the propensity of fully hydrated PC to retain a lamellar structure and that of PE to form non-bilayer phases depending on the temperature and acyl chain composition,[135,136] and the ability of PE bilayers to approach each other closer than PC bilayers due to differences in repulsive hydration forces.[137]

Divalent cation-induced lateral phase separation of acidic and neutral lipids has been observed in SUV composed of certain mixed membranes that undergo fusion and was thought to be an essential determinant of membrane fusion.[14] However, in PS dipalmitoylphosphatidyl choline (DPPC) LUV, Ca^{2+}-induced fusion was observed within a limited temperature range (see below) in the absence of a major phase separation of the two lipid components.[138] Ca^{2+}-induced phase separation in PS/PC mixtures appears to depend on the acyl chain species of the phospholipids and on the mole fraction of PS, with membranes containing ≥ 70 or ≤ 40 mol% PS not phase separating in the presence of Ca^{2+}.[139] In the case of PS dimyristoylphosphatidyl ethanolamine (DMPE) mixtures, Ca^{2+} induces phase separation while Mg^{2+} does not, although both ions induce membrane fusion, suggesting that phase separation is not required for fusion.[138] Likewise, in liposomes containing a low mole fraction of PS in PE, no phase separation is observed in the presence of Ca^{2+}, while fusion still takes place.[140]

D. FUSION OF CATIONIC LIPOSOMES

The fusion behavior of liposomes composed of or containing cationic lipids, although they do not occur in nature, have attracted attention not only because of their intrinsic physicochemical properties but also because of their application in the intracellular delivery of nucleic acids or proteins.[141,142] Liposomes composed of N-[1-(2,3-dioleyloxy)propyl]-N,N,N-trimethylammonium (DOTMA) and PE (1:1) can undergo fusion in the presence of polyvalent anions, with the sequence of effectiveness, citrate > EDTA > phosphate, while sulfate and dipicolinate are ineffective[55] (Figure 8). DOTMA liposomes containing 50 mol% PC are refractory to fusion by these multivalent anions, similar to the inhibitory effect of PC in divalent cation-induced fusion of negatively charged liposomes. However, DOTMA/PE liposomes can be induced to fuse with DOTMA/PC liposomes by high concentrations of citrate but not phosphate. DOTMA/PE liposomes can also fuse with negatively charged PS/PE or PS/PC (1:1) liposomes in the absence of multivalent anions, while the extent of fusion of DOTMA/PC liposomes with PS/PC liposomes is diminished substantially compared to fusion with PS/PE liposomes.[55] Clearly, the interaction of the negatively charged carboxyl and phosphate groups on PS with the positively charged trimethylammonium on DOTMA is sufficient to mediate fusion. It is likely that the polyanionic nature of the membrane surface of PS-containing liposomes also contributes to the induction of fusion.

Liposomes composed of diacyl cationic surfactants have also been shown to aggregate and fuse in the presence of anions. Sonicated dioctadecyldimethylammonium chloride (DODAC) liposomes prepared in glucose solutions aggregate in the presence of chloride ions.[143] Large (~ 300 nm in diameter) liposomes composed of didodecyldimethylammonium bromide prepared in HEPES buffer without NaCl can be induced to fuse with dipicolinic acid, sulfate of p-toluenesulfonic acid.[144,145] Curiously, smaller liposomes (~ 200 nm) of the same composition are resistant to fusion in the presence of dipicolinic acid, although they aggregate.

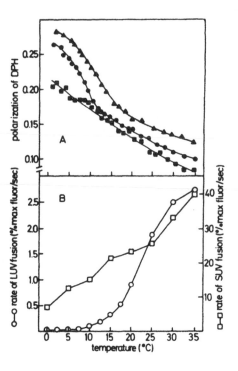

Figure 9 The dependence of the initial rate of Ca²⁺-induced fusion of PS LUV and SUV on membrane fluidity. **(A)** The fluorescence polarization of diphenylhexatriene as a function of temperature. (■) SUV without Ca²⁺; (●) LUV without Ca²⁺; (▲) LUV in the presence of 0.5 mM Ca²⁺. **(B)** Initial rate of fusion of SUV in the presence of 2 mM Ca²⁺ (□) and LUV in the presence of 5 mM Ca²⁺ (○), as a function of temperature. (From Wilschut, J., Düzgüneş, N., Hoekstra, D., and Papahadjopoulos, D., *Biochemistry*, 24, 8, 1985. With permission.)

E. ROLE OF MEMBRANE PHASE STATE

Calcium-induced fusion of PS LUV is inhibited when the membrane is below the gel-liquid crystalline phase transition temperature, although the liposomes can aggregate.[146] Above 15°C, the rate of fusion increases rapidly with temperature. This temperature coincides with the end of the phase transition of PS LUV in the presence of non-aggregating concentrations of Ca²⁺, that is before inter-membrane contact has been established (Figure 9). SUV composed of PS, however, do not reveal a distinct phase transition down to 0°C, i.e., they are still in a fluid state, and the fusion of such liposomes is not inhibited even at this temperature, although the rate of fusion does decrease with temperature. It appears, therefore, that a prerequisite for fusion is that the membrane be in a liquid crystalline state when the fusion-inducing cation binds to the membrane surface.

F. ROLE OF TEMPERATURE-INDUCED LATERAL PHASE SEPARATIONS

Although fusion is favored by a fluid phase state of the membrane, in certain phospholipid mixtures the liquid crystalline state can be inhibitory to fusion. This situation arises when the thoroughly mixed phospholipids constitute a phospholipid composition that is inhibitory to membrane fusion. As described above, LUV composed of an equimolar mixture of PS and PC are refractory to fusion in the presence of Ca²⁺. When the PC component is replaced by its saturated analog DPPC, however, the phase behavior of the membrane, as well as the lateral distribution of the individual phospholipids is altered within a particular temperature range.[138,147] Figure 10 shows the differential scanning calorimetry thermogram of LUV composed of a 1:1 mixture of PS and DPPC, and the temperature dependence of the fusion of the liposomes. When the membrane is in a liquid crystalline (fluid) state the membrane behaves like a PS/PC mixture, and fusion is inhibited. Fluid and solid phases are expected to co-exist within the temperature range of the transition endotherm, and at about 20°C fluid PS micro-domains are likely to be most abundant, overcoming the inhibitory effect of the DPPC on fusion. This observation, as well as similar results obtained in the CL/DPPC system,[119] suggest that micro-domains of phospholipid in membranes induced by temperature, or by other factors such as membrane proteins, can confer a different fusion behavior on the membrane than expected from its overall lipid composition.

G. INVERTED MICELLES OR POINT DEFECTS AS INTERMEDIATE LIPID STRUCTURES DURING MEMBRANE FUSION

The intermediate molecular structure(s) attained by phospholipids in the region of inter-membrane adhesion of fusing liposomes has not been identified unequivocally and has been the subject of extensive debate in the literature. Early studies on Ca²⁺-induced fusion of PS liposomes pointed to the role of a dehydrated interbilayer Ca/PS complex and the induction of a phase transition in the bilayer from the

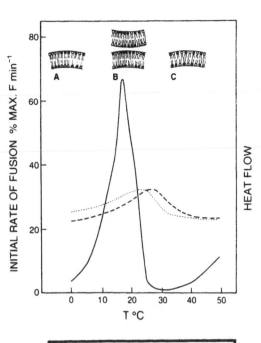

Figure 10 The initial rate of fusion of PS/DPPC (1:1) LUV induced by 5 (open circles) or 10 mM Ca²⁺ (closed circles) as a function of temperature and the differential scanning calorimetry thermograms in the absence (solid line) and presence of 10 mM Ca²⁺ (dashed line). The schematic diagram at the top depicts the formation of PS-rich fluid micro-domains surrounded by DPPC domains within the transition endotherm and a well-mixed fluid membrane above the endotherm. (From Düzgüneş, N., Paiement, J., Freeman, K.B., Lopez, N.G., Wilschut, J., and Papahadjopoulos, D., *Biochemistry*, 23, 3486, 1984. With permission.)

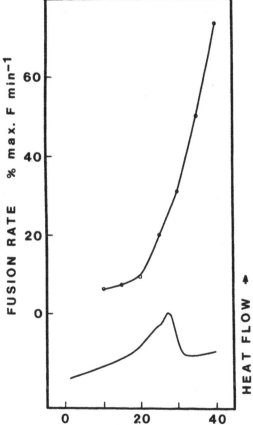

Figure 11 The temperature-dependence of the initial rate of Sr²⁺-induced fusion of PS LUV (upper curve) and its relationship to the phase state of the Sr/PS complex (lower curve). Fusion was monitored by the Tb/DPA assay, and the phase transition was determined by differential scannic calorimetry. (From Düzgüneş, N., Paiement, J., Freeman, K.B., Lopez, N.G., Wilschut, J., and Papahadjopoulos, D., *Biochemistry*, 23, 3486, 1984. With permission.)

liquid crystalline into a special gel phase.[17,75] However, subsequent studies indicated that Sr²⁺ and Ba²⁺ could induce fusion of PS LUV more rapidly when the divalent cation/PS complex was in a liquid crystalline state (Figure 11), obviating the requirement for a divalent cation-induced isothermal phase transition in the fusion reaction.[138,148] The latter observations, together with the concepts of contact-

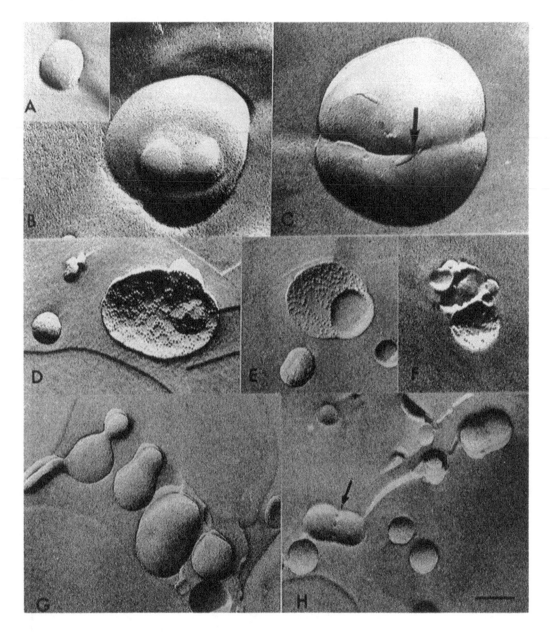

Figure 12 Freeze-fracture electron micrographs of quick-frozen liposomes. (**A**) A liposome before stimulation with Ca²⁺. (**B**) CL/PC (1:1) liposomes 1 s after the addition of 5 mM Ca²⁺. Several liposomes have fused, but no lipidic particles are apparent. (**C**) PS/PE (1:1) liposomes quick-frozen after a 1 s stimulation with Ca²⁺. Lipidic particles are not observed under these conditions that mediate membrane fusion. (**D,E**) CL/PC liposomes incubated for 2 h in Ca²⁺ and glycerol. Abundant lipidic particles are observed. (**F**) CL/PC liposomes incubated with Ca²⁺ alone for 2 h. Fewer lipidic particles are apparent in the absence of glycerol. (**G**) CL/PC liposomes incubated for 2 h in glycerol and treated with Ca²⁺ 1 s prior to quick freezing. The two vesicles at the lower right corner are aggregated, while at the top left three fused liposomes remain separated by narrow necks. The neck appears to have widened in the two central fusion products. (**H**) Same as in G but showing that lipidic particles are observed occasionally in glycerol-treated samples even in early times. Bar = 0.2 μm. (From Bearer, E.L., Düzgüneş, N., Friend, D.S., and Papahadjopoulos, D., *Biochim. Biophys. Acta, 693*, 93, 1982. With permission.)

induced dehydration of the interbilayer space[75] and the increased hydrophobicity and surface energy associated with membrane fusion,[122,123] have led to molecular models of fusion involving contact-induced packing defects in apposed monolayers that subsequently coalesce.[111,119]

Alternative models have been developed from observations of the hexagonal H_{II} phase and of lipidic particles following fusion of certain types of liposomes, detected by freeze-fracture electron micrographs and ^{31}P-NMR.[49,149-154] Rapid-freezing, freeze-fracture electron microscopy has shown, however, (Figure 12) that lipidic particles are not observed at early times following the induction of fusion of PS/PE or CL/PC (1:1) liposomes by Ca^{2+}.[51,152] Studies on the Ca^{2+}-induced fusion of PS/PE (1:3) liposomes monitored by the same technique have also produced no evidence of lipidic particles during early stages of membrane fusion.[155] The occurrence of lipidic particles in certain liposomes following fusion may be indicative of the propensity of the particular phospholipids to adopt transient non-bilayer structures during the membrane fusion reaction. However, as pointed out previously,[51,111,114] lipidic particles as defined by their morphology in electron micrographs are probably not involved in the fusion reaction. It is possible that non-bilayer intermediates in these liposome systems are too rapid to be detected by morphological techniques and may be localized to a small area within the region of inter-bilayer contact. The half-life of such non-bilayer intermediates have been predicted to be on the order of 1 ms.[156] In liposomes composed of N-methylated PE derivatives, a stable cubic phase has been observed at mildly acidic pH at temperatures below the bilayer-H_{II} transition. Membrane fusion without the loss of aqueous contents is observed under similar conditions, suggesting that the formation of the cubic phase may be involved in the fusion reaction.[157] Liposomes composed of natural PE species undergo fusion below the bilayer to H_{II} phase transition temperature,[158] but the involvement of a cubic phase in fusion has not been demonstrated in this case. Interlamellar attachment sites, or points of fusion, in frozen and thawed PC/PE liposomes have been interpreted to be point defects in the bilayer structure.[159] Thermal fluctuations of non-bilayer forming lipids out of the plane of a monolayer in contact with another, and the formation of a stalk between the apposed membranes, has been proposed to depend on the headgroup/acyl chain size ratio of the phospholipids in the membrane.[160,161] Studies on bilayers supported by mica surfaces have suggested that the major force that leads to fusion is hydrophobic attraction between acyl chains exposed to each other across the aqueous phase.[162] The progress of the membrane fusion reaction from a contact-induced defect (CID) to a curved bilayer annulus (CBA) via different possible pathways has been discussed by Papahadjopoulos et al..[111] Recent observations by Chernomordik et al.[163] indicating that lyso-PC inhibits the fusion of intracellular organelles, virus-cell fusion, and exocytosis has been interpreted to support the notion of a fusion stalk, a highly bent intermediate between the two fusing membranes.

V. MEMBRANE FUSION CAN BE INDUCED OR MODULATED BY PROTEINS, PEPTIDES AND POLYAMINES

A. PROTEINS

Reconstitution of viral envelope glycoproteins in liposomes has been instrumental in demonstrating the fusion activity of these proteins and the dependence of this activity on environmental factors such as pH, the relative surface density of the viral proteins, and lipid requirements in the membrane.[2,164-166] Addition of lectins to glycolipid-containing liposomes can mediate intermembrane attachment or lateral phase separation of the inhibitory glycolipids, thereby enhancing the rate of fusion in some cases and decreasing the threshold concentration of Ca^{2+} required to induce fusion.[80,167-171] The incorporation of the transmembrane protein glycophorin in PS membranes is inhibitory to the Ca^{2+}-induced fusion of these vesicles; incubation of the glycophorin vesicles with a lectin, however, facilitates rapid fusion in the presence of Ca^{2+}.[172] These observations point to the significance of membrane aggregating proteins that may facilitate the fusion activity of Ca^{2+}, whose binding to membranes is enhanced drastically upon intermembrane contact.[173-176]

The adrenal medullary and liver cytosolic protein synexin (annexin VII) also enhances the rate of fusion of liposomes and reduces the threshold Ca^{2+} concentration for fusion,[22,23,116] in some cases from millimolar down to 10 μM,[24] most likely due to the enhancement of the aggregation step of the overall fusion reaction[29] (Figure 13). The effect of synexin was found to be dependent on both the acidic and neutral phospholipid species in the liposome membrane[23,24] (Figure 13). Liposome fusion studies, as well as experiments on the fusion of liposomes with neutrophil secretory granules, have also revealed the

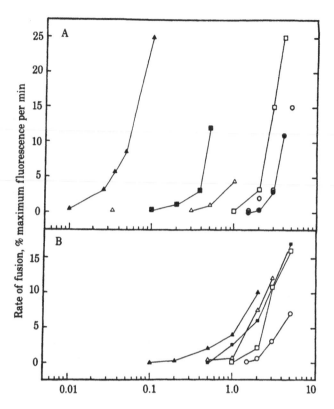

Figure 13 The effect of synexin on the initial rate of fusion of PA/PE (1:3) (Panel **A**), and PA/PE/PC (1:2:1) (Panel **B**) LUV, as a function of the Ca^{2+} or Mg^{2+} concentration. (○) Mg^{2+} only, at concentrations indicated numerically on the abscissa. (●) Synexin and Mg^{2+}, with Mg^{2+} concentrations indicated numerically on the abscissa. (□) Ca^{2+} only. (■) Synexin and Ca^{2+}. (△) 1.5 mM Mg^{2+} and Ca^{2+}. (▲) Synexin, 1.5 mM Mg^{2+} and Ca^{2+}. (From Hong, K., Düzgüneş, N., Ekerdt, R., and Papahadjopoulos, D., *Proc. Natl. Acad. Sci. USA*, 70, 4942, 1982.

presence of synexin-like molecules in neutrophils.[31] Neutrophil annexin I has been shown to enhance the Ca^{2+}-dependent fusion of PS liposomes[177] and to reduce the threshold Ca^{2+} concentration required for the fusion of PA/PE (1:4) liposomes.[178]

Studies with another cytoplasmic protein, clathrin, have indicated that the protein can penetrate and destabilize liposome membranes containing PS at neutral pH,[26,179] while fusion is observed only at acidic pH.[26] The membrane binding and conformational changes of the protein are involved in fusion of PS-containing vesicles in the pH range 5 to 6.[27,28] The regions of clathrin involved in membrane fusion have also been identified.[180]

A cationic protein from the acrosome granule of abalone spermatozoa has been shown to induce the fusion of negatively charged liposomes in the absence of Ca^{2+}.[181] The lung surfactant-associated proteins SP 5-18 induces the fusion of PG LUV in the presence of 3 mM Ca, which does not cause fusion by itself.[83] The purified lung surfactant protein SP-B could also induce the fusion of DPPC/PG (7:3) LUV.[182] These studies show the diversity of proteins that can induce or modulate the fusion of liposomes.

B. PEPTIDES

Various amphipathic or poly-ionic peptides can induce the fusion of liposomes. Polylysine induces the fusion of PS-containing liposomes at an optimal peptide/lipid ratio, suggesting that it mediates fusion by cross-linking membranes, but that it also interferes with fusion by steric hinderance.[61,183] In PA-containing membranes, however, it appears to induce membrane lipid mixing in the absence of any contents mixing.[184] The amphipathic peptide mellitin can also induce liposome fusion, most likely by partitioning into one liposome and electrostatically attracting another liposome to the surface.[185] Positively charged liposomes composed of PC, cholesterol, and a cationic detergent could undergo fusion in the presence of poly(aspartic acid).[186,187] Twelve-amino acid, basic, amphipathic helical peptides with a high hydrophobic moment were found to induce the fusion of PC or PA/PC SUV, accompanied by extensive leakage of aqueous contents.[188] The fusion activity of a 51-amino acid cationic peptide toward PS or PC LUV correlated with its hydrophobicity rather than the formation of an amphiphilic α-helix.[189] Rapaport et al.[190] demonstrated that the neurotoxic peptide pardaxin and several of its analogs could induce the fusion of SUV composed of PC and PS. Analogues of the peptide containing D-amino acids were not fusogenic.

Various peptides corresponding to the N-terminal "fusion domain" of viral envelope glycoproteins[10,191] can induce the fusion of certain types of liposomes. A 20-amino acid N-terminal peptide from the HA2 subunit of influenza hemagglutinin induces the fusion of SUV composed of palmitoyloleoyl-phosphatidyl choline (POPC) or egg PC,[192,193] but no fusion of LUV.[193] However, extensive pH-dependent fusion of large PE/PC/cholesterol (1:1:1) vesicles was observed, suggesting that the peptide may be inducing inverted micellar fusion intermediates, since certain species of PE can form non-bilayer structures. On the other hand, a 17-amino acid HA2 peptide induces neither aggregation nor fusion of PC or PS LUV, although it causes the destabilization of PC LUV, suggesting that the fusion activity may be related to the length of the peptide.[194,195] Mutant peptides corresponding to hemagglutinin mutants with impaired or no fusion activity in cell-cell fusion assays were found to have a corresponding impairment in their ability to interact with phospholipid membranes[193,194,196] and to mediate the fusion of POPC SUV at neutral pH.[197] It should be noted that SUV, because of their inherent lipid packing defects, may be susceptible to fusion under conditions that do not affect the fusion of LUV, which have a bilayer structure closer to that of biological membranes.[18,54,198] Thus, it is essential to examine the fusion activity of viral fusion peptides, or other synthetic peptides, using LUV rather than SUV.

Anionic derivatives of the 20-amino acid HA2 (influenza A/PR/8) N-terminal peptide were found to induce fusion of PC LUV at acidic pH, where the peptides become more hydrophobic due to protonation of the carboxyl groups, and assume an α-helical structure.[199] In contrast, a cationic peptide derivative became more hydrophobic and induced fusion at alkaline pH. Modification of the N-terminus of the anionic derivatives caused decreased membrane fusion, emphasizing the importance of the N-terminus in the fusion activity of the peptides.[200] Interestingly, a mixture of these anionic and cationic peptide derivatives was found to mediate the fusion of PC LUV at neutral pH.[201]

A 23 amino acid peptide corresponding to the N-terminus of the HIV-1 envelope transmembrane glycoprotein gp41 was also found to induce the fusion of LUV composed of DOPC/dioleoyl PE/cholesterol (1:1:1) and of palmitoyloleoyl phosphatidylglycerol (POPG) vesicles aggregated by Ca^{2+} or Mg^{2+} under conditions where the divalent cations alone did not induce fusion.[56,202] The peptide was found to be in a predominantly extended antiparallel β-structure under fusion conditions in this system, while pore formation was promoted by an α-helical structure. The ability of several gp41 N-terminal peptides to penetrate phospholipid bilayers was also found to correlate with their ability to form α-helices in non-polar media.[203] However, pore formation by the peptide in liposomes correlated with an α-helical structure. A similar correlation was found for a 16-amino acid N-terminal peptide from gp41, which also induced the fusion of LUV composed of PC/PE (1:1) or PC/PE/sphingomyelin/cholesterol (1:1:1:1.5) but not of LUV without PE in the membrane.[204]

Similarly, peptides of varying length corresponding to the N-terminus of the SIV envelope glycoprotein gp32 were shown to induce fusion of SUV of various compositions,[205] while fusion of LUV was observed only if the membranes contained PE. In this system, the shorter peptides (12 amino acids) induced more extensive fusion than the longer ones (16 or 24 amino acids), in contrast to the conclusion reached above for the influenza virus HA2 peptides. A 23-amino acid "fusion peptide" from the GP2 protein of the Lassa arenavirus was also shown to induce the fusion of DOPC SUV; fusion was enhanced at mildly acidic pH and inhibited by Ca^{2+} or high concentrations of NaCl.[206]

For peptide-induced membrane fusion, the initial aggregation step must be considered as a critical factor. It has been pointed out that the insertion of viral fusogenic peptides into the target membrane does not (necessarily) crosslink two membranes (liposomes), whereas insertion of a segment of a viral envelope protein into the target membrane does crosslink the viral and target membranes.[207] Clearly, vesicle aggregation at a primary minimum becomes more difficult with an increase in vesicle size.[112] Thus, the peptide GALA can induce the fusion of PC SUV but cannot agregate PC LUV,[54] although it can induce leakage of encapsulated fluorophores from LUV.[208] When PC or PG LUV are aggregated by a lectin or Ca^{2+}, respectively, the peptide GALA can induce liposome fusion.[209] It should also be noted that liposome fusion induced by viral "fusion peptides" may not necessarily reflect the fusogenic activity of the envelope protein as a whole, since the remainder of the protein may also participate in various stages of the fusion reaction.[195]

C. POLYAMINES

The polyamines spermine and spermidine are naturally occurring polycations that can induce the aggregation of negatively charged liposomes composed of PS or PA or their mixtures with PC or PE.[21] This property of polyamines was found to facilitate the fusion of such membranes induced by Ca^{2+} by reducing the minimum concentration of the divalent cation required for fusion and drastically increasing the rate of fusion.[20,21] The larger effect of spermine on the Ca^{2+}-induced fusion of PA liposomes compared to that of PS liposomes has been attributed to the enhancement of Ca^{2+} binding to PA in the presence of spermine.[210] When liposome membranes contain an acidic phospholipid, cholesterol, and a relatively high content of PE, spermine and spermidine can induce fusion in the absence of divalent cations. The concentration of spermine needed to induce the fusion of PS/PE/cholesterol (3:7:5) is about an order of magnitude lower than that of Ca^{2+}.[21]

VI. THE RATE CONSTANTS OF MEMBRANE ADHESION AND FUSION CAN BE DETERMINED BY A MASS-ACTION KINETIC ANALYSIS

Membrane fusion may be induced by a variety of factors, such as Ca^{2+} and other divalent or multivalent cations, peptides, viral glycoproteins, or other proteins.[2,3,10,12,111,112,114,116,119,211-213] Membrane fusion induced by all these processes can be described by a mass-action model and are amenable for mathematical analysis. The basis for the analysis is that the process of membrane fusion consists of three main steps[112,214,215]

1. Close approach or point contact between two membranes. In a system of vesicles, this is the aggregation step.
2. Membrane destabilization. The two apposed membranes that are ordered structures must undergo a transient disturbance in order that a new structure be formed later.
3. Membrane merging. The destabilized membranes fuse to form a single continuous membrane.

The mathematical procedure based on a mass action model has enabled us to distinguish between two steps in the overall fusion reaction, the close approach or aggregation, which involves a collision between particles and is of second order in particle concentrations, and the subsequent first order fusion reaction, which involves membrane destabilization and merging.

Ordinarily, the minimal set of rate constants required to describe the kinetics of fusion includes $f(s^{-1})$, the rate constant of fusion per se; $C(M^{-1}s^{-1})$, the rate constant of aggregation or adhesion; and $D(s^{-1})$, the rate constant of dissociation. The programs[216,217] enable the user to employ different rate constants for higher order aggregation-fusion products (see Table 2), but it is always desirable to employ a minimal number of parameters in describing any physical process. By focusing on the initial stages where only an aggregate of two particles, or dimer $[A = A(1,1)]$, is formed and is fused to a doublet $[F = F(1,1)]$, it is possible to determine the rate constants systematically, as will be explained below. In this case, the chain of reactions is given schematically by

$$L + V \rightleftharpoons A(1,1) \rightarrow F(1,1) \tag{1}$$

In Eq (1) L and V may denote liposomes and virions, or labeled and unlabeled liposomes. The indices in the matrices A and F denote the numbers of particles of types 1 and 2 in the aggregation or fusion products, respectively. (See References 216 and 218 for a detailed description.)

The kinetic equations are

$$dL/dt = -C \cdot L \cdot V + D \cdot A \tag{2}$$

$$dV/dt = -C \cdot L \cdot V + D \cdot A \tag{3}$$

$$dA/dt = C \cdot L \cdot V - f \cdot A - D \cdot A \tag{4}$$

$$dF/dt = f \cdot A$$

In Equation 4 the term $C \cdot L \cdot V$ is of second order in particle concentration and gives the rate of generation of aggregates, A, whereas the first order terms, F and D give the rate of annihilation of A due to fusion and dissociation.

By focusing on a very dilute system, where the aggregation will be the rate limiting step and fusion will occur relatively fast, the analysis of kinetics of fusion yields the rate constant of aggregation, C. From the results obtained with a more concentrated suspension and the knowledge of C, it is possible to determine the fusion rate constant, f, by relying mostly but not exclusively on the initial stages of the

Table 2 Rate constants of vesicle-vesicle aggregation and fusion[a]

Composition, Size, and References	Fusion Inducing Agents and Conditions	Rate Constant of Aggregation C (M^{-1}·s^{-1})	Rate Constant of Fusion f (s^{-1})	Rate Constant of De-aggregation[b] D (s^{-1})	Fusion Assay and Comments
PS LUV[84,30,218]	2 mM Ca^{2+}	2.3 × 10^6	0.004–0.01	S	Contents mixing
	5 mM Ca^{2+}	6.0 × 10^7	0.08	S	Contents mixing
PS/cholesterol (2:1) LUV[30]	2 mM Ca^{2+}	10^6	0.01–0.05	S	Contents mixing
	5 mM Ca^{2+}	3 × 10^7	0.2	S	Contents mixing
Chromaffin granule ghosts[217]	Synexin and pH 6, 37°C	5 × 10^9	>10	0.5	Membrane mixing
		3.2 × 10^9	1.3	0.5	Contents mixing
PS/PE (1:3) LUV[155]	4 mM Ca^{2+}	4 × 10^7	0.04^c	0	Contents mixing Size distribution determined by fast-freezing EM
POPC SUV[54]	GALA peptide at pH 5 and lipid/ peptide ratio 50/1	2.5 × 10^6	0.08	≤ 0.005	Membrane mixing
	lipid/peptide 100/1	1.3 × 10^6	0.03	≤ 0.005	Vesicle size determined by dynamic light scattering

[a]The uncertainties in the estimates of the rate constants are about 20, 25, and 50% for C, f, and D, respectively. More details can be found in the cited references. Whenever not stated explicitly, the measurements were performed at room temperature, and the medium included 100 mM NaCl, pH 7.4.

[b]The value 0 indicates that D << f. The symbol S indicates that D < f/2.

[c]$f_{ij} = f_{11}/6$

fusion process. Initially, the value for D affects the results minimally, since most of the particles are not aggregated.

In certain cases the rate constant f can be determined directly by pre-aggregating vesicles, e.g., by NaCl[219] or by a protein,[29,183] and then initiating fusion by adding Ca^{2+}. Similarly, influenza virus can adhere at neutral pH to a target membrane, such as that of a suspension cell with minimal endocytotic activity, an erythrocyte ghost, or a liposome, and then fusion is initiated upon lowering the pH of the suspension.[5]

From the simulation of the kinetics of fusion as monitored by a fluorescence assay, it is possible to determine the dependence of the rate constants on pH, Ca^{2+} concentration, temperature, peptide/lipid ratio, etc., and test a variety of hypotheses. A combination of calculations of cation binding to liposomes with fusion studies demonstrated that the rate constant of fusion of PS liposomes depends critically on the fraction of Ca^{2+} bound per PS molecule rather than on the concentration of Ca^{2+}.[78,112] The analysis revealed that certain proteins known as promoters of vesicle fusion, such as synexin,[22,23,211,220,221] only promoted the aggregation step by increasing the value of C and reducing the required level of Ca^{2+}.[29] Furthermore, an excess of polylysine[183] or synexin[29] resulted in a reduction of the fusion rate constant. The analysis of the effect of synexin on fusion of chromaffin granule ghosts at pH 6 (see Table 2) showed that the main activity of synexin was an enhancement of the rate of aggregation. At intermediate or excessive concentrations, synexin promoted moderately or inhibited the actual fusion step, respectively.[217]

The analysis of the results of Ca^{2+}-induced fusion of large unilamellar PS/cholesterol vesicles[222] showed that the effect of cholesterol was to increase slightly (up to twofold) the rate constant of fusion and reduce slightly the rate constant of aggregation (see Table 2). This could explain the apparent discrepancy that in certain cases cholesterol promoted the overall rate of vesicle fusion, whereas in other cases it inhibited it.[125]

The analysis also made possible a comparison between the results obtained by the Tb/DPA assay for volume mixing and by the resonance energy transfer assay, which monitors membrane mixing (see Section III). It was found that for Ca^{2+}-induced fusion of CL/DOPC (1:1) LUV at 25°C, the rate constants obtained from analysis of the lipid mixing assay were identical with those obtained with the Tb/DPA assay, but at 37°C the lipid mixing parameters were larger.[223] A similar study with PS vesicles showed that at 15°C the rate constants for either assay were the same, but at 35°C the lipid mixing assay aggregation rate constants were higher than the Tb/DPA ones, indicating the occurrence of events that result in mixing of bilayer lipids but not in mixing of vesicle contents.[84] The lipid mixing fusion rate constants are considerably higher than the Tb/DPA fusion rate constants, demonstrating the higher tendency of the vesicles, once aggregated, to mix lipids than to mix aqueous contents. This raises the possibility that the external monolayers are intermixing partially, since the lipid mixing probes are essentially unexhangeable. Indeed, Rosenberg et al.[61] and Ellens et al.[70] found evidence for such "semi-fusion" phenomena.

A table of rate constants obtained prior to 1987 is given in Nir.[218] Subsequently obtained rate constants of vesicle-vesicle fusion are shown in Table 2. One of the issues still under study is the effect of size on the tendency of liposomes to fuse.[224] For Ca^{2+}-induced fusion of PS vesicles, it was demonstrated that the fusion rate constants decrease dramatically with increasing vesicle size[112,126,215,225], whereas the Mg^{2+}-induced fusion of small unilamellar vesicles (SUV) terminates when they reach the size of LUV.[18] In parallel, it was shown that LUV can be aggregated but not fused by Mg^{2+}. In contrast, the effect of the amphipathic peptide GALA[54] on the fusion of vesicles exhibits different characteristics, despite the same trend that GALA can induce the fusion of PC SUV but cannot induce the fusion of LUV. In this case, the peptide cannot induce the aggregation of PC (or PG) LUV, although it destabilizes the vesicles as exhibited by their leakage, which occurs by means of a pore formation.[208] As illustrated in Table 2, for GALA-induced fusion of PC SUV at pH 5, the rate constant of aggregation is relatively small, whereas the rate constant of fusion is comparable to that observed in certain cases of fusion involving liposomes or viruses.

This survey illustrates that in certain cases vesicle fusion terminates after a small number of fusion rounds. From the final extents of fusion, obtained with a fluorescence assay for membrane mixing, it is possible to estimate the average number of vesicles in a fusion product by means of Equation 6 below.[217]

In a population of labeled and unlabeled particles whose number ratio is L/k, complete fusion to products consisting of n particles yields a fluorescence intensity, I, of

$$I = 100 \ [k/(k + L)] \cdot (n - 1)/n \tag{6}$$

As n becomes large, Equation 6 coincides with the expression for a homogeneous mixture. For fusion up to doublets (n = 2), the cases L/k = 1 and L/k = 1/9 yield I = 25 and 45, respectively.

Employing this analysis indicated that in GALA-induced fusion of PC SUV the number of fusion rounds varied between 2 and 4, as was also confirmed by dynamic light scattering measurements of the size distribution of the fusion products.[54]

These examples illustrate the usefulness of the application of mass-action model calculations as a means of comparing fusion assays and extracting rate constants, which are independent of vesicle concentration. The rate constants express the effect of various factors on aggregation and fusion as separate processes. This information on its own is not sufficient for a detailed description of the fusion process in molecular terms, but it provides mechanistic answers. On the other hand, one of the tests of any scientific theory is ultimately its ability to yield quantitative predictions. In this respect, the power of our calculations has been demonstrated in many cases, despite the fact that some assumptions and parameters are employed. The employment of fast-freezing electron microscopy to visualize the progress of the Ca^{2+}-induced fusion of PS/PE (1:3) liposomes, demonstrated the ability of the calculations to yield predictions for the vesicle sizes and the distribution of aggregation-fusion products.[155] This distribution has an interesting development in time. Initially, most of the vesicles are in monomers. One minute after the addition of Ca^{2+}, the fraction of monomers is significantly reduced, and dimers, trimers, tetramaers, etc., are observed. However, after 15 min the monomers become once again the predominant component, whereas the fraction of oligomers drops significantly, due to fusion of aggregated particles resulting in the formation of larger monomers, in accord with the calculations.

VII. LIPOSOMES CAN BE USED TO MONITOR THE FUSION ACTIVITY OF BIOLOGICAL MEMBRANES

Liposomes have been utilized in monitoring the membrane fusion activity of neutrophil secretory granules, neutrophil plasma membranes, cortical granules, chromaffin granules, and Golgi membranes. The specific granules of neutrophils have been shown to fuse with PA/PE (1:3) liposomes in the presence of synexin, arachidonic acid, and Ca^{2+}, using the fluorescence probe dilution and the R-18 dequenching methods described above.[31] Similarly, PS/PE (1:3) liposomes were shown to undergo Ca^{2+}-induced fusion with the cytoplasmic side of plasma membranes isolated from neutrophils.[226,227] In this system, neutrophil cytosol and annexin I enhanced fusion, while annexin V was inhibitory, and the threshold Ca^{2+} concentration for the induction of fusion was reduced to 5 μM in the presence of des(1–9)-annexin I (which lacks nine amino acids from the N-terminus of the parent protein).

Liposomes labeled with fluorescence probes have also been used to study the fusion characteristics of other intracellular membranes such as microsomes and the Golgi apparatus. Phosphatidylcholine (egg PC, DOPC, or POPC) liposomes can fuse with Golgi membranes in semi-intact or perforated fibroblasts in an ATP-dependent manner.[32] In this system, liposomes containing equimolar PC and either PS or sphingomyelin were completely resistant to fusion. In contrast, isolated Golgi membranes were observed to fuse with DOPC liposomes independent of ATP or cytosolic factors and dependent on intact integral membrane proteins in the Golgi membranes.[228] Liposomes composed of PC/PE/PA (10:5:1) were shown to undergo fusion with fungal microsomal membranes, and fusion was dependent on the proteins of microsomal membranes and was promoted by cytosolic proteins.[229] Sea urchin egg secretory granules were found to undergo fusion with liposomes composed of lipids extracted from egg cortices or of synthetic phospholipids and cholesterol.[230] This fusion reaction was not dependent on cytoplasmic proteins and was inhibited by N-ethylmaleimide. Fusion of chromaffin granule membranes with CL/PC/cholesterol (3:2:5) LUV was found to proceed efficiently in the absence of Ca^{2+} and was dependent on the intactness of the granule membrane proteins.[30] Coated vesicles from which the clathrin coat had been removed appeared to undergo Ca^{2+}-independent fusion with PS or PA SUV but not with neutral PC liposomes.[231]

VIII. LIPID-ENVELOPED VIRUSES CAN FUSE WITH LIPOSOMES

Lipid-enveloped viruses infect their host cell by fusing either with the plasma membrane or with the endosome membrane following endocytosis. To understand the molecular mechanisms of membrane fusion induced by the viral envelope glycoproteins, it is essential to know how the composition of the target membrane affects the fusion activity of such viruses. Liposomes have been used as convenient target membranes whose composition can be altered readily and in a controlled fashion. For example, that

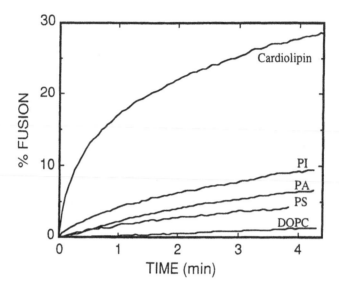

Figure 14 The time-course of fusion of HIV-1 at 37°C with LUV composed of the indicated phospholipids in a medium containing 150 mM NaCl, 0.1 mM EDTA and 10 mM TES at pH 7.5. Fusion is indicated as the fluorescence dequenching of R-18 incorporated in the virus membrane, as a percentage of maximal dequenching obtained by dispersing the virus membrane with the detergent octaethyleneglycol dodecyl ether. (From Larsen, C.E., Nir, S., Alford, D.R., Jennings, M., Lee, K.-D., and Düzgüneş, N., *Biochim. Biophys. Acta*, 1147, 223, 1993. With permission.)

the fusion activity of Semliki Forest virus required the presence of cholesterol in the target membrane was first established in experiments utilizing liposomes as target membranes.[33,164,232] The modulation of the fusion activity of influenza and Sendai viruses by the composition of the target membrane is described by de Lima and Hoekstra in the next chapter.[13]

Similar studies were performed recently with human and simian immunodeficiency viruses (HIV-1 and SIV, respectively). Although they appear to require the presence of the CD4 molecule on cell membranes for efficient infection, these viruses can fuse with liposomes without this receptor.[100,101,104,105] The rate of fusion of HIV-1 with liposomes, monitored by the R-18 dequenching assay, decreases according to the sequence: CL >> PI > CL/DOPC (3:7), PA > PS, PS/cholesterol (2:1) > PS/PC (1:1), PS/PE (1:1) > DOPC (Figure 14). Fusion is also observed by both negative stain and freeze-fracture electron microscopy (Figure 15). The fusion of immunodeficiency viruses with several types of liposomes is enhanced by the presence of Ca^{2+}, unlike influenza or Sendai viruses. A single HIV-1 virion appears to fuse with several liposomes. Compared to intact virions, the fusion products of the virus and liposomes have a reduced ability to undergo fusion. The fusion of HIV-1 with membranes containing its own envelope glycoproteins (i.e., fusion products) is strongly inhibited. This inhibition may result from (1) the limited capacity of the liposome membrane to accommodate the viral glycoproteins and (2) the inhibitory effect of viral glycoproteins in the fusion product membrane to the close approach and fusion of another virion. The fusion of HIV-1 with liposomes also inhibits the ability of virions to infect cultured cells, with the sequence of effectiveness CL >> PI > PS.[233,234]

IX. VIRAL MEMBRANE FUSION MECHANISMS CAN BE STUDIED USING LIPOSOMES

By using liposomes, it is possible to characterize the effect of lipids and glycolipids in the target membrane on the binding of the virus to and fusion with membranes as a function of pH, temperature, cations, etc. The kinetics and final extents of fusion can be monitored continuously by the employment of fluorescence assays for membrane mixing, such as the resonance energy transfer assay or the R-18 fluorescence dequenching assay. Studies on influenza virus, Sendai virus, and HIV-1, employing analysis by the mass action model, provide information on the effect of the liposome membrane composition, pH, and temperature on final extents and rate constants of fusion. Table 3 presents a few selected cases for virus-liposome and virus-cell fusion.

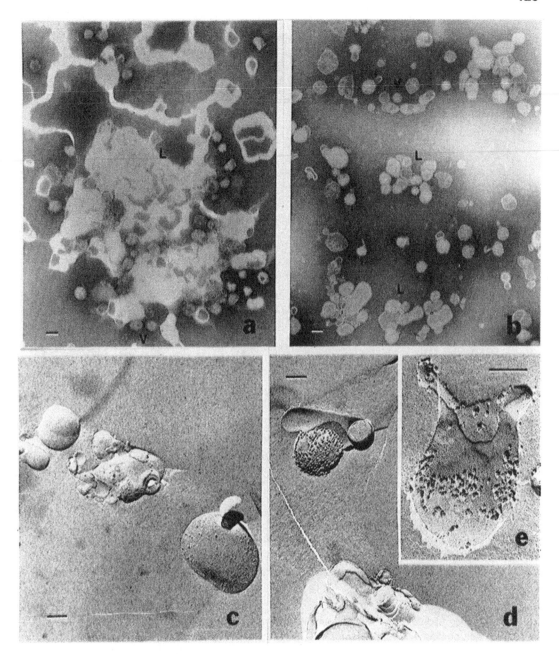

Figure 15 Fusion of HIV-1 with liposomes. (**a**) Negative strain electron micrograph of HIV-1 and CL liposomes incubated at 37°C for 5 min and fixed prior to negative staining. (V) Virus (L) liposomes. (**b**) Same as in a, except that DOPC liposomes were used. (**c,d**) Freeze-fracture electron micrograph of CL liposomes incubated with HIV-1 for 5 min at 37°C and then fixed. (**e**) Freeze-fracture electron micrograph of DOPC liposomes incubated with HIV-1 for 4 h at 37°C and then fixed. Bars = 0.1 μm. (From Larsen, C.E., Nir, S., Alford, D.R., Jennings, M., Lee, K.-D., and Düzgüneş, N., *Biochim. Biophys. Acta*, 1147, 223, 1993. With permission.)

This comparison indicates that in general the fusion rate constants of virus-liposome fusion are larger than those found for virus–cell membrane fusion. One possible interpretation of this observation may be that cell membrane proteins and glycoproteins have some inhibitory effect on viral membrane fusion. However, Table 3 indicates that an increase in the fraction of PC in the target membrane results in lower fusion activity, similar to that observed in vesicle-vesicle fusion. It is known that the external monolayers of plasma membranes are rich in PC and contain little PE and PS. Hence, it is likely that for viruses fusing

Table 3 **The effect of the target membrane on the percentage of virions capable of fusing and the fusion rate constant f at 37°C[a]**

Virus and Target Membrane	% of Fusion-Active Virus	f (s⁻¹)	Ref.
Influenza virus (pH 5)			
CL	100	0.9–1.7	216, 35
PS	100	0.15	35
40% CL, 60% PC/PE (2:1)	80	0.25	35
20% CL, 80% PC/PE (2:1)	70	0.20	35
40% PS, 60% PC/PE (2:1)	80	0.1	35
5% GD1a, 39% chol, 56% PC/PE (2:1)	20–45	0.5	35
5% GD1a, 95% PC/PE (2:1)	20–40	0.2	35
PC/PE (2:1)	20–40	0.25	35
Erythrocyte ghosts	100	0.1	90, 91
HL-60 cells	100	0.01–0.1	91
S49 and CEM cells	100	0.005–0.05	91
PC-12 cells	100	0.03–0.07	92
Sendai virus (neutral pH)			
CL	60	0.02	242
CL/DOPC (1:1)	30	0.009	242
PS	25	0.005	242
Erythrocyte ghosts	100	0.008–0.07	93
HL-60 cells	100	0.001–0.1	95
PC-12 cells	100	0.03	237
HIV-1 (neutral pH)			
CL	70	0.1	105
PS/PE (1:1)	40	0.001	105
DOPC	50	0.003	105
PS/PC (1:1)	50	0.0007	105

[a]Uncertainties in the percentage of fusion active virus and f are 10–15% and 30%, respectively. The X-47 strain was used in influenza virus-liposome fusion studies, while in virus-cell fusion experiments the A/PR/8/34 and X-87 strains were employed. The X-87 strain gave lower f values. The Z strain of Sendai was employed in the virus fusion studies. The Z/SF isolate gave the lower f values with the HL-60 cells.

with liposomes composed of lipids in the same ratio as in the external monolayers of plasma membranes, the fusion rate constants would be similar to, or perhaps even lower than, those found in virus-cell membrane fusion. Further studies are necessary to verify this speculation. We have initiated studies to analyze the effect on the fusion rate constant of enzymatic removal of cellular proteins and glycoproteins from the plasma membrane.[235]

The effect on viral fusion of glycolipids in target liposomes has been studied in only a few cases. For influenza virus fusing with liposomes at 37°C, it was found that the ganglioside G_{D1a} has little effect on the fusion rate constants or on the final extents of fusion but enhances threefold the rate constant of aggregation.[35] Similarly, preliminary results do not indicate very dramatic effects of this ganglioside on the fusion of Sendai virus with liposomes at 37°C.[236]

Table 3 illustrates that in the case of virus fusion with neutral liposomes only a certain fraction of the virus can fuse, even if the liposomes are in excess. In the case of Sendai virus, only 25% of the virions undergo fusion with PS liposomes. In contrast, all the virions are capable of fusing with erythrocyte ghosts and several suspension and adherent cells in the case of influenza virus[90-92] and Sendai virus,[95,237] provided that the ratio of the number of virions per cell is not too large. In the case of immunodeficiency viruses such as HIV-1, HIV-2, and SIV, this issue is still under investigation. The question of partial fusion activity has been discussed previously,[238] and a model has been proposed by Nir et al.[90]

In the case of Sendai virus, fusion inactive virions bind essentially irreversibly to liposomes. Yet, preliminary results have revealed that such bound, unfused virions can be released by sucrose gradient centrifugation. The separated unfused virions subsequently fuse when incubated with a "fresh" batch of liposomes. We have concluded, therefore, that the fraction of initially bound, unfused virions does not consist of defective particles, but rather of particles bound to liposomes via "inactive" sites.[90] Recently, a similar phenomenon was also observed in the case of influenza virus. The released virus could fuse again with liposomes despite several hours of incubation at pH 5, under which conditions the virus is "inactivated."[239]

Why the virions have complete fusion activity toward cell membranes and incomplete activity toward liposomes is not clear. It is possible that when the virus binds to the cell surface via an inactive site on the virus, it has, geometrically, the opportunity to establish contact with the cell membrane via another active site. Another possibility is that dissociation of the virus from the cell surface occurs more easily, perhaps via the action of the viral neuraminidase.

Liposomes also constitute a convenient system for studies on the insertion of viral envelope proteins into the target membrane, particularly since the photoaffinity probes used for these experiments only label viral proteins, as no other proteins are present. Using this technique, Novick and Hoekstra[240] have shown that the F-protein of Sendai virus is the first envelope protein to insert into the target membrane. Harter et al.[241] have demonstrated that the N-terminal "fusion peptide" of the cleaved influenza virus hemagglutinin (HA2) inserts into target liposomes in an α-helical configuration.

Studies of virus-liposome fusion have also been helpful in attempts to inhibit infection of cells by HIV-1. We have found that pre-incubation of HIV-1 with CL liposomes inhibits the infectivity of the virus.[101,233,234] Liposomes composed of other negatively charged phospholipids, such as PS or PI, were ineffective under similar conditions but were inhibitory to infection following a longer (2–24 h) preincubation with the virus.[234] Our explanation of these findings stems from our HIV-1 liposome fusion studies,[105] where it was shown that CL liposomes fuse significantly faster with HIV-1 than PS or PI liposomes do (Figure 14 and Table 3). Thus, liposomes can effectively compete with the cells for binding and fusion with the virus. In addition, the fusion products were shown to lose their ability to fuse with additional liposomes,[105] and hence presumably also with cells.

X. CONCLUDING REMARKS

The use of liposomes in studies of membrane fusion has generated much insight into the biophysical behavior of individual membrane components under experimental conditions that induce fusion. Although liposomes are a simple model of biological membranes, they have proved to be essential in certain investigations, such as for the photoaffinity labeling of viral proteins during fusion, studying the role of target membrane components in viral fusion, and the establishment of rigorous criteria for the reliability of membrane fusion assays. Clearly, the insights gained with the utilization of the liposome system, must then be extended to the biological phenomena of interest. It is unfortunate that there is a general sense of distrust of liposomes among cell biologists and virologists studying membrane fusion. Yet, many studies in cell biology and virology that have led to the establishment of essential concepts of the molecular mechanisms of membrane fusion in different experimental systems have utilized liposomes. Hopefully this review will help future investigators place the utility of liposomes in membrane fusion studies into proper perspective, balancing the advantages and limitations of the system.

REFERENCES

1. Poste, G. and Allison, A.C., Membrane fusion, *Biochim. Biophys. Acta*, 300, 421, 1973.
2. Düzgüneş, N., Membrane fusion, in *Subcellular Biochemistry*, Vol. 11, Roodyn, D.B., Ed., Plenum, New York, 1985, 195.
3. Blumenthal, R., Membrane fusion, *Curr. Top. Membr. Transp.* 29, 203, 1987.
4. Zimmerberg, J., Vogel, S.S., and Chernomordik, L.V., Mechanisms of membrane fusion, *Annu. Rev. Biophys. Biomol. Struct.*, 22, 433, 1993.
5. Wilschut, J., Intracellular membrane fusion, *Curr. Opin. Cell Biol.*, 1, 639, 1989.
6. Wilson, D.W., Whiteheart, S.W., Orci, L., and Rothman, J.E., Intracellular membrane fusion, *Trends Biochem. Sci.*, 16, 334, 1991.
7. Rothman, J.E., and Orci, L., Molecular dissection of the secretory pathway, *Nature*, 355, 409, 1992.

8. Wakelam, M.J.O., Myoblast fusion-a mechanistic analysis, in *Membrane Fusion in Fertilization, Cellular Transport and Viral Infection*, Düzgüneş, N. and Bronner, F., Eds., Academic Press, San Diego, 1988, 87.

9. Yanagimachi, R., Sperm-egg fusion, in *Membrane Fusion in Fertilization, Cellular Transport and Viral Infection*, Düzgüneş, N. and Bronner, F., Eds., Academic Press, San Diego, 1988, 3.

10. White, J., Kielian, M., and Helenius, A., Membrane fusion proteins of enveloped animal viruses, *Q. Rev. Biophys.*, 16, 151, 1983.

11. White, J.M., Membrane fusion, *Science*, 258, 917, 1992.

12. Hoekstra, D. and Kok, J.W., Entry mechanisms of enveloped viruses: implications for fusion of intracellular membranes, *Biosci. Rep.*, 9, 273, 1989.

13. de Lima, M.C.P. and Hoekstra, D., Liposomes, viruses and membrane fusion, in this volume.

14. Papahadjopoulos, D., Poste, G., Schaeffer, B.E., and Vail, W.J., Membrane fusion and molecular segregation in phospholipid vesicles, *Biochim. Biophys. Acta*, 352, 10, 1974.

15. Papahadjopoulos, D., Vail, W.J., Jakobson, K., and Poste, G., Cochleate lipid cylinders: formation by fusion of unilamellar lipid vesicles, *Biochim. Biophys. Acta*, 394, 483, 1975.

16. Papahadjopoulos, D., Vail, W.J., Pangborn, W.A., and Poste, G., Studies on membrane fusion. II. Induction of fusion in pure phospholipid membranes by calcium and other divalent metals, *Biochim. Biophys. Acta*, 448, 265, 1976.

17. Papahadjopoulos, D., Vail, W.J., Newton, C., Nir, S., Jacobson, K., Poste, G. and Lazo, R., Studies on membrane fusion. III. The role of calcium-induced phase changes, *Biochim. Biophys. Acta*, 465, 579, 1977.

18. Wilschut, J., Düzgüneş, N., and Papahadjopoulos, D., Calcium/magnesium specificity in membrane fusion: kinetics of aggregation and fusion of phosphatidylserine vesicles and the role of bilayer curvature, *Biochemistry*, 20, 3126, 1981.

19. Fraley, R., Wilschut, J., Düzgüneş, N., Smith, C., and Papahadjopoulos, D., Studies on the mechanism of membrane fusion: the role of phosphate in promoting calcium-induced fusion of phospholipid vesicles, *Biochemistry*, 19, 6021, 1980.

20. Hong, K., Schuber, F., and Papahadjopoulos, D., Polyamines: biological modulators of membrane fusion. *Biochim. Biophys. Acta*, 732, 469, 1983.

21. Schuber, F., Hong, K., Düzgüneş, N., and Papahadjopoulos, D., Polyamines as modulators of membrane fusion: aggregation and fusion of liposomes, *Biochemistry*, 22, 6134, 1983.

22. Hong, K., Düzgüneş, N., and Papahadjopoulos, D., Role of synexin in membrane fusion, *J. Biol. Chem.*, 256, 3651, 1981.

23. Hong, K., Düzgüneş, N., and Papahadjopoulos, D., Modulation of membrane fusion by calcium-binding proteins, *Biophys. J.*, 37, 296, 1982.

24. Hong, K., Düzgüneş, N., Ekerdt, R., and Papahadjopoulos, D., Synexin facilitates fusion of specific phospholipid vesicles at divalent cation concentrations found intracellularly, *Proc. Natl. Acad. Sci. USA*, 70, 4942, 1982.

25. Blumenthal, R., Henkart, M., and Steer, C.J., Clathrin-induced pH dependent fusion of phosphatidylcholine vesicles, *J. Biol. Chem.*, 258, 3409, 1983.

26. Hong, K., Yoshimura, T., and Papahadjopoulos, D., Interaction of clathrin with liposomes: pH-dependent fusion of phospholipid membranes induced by clathrin, *FEBS Lett.*, 191, 17, 1985.

27. Yoshimura, T., Maezawa, S., and Hong, K., Exposure of hydrophobic domains of clathrin in its membrane fusion-inducible pH region, *J. Biochem.*, 101, 1265, 1987.

28. Maezawa, S., Yoshimura, T., Hong, K., Düzgüneş, N., and Papahadjopoulos, D., Clathrin-induced fusion of phospholipid vesicles associated with its membrane binding and conformational change, *Biochemistry*, 28, 1422, 1989.

29. Meers, P., Bentz, J., Alford, D., Nir. S., Papahadjopoulos, D., and Hong, K., Synexin enhances the aggregation rate but not the fusion rate of liposomes, *Biochemistry*, 27, 4430, 1988.

30. Bental, M., Lelkes, P.I., Scholma, J., Hoekstra, D., and Wilschut, J., Ca^{2+}-independent, protein-mediated fusion of chromaffin granule ghosts with liposomes, *Biochim. Biophys. Acta*, 774, 296, 1984.

31. Meers, P., Ernst, J.D., Düzgüneş, N., Hong, K., Fedor, J., Goldstein, I.M., and Papahadjopoulos, D., Synexin-like proteins from human polymorphonuclear leukocytes. Identification and characterization of granule-aggregating and membrane-fusing activities, *J. Biol. Chem.*, 262, 7850, 1987.

32. Kobayashi, T. and Pagano, R.E., ATP-dependent fusion of liposomes with the Golgi apparatus of perforated cells, *Cell*, 55, 797, 1988.

33. White, J. and Helenius, A., pH-dependent fusion between the Semliki Forest virus membrane and liposomes, *Proc. Natl. Acad. Sci. USA*, 77, 3273, 1980.
34. Stegmann, T., Hoekstra, D., Scherphof, G., and Wilschut, J., Kinetics of pH-dependent fusion between influenza virus and liposomes, *Biochemistry*, 24, 3107, 1985.
35. Stegmann, T., Nir, S., and Wilschut, J., Membrane fusion activity of influenza virus. Effects of gangliosides and negatively charged phospholipids in target liposomes, *Biochemistry*, 28, 1698, 1989.
36. Citovsky, V., Blumenthal, R., and Loyter, A., Fusion of Sendai virions with phosphatidylcholine-cholesterol liposomes reflects the viral activity required for fusion with biological membranes, *FEBS Lett.*, 193, 135, 1985.
37. Hosaka, Y. and Shimizu, Y.K., Artificial assembly of envelope particles of HVJ (Sendai virus) II. Lipid components for formation of the active hemolysin, *Virology*, 49, 640, 1972.
38. Hsu, M.C., Scheid, A., and Choppin, P.W., Reconstitution of membranes with individual paramyxovirus glycoproteins and phospholipid in cholate solution, *Virology*, 95, 476, 1979.
39. Nakanishi, M., Uchida, T., Kim, J., and Okada, Y., Glycoproteins of Sendai virus (HVJ) have a critical ratio for fusion between virus envelopes and cell membranes, *Exp. Cell Res.*, 142, 95, 1982.
40. Harmsen, M.C., Wilschut, J., Scherphof, G., Hulstaert, C., and Hoekstra, D., Reconstitution and fusogenic properties of Sendai virus envelopes, *Eur. J. Biochem.*, 149, 591, 1985.
41. Chejanovsky, N., Zakai, N., Amselem, S., Barenholz, Y., and Loyter, A., Membrane vesicles containing the Sendai virus binding glycoprotein, but not the viral fusion protein, fuse with phosphatidylserine liposomes at low pH., *Biochemistry*, 25, 4810, 1986.
42. Petri, W.A., Pal, R., Barenholz, Y., and Wagner, R.R., Fluorescence studies of dipalmitoyl-phosphatidylcholine vesicles reconstituted with the glycoprotein of vesicular stomatitis virus, *Biochemistry*, 20, 2796, 1981.
43. Metsikkö, K., van Meer, G., and Simons, K., Reconstitution of the fusogenic activity of vesicular stomatitis virus, *EMBO J.*, 5, 3429, 1986.
44. Kawasaki, K., Sato, S.B., and Ohnishi, S.-I., Membrane fusion activity of reconstituted vesicles of influenza virus hemagglutinin glycoproteins, *Biochim. Biophys. Acta*, 733, 286, 1983.
45. Sizer, P.J.H., Miller, A., and Watts, A., Functional reconstitution of the integral membrane proteins of influenza virus into phospholipid liposomes, *Biochemistry*, 26, 5106, 1987.
46. Nussbaum, O., Lapidot, M., and Loyter, A., Reconstitution of functional influenza virus envelopes and fusion with membranes and liposomes lacking virus receptors, *J. Virol.* 61, 2245, 1987.
47. Stegmann, T., Morselt, H.W.M., Booy, F.P., van Breemen, J.F.L., Scherphof, G., and Wilschut, J., Functional reconstitution of influenza virus envelopes, *EMBO J.*, 6, 2651, 1987.
48. Lapidot, M., and Loyter, A., Fusogenic properties of reconstited hybrid vesicles containing Sendai and influenza envelope glycoproteins: fluorescence dequenching and fluorescence microscopy studies, *Biochim. Biophys. Acta*, 980, 281, 1989.
49. Verkleij, A.J., Mombers, C., Gerritsen, W.J., Leunissen-Bijvelt, L. and Cullis, P.R., Fusion of phospholipid vesicles in association with the appearance of lipidic particles as visualized by freeze-fracturing, *Biochim. Biophys. Acta*, 555, 358, 1979.
50. Miller, D.C. and Dahl, G.P., Early events in calcium-induced liposome fusion, *Biochim. Biophys. Acta*, 689, 165, 1982.
51. Bearer, E.L., Düzgüneş, N., Friend, D.S., and Papahadjopoulos, D., Fusion of phospholipid vesicles arrested by quick freezing. The question of lipidic particles as intermediates in membrane fusion, *Biochim. Biophys. Acta*, 693, 93, 1982.
52. Maeda, T. and Ohnishi, S.I., Membrane fusion. Transfer of phospholipid molecules between phospholipid bilayer membranes, *Biochem. Biophys. Res. Commun.*, 60, 1509, 1974.
53. Day, E.P., Ho, J.T., Kunze, R.K. Jr., and Sun, S.T., Dynamic light scattering study of calcium-induced fusion in phospholipid vesicles, *Biochim. Biophys. Acta*, 470, 503, 1977.
54. Parente, R.A., Nir, S., and Szoka, F.C. Jr., pH-dependent fusion of phosphatidylserine small vesicles, *J. Biol. Chem.*, 263, 4724, 1988.
55. Düzgüneş, N., Goldstein, J.A., Friend, D.S., and Felgner, P.L., Fusion of liposomes containing a novel cationic lipid, N-[2,3-(dioleyloxy)propyl]-N,N,N-trimethylammonium: induction by multivalent anions and asymmetric fusion with acidic phospholipid vesicles, *Biochemistry*, 28, 9179, 1989.
56. Nieva, J., Nir, S., Muga, A., Goñi, F.M., and Wilschut, J., Interaction of the HIV-1 fusion peptide with negatively charged vesicles: different structural requirements for fusion and leakage, *Biochemistry*, 33, 3201, 1994.

57. Papahadjopoulos, D., Hui, S., Vail, W.J., and Poste, G., Studies on membrane fusion I. Interaction of pure phospholipid membranes and the effect of myristic acid, lysolecithin, proteins and DMSO, *Biochim. Biophys. Acta*, 448, 245, 1976.

58. Uster, P.S. and Deamer, D.W., 1981, Fusion competence of phosphatidylserine-containing liposomes quantitatively measured by a fluorescence resonance energy transfer assay, *Arch. Biochem. Biophys.*, 209, 385, 1981.

59. Vanderwerf, P. and Ullman, E.F., Monitoring of phospholipid vesicle fusion by fluorescence energy transfer between membrane-bound dye labels, *Biochim. Biophys. Acta*, 596, 302, 1980.

60. Struck, D. K., Hoekstra, D., and Pagano, R. E., Use of resonance energy transfer to monitor membrane fusion, *Biochemistry*, 20, 4093, 1981.

61. Rosenberg, J., Düzgüneş, N., and Kayalar, C., Comparison of two liposome fusion assays monitoring the intermixing of aqueous contents and of membrane components, *Biochim. Biophys. Acta*, 735, 173, 1983.

61a. Walter, A., Steer, C.J., and Blumenthal, R., Polylysine induces pH-dependent fusion of acidic phospholipid vesicles: a model for polycation-induced fusion, *Biochim. Biophys. Acta*, 861, 319, 1986.

62. Düzgüneş, N., Allen, T.M., Fedor, J., and Papahadjopoulos, D., Lipid mixing during membrane aggregation and fusion. Why fusion assays disagree, *Biochemistry*, 26, 8435, 1987.

63. Silvius, J.R., Leventis, R., Brown, P.M., and Zuckerman M., Novel fluorescent phospholipids for assays of lipid mixing between membranes, *Biochemistry*, 26, 4279, 1987.

64. Leventis, R., and Silvius, J.R., Intermembrane lipid-mixing assays using acyl chain-labeled coumarinyl phospholipids, in *Membrane Fusion Techniques, Methods in Enzymology*, Vol. 220, Düzgüneş, N., Ed., Academic Press, San Diego, 1993, 32.

65. Hoekstra, D. and Düzgüneş, N., Lipid mixing assays to determine fusion in liposome systems, in *Membrane Fusion Techniques, Methods in Enzymology*, Vol. 220, Düzgüneş, N., Ed., Academic Press, San Diego, 1993, 15.

66. Düzgüneş, N. and Bentz, J., Fluorescence assays for membrane fusion, in *Spectroscopic Membrane Probes*, L.M. Loew, Ed., CRC Press, Boca Raton, Florida, 1988, 117.

67. Hoekstra, D., Yaron, A., Carmel, A., and Scherphof, G., Fusion of phospholipid vesicles containing a trypsin-sensitive fluorogenic substrate and trypsin, *FEBS Lett.*, 106, 176, 1979.

68. Wilschut, J. and Papahadjopoulos, D., Ca^{2+}-induced fusion of phospholipid vesicles monitored by mixing of aqueous contents, *Nature*, 281, 690, 1979.

69. Wilschut, J., Düzgüneş, N., Fraley, R., and Papahadjopoulos, D., Studies on the mechanism of membrane fusion: kinetics of Ca^{2+}-induced fusion of phosphatidylserine vesicles followed by a new assay for mixing of aqueous vesicle contents, *Biochemistry*, 19, 6011, 1980.

70. Ellens, H., Bentz, J., and Szoka, F. C., H$^+$- and Ca^{2+}-induced fusion and destabilization of liposomes, *Biochemistry*, 24, 3099, 1985.

71. Düzgüneş, N., Straubinger, R. M., Baldwin, P. A., Friend, D. S., and Papahadjopoulos, D., Proton-induced fusion of oleic/acid-phosphatidylethanolamine liposomes, *Biochemistry*, 24, 3091, 1985.

72. Stutzin, A., A fluorescence assay for monitoring and analyzing fusion of biological membrane vesicles in vitro, *FEBS Lett.*, 197, 274, 1986.

73. Düzgüneş, N. and Wilschut, J., Fusion assays monitoring intermixing of aqueous contents, in *Membrane Fusion Techniques, Methods in Enzymology*, Vol. 220, Düzgüneş, N., Ed., Academic Press, San Diego, 1993, 3.

74. Düzgüneş, N., Allen, T.M., Fedor, J. and Papahadjopoulos, D., 1988, Why fusion assays disagree, in *Molecular Mechanisms of Membrane Fusion*, Ohki, S., Doyle, D., Flanagan, T., Hui, S.W., and Mayhew, E., Eds., Plenum Press, New York, 1988, 543.

75. Portis, A., Newton, C., Pangborn, W., and Papahadjopoulos, D., Studies on the mechanism of membrane fusion: evidence for an intermembrane Ca^{2+}-phospholipid complex, synergism with Mg^{2+} and inhibition by spectrin, *Biochemistry*, 18, 780, 1979.

76. Düzgüneş, N., Wilschut, J., Fraley, R., and Papahadjopoulos, D., Studies on the mechanism of membrane fusion: role of head-group composition in calcium- and magnesium-induced fusion of mixed phospholipid vesicles, *Biochim. Biophys. Acta*, 642, 182, 1981.

77. Wilschut, J., Düzgüneş, N., Hong, K., Hoekstra, D., and Papahadjopoulos, D., Retention of aqueous contents during divalent cation-induced fusion of phospholipid vesicles, *Biochim. Biophys. Acta*, 734, 309, 1983.

78. Bentz, J., Düzgüneş N., and Nir, S., Kinetics of divalent cation induced fusion of phosphatidylserine vesicles: correlation between fusogenic capacities and binding affinities, *Biochemistry*, 22, 3320, 1983.

79. Nir, S., Düzgüneş, N. and Bentz, J., Binding of monovalent cations to phosphatidylserine and modulation of Ca^{2+}- and Mg^{2+}-induced vesicle fusion, *Biochim. Biophys. Acta*, 735, 160, 1983.

80. Hoekstra, D., Düzgüneş, N., and Wilschut, J., Agglutination and fusion of globoside GL-4 containing phospholipid vesicles mediated by lectins and Ca^{2+}, *Biochemistry*, 24, 565, 1985.

81. Ellens, H., Bentz, J., and Szoka, F.C., pH-induced destabilization of phosphatidylethanolamine-containing liposomes: role of inter-bilayer contact, *Biochemistry*, 23, 1532, 1984.

82. Sundler, R. and Papahadjopoulos, D., Control of membrane fusion by phospholipid head groups. I. Phosphatidate/phosphatidylinositol specificity, *Biochim. Biophys. Acta*, 649, 743, 1981.

83. Shiffer, K., Hawgood, S., Düzgüneş, N., and Goerke, J., The interactions of the low molecular weight group of surfactant-associated proteins (SP 5–18) with pulmonary surfactant lipids, *Biochemistry* 27, 2689, 1988.

84. Wilschut, J., Scholma, J., Bental, M., Hoekstra, D., and Nir, S., Ca^{2+}-induced fusion of phosphatidylserine vesicles: mass action kinetic analysis of membrane lipid mixing and aqueous contents mixing, *Biochim. Biophys. Acta*, 821, 45, 1985.

85. Walter, A. and Siegel, D.P., Divalent cation-induced lipid mixing between phosphatidylserine liposomes studied by stopped-flow fluorescence measurements: effect of temperature, comparison of barium and calcium, and perturbation by DPX, *Biochemistry*, 32, 3271, 1993.

86. Düzgüneş, N., Newton, C., Fisher, K., Fedor, J., and Papahadjopoulos, D., Monolayer coupling in phosphatidylserine bilayers: distinct phase transitions induced by magnesium interacting with one or both monolayers, *Biochim. Biophys. Acta*, 944, 391, 1988.

87. Hoekstra, D., de Boer, T., Klappe, K., and Wilschut, J., Fluorescence method for measuring the kinetics of fusion between biological membranes, *Biochemistry*, 23, 5675, 1984.

88. Hoekstra, D. and Klappe, K., Fluorescence assays to monitor fusion of enveloped viruses, in *Membrane Fusion Techniques, Methods in Enzymology*, Vol. 220, Düzgüneş, N., Ed., Academic Press, San Diego, 1993, 261.

89. Stegmann, T., Hoekstra, D., Scherphof, G., and Wilschut, J., Fusion activity of influenza virus. A comparison between biological and artificial target membrane vesicles, *J. Biol. Chem.*, 261, 10966, 1986.

90. Nir, S., Düzgüneş, N., de Lima, M.C.P., and Hoekstra, D., Fusion of enveloped viruses with cells and liposomes: activation and inactivation, *Cell Biophys.*, 17, 181, 1990.

91. Düzgüneş, N., Pedroso de Lima, M.C., Stamatatos, L., Flasher, D., Alford, D., Friend, D.S., and Nir, S., Fusion activity and inactivation of influenza virus: kinetics of low-pH induced fusion with cultured cells. *J. Gen Virol.*, 73, 27, 1992.

92. Ramalho-Santos, J., Nir, S., Düzgüneş, N., Carvalho, A.P., and de Lima, M.C.P., A common mechanism for influenza virus fusion activity and inactivation. *Biochemistry*, 32, 2771, 1993.

93. Nir, S., Klappe, K., and Hoekstra, D., Kinetics and extent of fusion between Sendai virus and erythrocyte ghosts: application of a mass action kinetic model, *Biochemistry* 25, 2155, 1986.

94. Klappe, K., Wilschut, J., Nir, S., and Hoekstra, D., Parameters affecting fusion between Sendai virus and liposomes. Role of viral proteins, liposome composition, and pH, *Biochemistry*, 25, 8252, 1986.

95. de Lima, M.C.P., Nir, S., Flasher, D., Klappe, K., Hoekstra, D., and Düzgüneş, N., Fusion of Sendai virus with human HL-60 and CEM cells: different kinetics of fusion for two isolates, *Biochim. Biophys. Acta*, 1070, 446, 1991.

96. Loyter, A. and Citovsky, V., The role of envelope glycoproteins in the fusion of Sendai virus with liposomes, in *Membrane Fusion*, Wilschut, J. and Hoekstra, D., Eds., Marcel Dekker, New York, 1991, 375.

97. Miller, N. and Hutt-Fletcher, L.M., A monoclonal antibody to glycoprotein gp85 inhibits fusion but not attachment of Epstein-Barr virus, *J. Virol.*, 62, 2366, 1988.

98. Gilbert, M.J., Mason, D., White, J.M., Fusion of Rous sarcoma virus with host cells does not require exposure to low pH, *J. Virol.*, 64, 5106, 1990.

99. Srinivasakumar, N., Ogra, P.L., and Flanagan, T.D., Characteristics of fusion of respiratory syncytial virus with HEp-2 cells as measured by R18 fluorescence dequenching assay, *J. Virol.*, 65, 4063, 1991.

100. Larsen, C.E., Alford, D.R., Young, L.J.T., McGraw, T.P., and Düzgüneş, N., Fusion of simian immunodeficiency virus with liposomes and erythrocyte ghost membranes: effects of lipid composition, pH and calcium. *J. Gen. Virol.*, 71, 1947, 1990.

101. Düzgüneş, N., Larsen, C.E., Konopka, K., Alford, D.R., Young, L.J.T., McGraw, T., Davis, B.R., Nir, S., and Jennings, M., Fusion of HIV-1 and SIV$_{mac}$ with liposomes and modulation of HIV-1 infectivity, in *Mechanisms and Specificity of HIV Entry into Host Cells*, Düzgüneş, N., Ed., Plenum Press, New York, 1991, 167.

102. Stamatatos, L. and Düzgüneş, N., Simian immunodeficiency virus (SIV$_{mac}$251) membrane lipid mixing with CD4$^+$ and CD4$^-$ cells *in vitro* does not necessarily result in internalization of the viral core proteins and productive infection, *J. Gen. Virol.*, 74, 1043, 1993.

103. Sinangil, F., Loyter, A., and Volsky, D.J., Quantitative measurement of fusion between human immunodeficiency virus and cultured cells using membrane fluorescence dequenching, *FEBS Lett.*, 239, 88, 1988.

104. Düzgüneş, N., Larsen, C.E., Stamatatos, L., and Konopka, K., Fusion of immunodeficiency viruses with liposomes and cells: inhibition of HIV-1 infectivity by cardiolipin liposomes, in *Membrane Interactions of Human Immunodeficiency Virus*, Aloia, R.C. and Curtain, C.C., Eds., Wiley-Liss, New York, 1992, 317.

105. Larsen, C.E., Nir, S., Alford, D.R., Jennings, M., Lee, K.-D., and Düzgüneş, N., Human immunodeficiency virus type 1 (HIV-1) fusion with model membranes: kinetic analysis and the effects of pH and divalent cations. *Biochim. Biophys. Acta*, 1147, 223, 1993.

106. Stamatatos, L., Nir, S., and Düzgüneş, N., Membrane lipid mixing and nucleocapsid entry are distinct events during human immunodeficiency virus infection, manuscript in preparation.

107. Amselem, S., Barenholz, Y., Loyter, A., Nir, S., and Lichtenberg, D., Fusion of Sendai virus with negatively charged liposomes as studied by pyrene-labelled phospholipid liposomes, *Biochim. Biophys. Acta*, 860, 301, 1986.

108. Barenholz, Y., Pal, R., and Wagner, R.R., Metabolic labeling of viral membrane lipids by fluorescent fatty acids: studying virus fusion with target membranes, in *Membrane Fusion Techniques, Methods in Enzymology*, Vol. 220, Düzgüneş, N., Ed., Academic Press, San Diego, 1993, 288.

109. Bron, R., Wahlberg, J.M., Garoff, H., and Wilschut, J., Membrane fusion of Semliki Forest virus in a model system: correlation between fusion kinetics and structural changes in the envelope glycoprotein, *EMBO J.*, 12, 693, 1993.

110. Papahadjopoulos, D., Nir, S., and Düzgüneş, N., Molecular mechanisms of calcium-induced membrane fusion, *J. Bioenerget. Biomembr.* 22, 157, 1990.

111. Papahadjopoulos, D., Poste, G., and Vail, W.J., Studies on membrane fusion with natural and model membranes, *Methods Membr. Biol.*, 10, 1, 1979.

112. Nir, S., Bentz, J., Wilschut, J., and Düzgüneş, N., Aggregation and fusion of phospholipid vesicles, *Prog. Surface Sci.*, 13, 1, 1983.

113. Sundler, R., Role of phospholipid head group structure and polarity in the control of membrane fusion, in *Membrane Fluidity, Biomembranes* Vol. 12, Kates, M. and Manson, L.A., Eds., Plenum Press, New York, 1984, 563.

114. Düzgüneş, N., Wilschut, J., and Papahadjopoulos, D., Control of membrane fusion by divalent cations, phospholipid head-groups and proteins, in *Physical Methods on Biological Membranes and their Model Systems*, Conti, F., Blumberg, W.E., DeGier, J., and Pocchiari, F., Eds., Plenum Press, New York, 1985, 193.

115. Düzgüneş, N., Hong, K., Baldwin, P.A., Bentz, J., Nir, S., and Papahadjopoulos, D., Fusion of phospholipid vesicles induced by divalent cations and protons. Modulation by phase transitions, free fatty acids, monovalent cations, and polyamines, in *Cell Fusion*, Sowers, A.E., Ed., Plenum Press, New York, 1987, 241.

116. Hong, K., Düzgüneş, N., Meers, P.R., and Papahadjopoulos, D., Protein modulation of liposome fusion, in *Cell Fusion*, Sowers, A.E., Ed., Plenum Press, New York, 1987, 269.

117. Prestegard, J.H. and O'Brien, M.P., Membrane and vesicle fusion, *Ann. Rev. Phys. Chem.*, 38, 383, 1987.

118. Nir, S., Van der Waals interactions between surfaces of biological interest, *Prog. Surf. Sci.*, 8, 1, 1977.

119. Wilschut, J., Membrane fusion in lipid vesicle systems. An overview, in *Membrane Fusion*, Wilschut, J. and Hoekstra, D., Eds., Marcel Dekker, New York, 1991, 89.

120. Nir, S., Bentz, J., and Düzgüneş, N., Two modes of reversible vesicle aggregation: particle size and the DLVO theory, *J. Colloid Interface Sci.*, 84, 266, 1981.

121. McLaughlin, S., Mulrine, N., Gresalfi, T., Vaio, G., and McLaughlin, A., The adsorption of divalent cations to bilayer membranes containing phosphatidylserine, *J. Gen. Physiol.*, 77, 445, 1981.

122. Ohki, S. and Düzgüneş, N., Divalent cation-induced interaction of phospholipid vesicle and monolayer membranes, *Biochim. Biophys. Acta*, 552, 438, 1979.

123. Ohki, S., A mechanism of divalent ion induced phosphatidylserine membrane fusion, *Biochim. Biophys. Acta*, 689, 1, 1982.

124. Ohki, S. and Ohshima, H., Divalent cation-induced surface tension increase in acidic phospholipid membranes. Ion binding and membrane fusion, *Biochim. Biophys. Acta*, 776, 177, 1984.

125. Shavnin, S.A., Pedroso de Lima, M.C., Fedor, J., Wood, P., Bentz, J., and Düzgüneş, N., Cholesterol affects divalent cation-induced fusion and isothermal phase transitions of phospholipid membranes, *Biochim. Biophys. Acta*, 946, 405, 1988.

126. Bentz, J. and Düzgüneş, N., Fusogenic capacities of divalent cations and the effect of liposome size, *Biochemistry*, 24, 5436, 1985.

127. Hammoudah, M.M., Nir, S., Bentz, J., Mayhew, E., Stewart, T.P., Hui, S.W., and Kurland, R.J., Interactions of La^{3+} with phosphatidylserine vesicles. Binding, phase transition, leakage, 31P-NMR and fusion, *Biochim. Biophys. Acta*, 645, 102, 1981.

128. Bentz, J., Alford, D., Cohen, J., and Düzgüneş, N., 1988, La^{3+}-induced fusion of phosphatidylserine liposomes. Close approach, intermembrane intermediates, and the electrostatic membrane potential, *Biophys. J.*, 53, 593.

129. Freeman, K.B., unpublished data.

130. Sundler, R., Düzgüneş, N., and Papahadjopoulos, D., Control of membrane fusion by phospholipid head groups. II. The role of phosphatidylethanolamine in mixtures with phosphatidate and phosphatidylinositol, *Biochim. Biophys. Acta*, 649, 751, 1981.

131. Düzgüneş, N., Nir, S., Wilschut, J., Bentz, J., Newton, C., Portis, A., and Papahadjopoulos, D., Calcium- and magnesium-induced fusion of mixed phosphatidylserine/phosphatidylcholine vesicles: Effect of ion binding, *J. Membr. Biol.*, 59, 115, 1981.

132. Liao, M.-J. and Prestegard, J.H., Ion specificity in fusion of phosphatidic acid-phosphatidylcholine mixed lipid vesicles, *Biochim. Biophys. Acta*, 601, 453, 1980.

133. Jendrasiak, G.L. and Hasty, J.H., The hydration of phospholipids, *Biochim. Biophys. Acta*, 337, 79, 1974.

134. Hauser, H., Pascher, I., Pearson, R.H., and Sundell, S., Preferred conformation and molecular packing of phosphatidylethanolamine and phosphatidylcholine, *Biochim. Biophys. Acta*, 650, 21, 1981.

135. Reiss-Husson, F., Structure des phases liquid-crystallines de différents phospholipides, monoglycerides, sphingolipides, anhydrides ou en présence d'eau, *J. Mol. Biol.*, 25, 363, 1967.

136. Cullis, P.R., Tilcock, C.P., and Hope, M.J., Lipid polymorphism, in *Membrane Fusion*, Wilschut, J. and D. Hoekstra, Eds., Marcel Dekker, New York, 1991, 35.

137. Parsegian, V.A. and Rand R.P., Forces governing lipid interaction and rearrangement, in *Membrane Fusion*, Wilschut, J. and D. Hoekstra, Eds., Marcel Dekker, New York, 1991, 65.

138. Düzgüneş, N., Paiement, J., Freeman, K.B., Lopez, N.G., Wilschut, J., and Papahadjopoulos, D., Modulation of membrane fusion by ionotropic and thermotropic phase transitions, *Biochemistry*, 23, 3486, 1984.

139. Silvius, J.R. and Gagné, J., Calcium-induced fusion and lateral phase separations in phosphatidylcholine-phosphatidylserine vesicles. Correlation by calorimetric and fusion measurements, *Biochemistry*, 23, 3241, 1984.

140. Silvius, J.R. and Gagné, J., Lipid phase behavior and calcium-induced fusion of phosphatidylethanolamine-phosphatidylserine vesicles. Calorimetric and fusion studies, *Biochemistry*, 23, 3232, 1984.

141. Felgner, P.L., Gadek, T. R., Holm, M., Roman, R., Chan, H.W., Wenz, M., Northrop, J.P., Ringold, G.M., and Danielsen, M., Lipofectin: a highly efficient, lipid-mediated DNA-transfection procedure, *Proc. Natl. Acad. Sci. USA*, 84, 7413, 1987.

142. Düzgüneş, N. and Felgner, P.L., Intracellular delivery of of nucleic acids and transcription factors by cationic liposomes, in *Membrane Fusion Techniques*, *Methods in Enzymology*, Vol. 221, Düzgüneş, N., Ed., Academic Press, San Diego, 1993, 303.

143. Carmona-Ribeiro, A.M., Yoshida, L.S., and Chaimovich, H., Salt effects on the stability of dioctadecyldimethylammonium chloride and sodium dihexadecyl phosphate vesicles, *J. Phys. Chem.*, 89, 2928, 1985.

144. Rupert, L.A.M., Hoekstra, D., and Engberts, J. B.F.N., Fusogenic behavior of didodecyl-dimethylammonium bromide bilayer vesicles, *J. Am. Chem. Soc.*, 107, 2628, 1985.

145. Rupert, L.A.M., Engberts, J.B.F.N., and Hoekstra, D., Role of membrane hydration and membrane fluidity in the mechanism of anion-induced fusion of didodecyldimethylammonium bromide vesicles, *J. Am. Chem. Soc.* 108, 3920, 1986.

146. Wilschut, J., Düzgüneş, N., Hoekstra, D., and Papahadjopoulos, D., Modulation of membrane fusion by membrane fluidity: temperature dependence of divalent cation-induced fusion of phosphatidylserine vesicles, *Biochemistry*, 24, 8, 1985.

147. Stewart, T.P., Hui, S.W., Portis, A.R., and Papahadjopoulos, D., Complex phase mixing of phosphatidylcholine and phosphatidylserine in multilamellar membrane vesicles, *Biochim. Biophys. Acta*, 556, 1, 1979.

148. Bentz, J., Düzgüneş, N., and Nir, S., Temperature dependence of divalent cation induced fusion of phosphatidylserine liposomes: evaluation of the kinetic rate constants, *Biochemistry*, 24, 1064, 1985.

149. Cullis, P.R. and Hope, M.J., Effects of fusogenic agent on membrane structure of erythrocyte ghosts and the mechanism of membrane fusion, *Nature*, 271, 672, 1978.

150. Cullis, P.R. and Verkleij, A.J., Modulation of membrane structure by Ca^{2+} and dibucaine as detected by ^{31}P NMR, *Biochim. Biophys. Acta*, 552, 546, 1979.

151. Verkleij, A.J., van Echteld, C.J.A., Gerritsen, W.J., Cullis, P.R., and de Kruijff, B., The lipidic particle as an intermediate structure in membrane fusion processes and bilayer to hexagonal H_{II} transitions, *Biochim. Biophys. Acta*, 600, 620, 1980.

152. Verkleij, A.J., Leunissen-Bijvelt, J., de Kruijff, B., Hope, M., and Cullis, P.R., Non-bilayer structures in membrane fusion, in *Cell Fusion, Ciba Foundation Symposium*, Vol. 103, Pitman Books, London, 1984, 45.

153. Hope, M.J., Walker, D.C., and Cullis, P.R., Calcium and pH-induced fusion of small unilamellar vesicles consisting of phosphatidylethanolamine and negatively charged phospholipids: a freeze-fracture study, *Biochem. Biophys. Res. Commun.*, 110, 15, 1983.

154. Verkleij, A.J., Role of nonbilayer lipids in membrane fusion, in *Membrane Fusion*, Wilschut, J. and Hoekstra, D., Eds., Marcel Dekker, New York, 1991, 155.

155. Hui, S., Nir, S., Stewart, T.P., Boni, L.T., and Huang, S.K., Kinetic measurements of fusion of phosphatidylserine-containing vesicles by electron microscopy and fluorometry, *Biochim. Biophys. Acta*, 941, 130, 1988.

156. Siegel, D.P., Inverted micellar intermediates and the transitions between lamellar, cubic and inverted hexagonal lipid phases. II. Implications for membrane-membrane interactions and membrane fusion, *Biophys. J.*, 49, 1171, 1986.

157. Ellens, H., Siegel, D.P., Alford, D., Yeagle, P.L., Boni, L., Lis, L.J., Quinn, P.J., and Bentz, J., Membrane fusion and inverted phases, *Biochemistry*, 28, 3692, 1989.

158. Allen, T.M., Hong, K. and Papahadjopoulos, D., Membrane contact, fusion and hexagonal H_{II} transitions in phosphatidylethanolamine liposomes, *Biochemistry*, 29, 2976, 1990.

159. Hui, S.W., Stewart, T.P., Boni, L.T., and Yeagle, P.L., Membrane fusion through point defects in bilayers, *Science*, 212, 921, 1981.

160. Markin, V.S., Kozlov, M.M., and Borovjagin, V.L., On the theory of membrane fusion. The stalk mechanism, *Gen. Physiol. Biophys.*, 5, 361, 1984.

161. Chernomordik, L.V., Kozlov, M.M., Melikyan, G.B., Abidor, I.G., Markin, V.S., and Chizmadzhev, Y.A., The shape of lipid molecules and monolayer membrane fusion, *Biochim. Biophys. Acta*, 812, 643, 1985.

162. Helm, C.A., Israelachvili, J.N., and McGuiggan, P.M., Molecular mechanisms and forces involved in the adhesion and fusion of amphiphilic bilayers, *Science*, 246, 919, 1989.

163. Chernomordik, L.V., Vogel, S.S., Sokoloff, A., Onaran, H.O., Leikina, E.A., and Zimmerberg, J., Lysolipids reversibly inhibit Ca^{2+}-, GTP- and pH-dependent fusion of biological membranes, *FEBS Lett.*, 318, 71, 1993.

164. Düzgüneş, N., Cholesterol and membrane fusion, in *Biology of Cholesterol*, Yeagle, P.L., Ed., CRC Press, Boca Raton, FL, 1988, 197.

165. Walter, A., Eidelman, O., Ollivon, M., and Blumenthal, R., Functional reconstitution of viral envelopes, in *Membrane Fusion*, Wilschut, J. and Hoekstra, D., Eds., Marcel Dekker, New York, 1991, 395.

166. Bron, R., Ortiz, A., Dijkstra, J., Stegmann, T., and Wilschut, J., Preparation, properties, and applications of reconstituted influenza virus envelopes (virosomes), in *Membrane Fusion Techniques, Methods in Enzymology*, Vol. 220, Düzgüneş, N., Ed., Academic Press, San Diego, 1993, 313.

167. Sundler, R. and Wijkander, J., Protein-mediated intermembrane contact specifically enhances Ca^{2+}-induced fusion of phosphatidate-containing membranes, *Biochim. Biophys. Acta*, 730, 391, 1983.

168. Düzgüneş, N., Hoekstra, D., Hong, K., and Papahadjopoulos, D., Lectins facilitate calcium-induced fusion of phospholipid vesicles containing glycosphingolipids, *FEBS Lett.*, 173, 80, 1984.

169. Hoekstra, D. and Düzgüneş, N., *Ricinus communis* agglutinin-mediated agglutination and fusion of glycolipid-containing phospholipid vesicles. Effect of carbohydrate headgroup size, calcium ions and spermine, *Biochemistry*, 25, 1321, 1986.

170. Düzgüneş, N. and Hoekstra, D., Agglutination and fusion of glycolipid-phospholipid vesicles mediated by lectins and calcium ions, *Studia Biophysica*, 111, 5, 1986.

171. Hoekstra, D. and Düzgüneş, N., Lectin-carbohydrate interactions in model and biological membrane systems, in *Subcellular Biochemistry*, Vol 14, Harris, J.R. and Etamadi, A.-H., Eds., Plenum Press, New York, 1989, 229.

172. deKruijff, B., de Gier, J., van Hoogevest, P., van der Steen, N., Taraschi, T.F., and de Kroon, T., Effects of an integral membrane glycoprotein on phospholipid vesicle fusion, in *Membrane Fusion*, Wilschut, J. and D. Hoekstra, Eds., Marcel Dekker, New York, 1991, 209.

173. Ekerdt, R. and Papahadjopoulos, D., Intermembrane contact affects calcium binding to phospholipid vesicles, *Proc. Natl. Acad. Sci. USA*, 79, 2273, 1982.

174. Düzgüneş, N. and Papahadjopoulos, D., Ionotropic effects on phospholipid membranes: Calcium-magnesium specificity in binding, fluidity, and fusion, in *Membrane Fluidity in Biology*, Vol. 2, Aloia, R.C., Ed., Academic Press, New York, 1983, 187.

175. Nir, S., A model for cation adsorption in closed systems: application to calcium binding to phospholipid vesicles, *J. Coll. Interface Sci.*, 102, 313, 1984.

176. Feigenson, G.W., On the nature of calcium ion binding between phosphatidylserine lamellae, *Biochemistry*, 25, 5819, 1986.

177. Blackwood, R.A. and Ernst, J.D., Characterization of Ca^{2+}-dependent phospholipid binding, vesicle aggregation and membrane fusion by annexins, *Biochem. J.*, 266, 195, 1990.

178. Francis, J.W., Balazovich, K.J., Smolen, J.E., Margolis, D.I., and Boxer, L.A., Human neutrophil annexin I promotes granule aggregation and modulates Ca^{2+}-dependent membrane fusion, *J. Clin. Invest.*, 90, 537, 1992.

179. Seppen, J., Ramalho-Santos, J., de Carvalho, A.P., ter Best, M., Kok, J.W., Pedroso de Lima, M.C., and Hoekstra, D., Interaction of clathrin with large unilamellar vesicles at neutral pH. Lipid dependence and protein penetration, *Biochim. Biophys. Acta*, 1106, 209, 1992.

180. Maezawa, S. and Yoshimura, T., Determination of the regions of the clathrin molecule inducing membrane fusion, *Biochemistry*, 29, 1813, 1990.

181. Hong, K. and Vacquier, V.D., Fusion of liposomes induced by a cationic protein from the acrosome granule of abalone spermatozoa, *Biochemistry*, 25, 543, 1986.

182. Poulain, F.R., Allen, L., Williams, M.C., Hamilton, R.L., and Hawgood, S., Effects of surfactant apolipoproteins on liposome structure: implications for tubular myelin formation, *Am. J. Physiol.*, 262, L730, 1992.

183. Gad, A.E., Bental, M., Elyashiv, G., Weinberg, H., and Nir, S., Promotion and inhibition of vesicle fusion by polylysine, *Biochemistry*, 24, 6277, 1985.

184. Bondeson, J. and Sundler, R., Lysine peptides induce lipid intermixing but not fusion of phosphatidic acid-containing vesicles, *FEBS Lett.*, 190, 283, 1985.

185. Eytan, G.D. and Almary, T., Mellitin-induced fusion of acidic liposomes, *FEBS Lett.*, 156, 29, 1983.

186. Beigel, M., Keren-Zur, M., Laster, Y., and Loyter, A., Poly(aspartic acid)-dependent fusion of liposomes bearing the quaternary ammonium detergent [[[(1,1,3,3-tetramethylbutyl) cresoxy]-ethoxy]ethyl]dimethylbenzylammonium hydroxide, *Biochemistry*, 27, 660, 1988.

187. Keren-Zur, M., Beigel, M., and Loyter, A., Induction of fusion in aggregated and nonaggregated liposomes bearing cationic detergents, *Biochim. Biophys. Acta*, 983, 253, 1989.

188. Suenaga, M., Lee, S., Park, N.G., Aoyagi, H., Kato, T., Umeda, A., and Amako, K., Basic amphipathic helical peptides induce destabilization and fusion of acidic and neutral liposomes, *Biochim. Biophys. Acta*, 981, 143, 1989.

189. Yoshimura, T., Goto, Y., and Aimoto, S., Fusion of phospholipid vesicles induced by an amphiphilic model peptide: close correlation between fusogenicity and hydrophobicity of the peptide in an α-helix, *Biochemistry*, 31, 6119, 1992.

190. Rapaport, D., Hague, G.R., Pouny, Y., and Shai, Y., pH- and ionic strength–dependent fusion of phospholipid vesicles induced by pardaxin analogues or by mixtures of charge-reversed peptides, *Biochemistry*, 32, 3291, 1993.

191. Ohnishi, S.-I., Fusion of viral envelopes with cellular membranes, in *Membrane Fusion in Fertilization, Cellular Transport and Viral Infection*, Düzgüneş, N. and Bronner, F., Eds., Academic Press, San Diego, 1988, 257.

192. Murata, M., Sugahara, Y., Takahashi, S., and Ohnishi, S.-I., pH-Dependent membrane fusion activity of a synthetic twenty amino acid peptide with the same sequence as that of the hydrophobic segment of influenza virus hemagglutinin, *J. Biochem.*, 102, 957, 1987.

193. Rafalski, M., Ortiz, A., Rockwell, A., van Ginkel, L.C., Lear, J., De Grado, W.F., and Wilschut, J., Membrane fusion activity of the influenza virus hemagglutinin: interaction of HA2 N-terminal peptides with phospholipid vesicles, *Biochemistry*, 30, 10211, 1991.

194. Düzgüneş, N. and Gambale, F., Membrane action of synthetic peptides from influenza virus hemagglutinin and its mutants, *FEBS Lett.*, 227, 110, 1988.

195. Düzgüneş, N. and Shavnin, S.A., Membrane destabilization by N-terminal peptides of viral envelope proteins. *J. Membrane Biol.*, 128, 71, 1992.

196. Burger, K.N.J., Wharton, S.A., Demel, R.A., and Verkleij, A.J., The interaction of synthetic analogs of the N-terminal fusion sequence of influenza virus with a lipid monolayer. Comparison of fusion-active and fusion-defective analogs, *Biochim. Biophys. Acta*, 1065, 121, 1991.

197. Wharton, S.A., Martin, S.R., Ruigrok, R.W.H., Skehel, J.J., and Wiley, D. C., Membrane fusion by peptide analogues of influenza virus hemagglutinin, *J. Gen. Virol.*, 69, 1847, 1988.

198. Stegmann, T., Doms, R.W., and Helenius, A., Protein-mediated membrane fusion, *Annu. Rev. Biophys. Chem.*, 18, 187, 1989.

199. Murata, M., Takahashi, S., Kagiwada, S., Suzuki, A., and Ohnishi, S.-I., pH-dependent membrane fusion and vesiculation of phospholipid large unilamellar vesicles induced by amphiphilic anionic and cationic peptides, *Biochemistry*, 31, 1986, 1992.

200. Murata, M., Kagiwada, S., Hishida, R., Ishiguro, R., Ohnishi, S.-I., and Takahashi, S., Modification of the N-terminus of membrane fusion-active peptides blocks the fusion activity, *Biochem. Biophys. Res. Commun.*, 179, 1050, 1991.

201. Murata, M., Kagiwada, S., Takahashi S., and Ohnishi, S., Membrane fusion induced by mutual interaction of the two charge-reversed amphiphilic peptides at neutral pH, *J Biol. Chem.*, 266, 14353, 1991.

202. Nieva, J., Nir, S., and Wilschut, J., unpublished data.

203. Slepushkin, V.A., Andreev, V.A., Sidorova, M.V., Melikyan, G.B., Grigoriev, V.B., Chumakov, V.M., Grinfeldt, A.E., Manukyan, R.A., and Karamov, E.V., Investigation of human immunodeficiency virus fusion peptides. Analysis of interrelations between their structure and function, *AIDS Res. Hum. Retroviruses*, 8, 9, 1992.

204. Martin, I., Defrise-Quertain, F., Decroly, E., Vandenbranden, M., Brasseur, R., and Ruysschaert, J.-M., Orientation and structure of the NH_2-terminal HIV-1 gp41 peptide in fused and aggregated liposomes, *Biochim. Biophys. Acta*, 1145, 124, 1993.

205. Martin, I., Defrise-Quertain, F., Mandieau, V., Saermark, T., Burny, A., Brasseur, R., Ruysschaert, J.-M., and Vandenbranden, M., Fusogenic activity of SIV (simian immunodeficiency virus) peptides located in the gp32 NH2 domain, *Biochem. Biophys. Res. Commun.*, 175, 872, 1991.

206. Glushakova, S.E., Omelyanenko, V.G., Lukashevitch, I.S., Bogdanov, A.A., Moshnikova, A.B., Kozytch, A.T., and Torchilin, V.P., The fusion of artificial lipid membranes induced by the synthetic arenavirus 'fusion peptide,' Biochim. Biophys. Acta, 1110, 202, 1992.

207. Epand, R.M., Cheetham, J.J., Epand, R.F., Yeagle, P.L., Richardson, C.D., Rockwell, A., and DeGrado, W.F., Peptide models for the membrane destabilizing actions of viral fusion proteins, *Biopolymers*, 32, 309, 1992.

208. Parente, R.A., Nir, S., and Szoka, F.C. Jr., Mechanism of leakage of phospholipid contents induced by the peptide GALA, *Biochemistry*, 29, 8720, 1990.

209. Parente, R.A. and Szoka, F., unpublished data.

210. Meers, P., Hong, K., Bentz, J., and Papahadjopoulos, D., Spermine as a modulator of membrane fusion: interaction with acidic phospholipids, *Biochemistry*, 25, 3109, 1986.

211. Pollard, H.B., Rojas, E., Burns, A.L. and Parra, C., Synexin, calcium and the hydrophobic bridge hypothesis for membrane fusion, in *Molecular Mechanisms of Membrane Fusion*, Ohki, S., Doyle, D., Flanagan, T., Hui, S.W. and Mayhew, E., Eds., Plenum, New York, 1988, 341.

212. Hoekstra, D., Membrane fusion of enveloped viruses: especially a matter of proteins, *J. Bioenerget. Biomembr.*, 22, 121, 1990.

213. Larsen, C., Ellens, H., and Bentz, J., Membrane fusion induced by the HIV env glycoprotein, in *Membrane Interactions of Human Immunodeficiency Virus*, Aloia, R.C and Curtain, C.C., Eds., Wiley-Liss, New York, 1992, 143.

214. Nir, S., Bentz, J., and Wilschut, J., Mass action kinetics of phosphatidylserine vesicle fusion as monitored by coalescence of internal vesicle volumes, *Biochemistry*, 19, 6030, 1980.

215. Bentz, J., Nir, S., and Wilschut, J., Mass action kinetics of vesicle aggregation and fusion, *Colloids Surfaces*, 6, 333, 1983.

216. Nir, S., Stegmann, T. and Wilschut, J., Fusion of influenza virus with cardiolipin liposomes at low pH: mass action analysis of kinetics and fusion, *Biochemistry* 25, 257, 1986.

217. Nir, S., Stutzin, A., and Pollard, H.B., Effect of synexin on aggregation and fusion of chromaffin granule ghosts at pH 6, *Biochim. Biophys. Acta*, 903, 309, 1987.

218. Nir, S., Modeling of aggregation and fusion of phospholipid vesicles, in *Membrane Fusion*, Wilschut, J. and Hoekstra, D., Eds., Marcel Dekker, New York, 1991, 127.

219. Braun, G., Lelkes, P.I., and Nir, S., Effect of cholesterol on Ca^{2+}-induced aggregation and fusion of sonicated phosphatidylserine/cholesterol vesicles, *Biochim. Biophys. Acta.*, 812, 688, 1985.

220. Pollard, H.B., Burns, A.L., and Rojas, E., Synexin (annexin VII)): a cytosolic calcium binding protein which promotes membrane fusion and forms calcium channels in artificial bilayer and natural membranes, *J. Membr. Biol.*, 117, 101, 1990.

221. Creutz, C.E., Cis-unsaturated fatty acids induce the fusion of chromaffin granules aggregated by synexin, *J. Cell Biol.*, 91, 247, 1981.

222. Bental, M., Wilschut, J., Scholma, J., and Nir, S., Ca^{2+}-induced fusion of large unilamellar phosphatidylserine/cholesterol vesicles, *Biochim. Biophys. Acta*, 898, 239, 1987.

223. Wilschut, J., Nir, S., Scholma, J., and Hoekstra, D., Kinetics of Ca^{2+}-induced fusion of cardiolipin-phosphatidylcholine vesicles: correlation between vesicle aggregation, bilayer destabilization, and fusion, *Biochemistry*, 24, 4630, 1985.

224. Wilschut, J., Nieva, J.L., and Nir, S., in preparation.

225. Nir, S., Wilschut, J., and Bentz, J., The rate of fusion of phospholipid vesicles and the role of bilayer curvature, *Biochim. Biophys. Acta*, 688, 275, 1982.

226. Oshry, L., Meers, P., Mealy, T., and Tauber, A.I., Annexin-mediated membrane fusion of human neutrophil plasma membranes and phospholipid vesicles, *Biochim. Biophys. Acta*, 1006, 239, 1991.

227. Meers, P., Mealy, T., Pavlotsky, N., and Tauber, A.I., Annexin I–mediated vesicular aggregation: mechanism and role in human neutrophils, *Biochemistry*, 31, 6472, 1992.

228. Kagiwada, S., Murata, M., Hishida, R., Tagaya, M., Yamasgina, S., and Ohnishi, S.-I., *In vitro* fusion of rabbit liver Golgi membranes with liposomes, *J. Biol. Chem.*, 268, 1430, 1993.

229. Martinez-Bazenet, C., Audigier-Petit, C., Frot-Coutaz, J., Got, R., Nicolau, C., and Létoublon, R., Protein-mediated fusion of liposomes with microsomal membranes of *Aspergillus niger*: evidence for a complex mechanism dealing with membranous and cytosolic fusogenic proteins, *Biochim. Biophys. Acta*, 943, 35, 1988.

230. Vogel, S.S., Chernomordik, L.V., and Zimmerberg, J., Calcium-triggered fusion of exocytotic granules requires proteins in only one membrane, *J. Biol. Chem.*, 267, 25640, 1992.

231. Lawaczeck, R., Gervais, M., Nandi, P.K., and Nicolau, C., Fusion of negatively charged liposomes with clathrin-uncoated vesicles, *Biochim. Biophys. Acta*, 903, 112, 1987.

232. Kielian, M. and Helenius, A., Role of cholesterol in fusion of Semliki Forest virus with liposomes, *J. Virol.*, 52, 281, 1984.

233. Konopka, K., Davis, B.R., Larsen, C.E., Alford, D.R., Debs, R.J., and Düzgüneş, N., Liposomes modulate human immunodeficiency virus infectivity. *J. Gen. Virol.*, 71, 2899, 1990.

234. Konopka, K., Davis, B.R., Larsen, C.E., and Düzgüneş, N., Anionic liposomes inhibit human immunodeficiency virus type 1 (HIV-1) infectivity in CD4+ A3.01 and H9 cells, *Antiviral Chem. Chemother.*, 4, 179, 1993.

235. de Lima, M.C.P., Nir, S., Flasher, D., and Düzgüneş, N., unpublished data.

236. van Gorkom, L.C.M., Cheetham, J.J., and Epand, R., personal communication.

237. de Lima, M.C.P., Ramalho-Santos, J., Martins, M.F., Carvalho, A.P, Bairos, V., and Nir, S., Kinetic modeling of Sendai virus fusion with PC-12 cells: effect of pH and temperature on fusion and viral inactivation, *Eur. J. Biochem.*, 205, 181, 1992.

238. Nir, S., Stegmann, T., Hoekstra, D., and Wilschut, J., Kinetics and extents of fusion of influenza virus and Sendai virus with liposomes, in *Molecular Mechanisms of Membrane Fusion*, Ohki, S., Doyle, D., Flanagan, T.D., Hui, S.W., and Mayhew, E., Eds., Plenum, New York, 1988, 451.

239. Ramalho-Santos, J., de Lima, M.C.P., and Nir, S., unpublished data.

240. Novick, S.L. and Hoekstra, D., Membrane penetration of Sendai virus glycoproteins during the early stages of fusion with liposomes as determined by hydrophobic photoaffinity labeling, *Proc. Natl. Acad. Sci. USA*, 85, 7433, 1988.

241. Harter, C., James, P., Bächi, T., Semenza, G., and Brunner, J., Hydrophobic binding of the ectodomain of influenza hemagglutinin to membranes occurs through the "fusion peptide." *J. Biol. Chem.*, 264, 6459, 1989.

242. Nir, S., Klappe, K., and Hoekstra, D., Mass action analysis of kinetics and extent of fusion between Sendai virus and liposomes, *Biochemistry*, 25, 8261, 1986.

Chapter 8

Liposomes, Viruses, and Membrane Fusion

Maria C.P. de Lima and Dick Hoekstra

CONTENTS

I. INTRODUCTION

Membrane-bounded or so-called enveloped viruses penetrate and infect their host cells by transfer of the genetic material contained within the viral nucleocapsid into the cytoplasm of the target cell. This transfer mechanism is accomplished by a process involving fusion of the viral envelope with a cellular target membrane.

Two different pathways have evolved that appear to mediate the entry, the route depending on the virus family. Paramyxoviruses (e.g., Sendai virus) enter the cell by fusing with the plasma membrane in the neutral pH range.[1,2] Most other enveloped viruses, such as togaviruses (e.g., Semliki Forest virus, SFV), rhabdoviruses (e.g., vesicular stomatitis virus, VSV), and myxoviruses (e.g., Influenza virus) are taken up by the host cells via receptor-mediated endocytosis and the fusion of cellular and viral membranes is triggered by the acidic pH within endosomal compartments.[1,2] Usually, these viruses display sharp pH-dependent fusion profiles with threshold pH values ranging from 5.0 to 6.5. Within this range, the pH at which the virus acquires optimal fusion competence depends on the virus involved. The route of infectious entry of the retroviruses, human immunodeficiency virus (HIV) and simian immunodeficiency virus from macaques (SIV_{mac}), the causative agents of the acquired immunodeficiency syndrome (AIDS) and simian AIDS respectively, is unclear. Although most reports point to fusion of these viruses with the plasma membrane, i.e., entry occurs in the neutral pH range,[3-5] endocytosis preceding viral-cell membrane fusion (i.e., at the endosomal level) may provide an additional entry pathway.[6-8]

Most enveloped viruses have a relatively simple membrane composition, consisting of the lipids derived from the host cell, which are acquired at the site of viral budding, and containing often as few as one or two specific, virally encoded glycoproteins. It has been firmly established that these specific envelope proteins are intimately involved in triggering the fusion reaction. The most convincing evidence that the fusion activity is confined to a single viral glycoprotein has been obtained from numerous studies in which viral glycoproteins from cloned copies of their genes were expressed in eukaryotic cells. After appearance of the protein on the plasma membrane, cell-cell fusion was observed to occur.[9-16] Additional evidence has been provided by demonstrating fusion activity of reconstituted purified viral fusion proteins into liposomes.[17-20]

Although considerable progress has been made in elucidating the role of various viral proteins in the fusion process, the molecular mechanism underlying this event remains elusive. During the last decade extensive research has been devoted to characterizing the parameters modulating the fusion activity of viruses, as an attempt to gain insight into the molecular details of virus-membrane interactions. The complexity of biological membranes has obviously prompted the search for more simple and controllable systems as targets for assessing virus fusion activity. Because of their simplicity and the ease of manipulating their composition, liposomes (phospholipid vesicles) provide a valuable target membrane system for *in vitro* characterization of the fusion properties of viruses. Obviously, such artificial membranes are by far too simple a model for clarifying the molecular details relevant to virus-cell interactions, where specific proteins must be involved too. However, it is equally clear that when care is taken to use conditions that are comparable to those occurring *in vivo*, striking similarities can be recognized between the characteristics of fusion of the viral membrane with liposomes and those observed upon virus fusion with biological targets.

This chapter will review studies on interaction of lipid enveloped viruses with liposomes, and the various approaches used, emphasizing the role of modulating factors, particularly the target membrane composition, on virus fusion activity. Some attention will be paid to kinetic studies on the interaction of viruses with liposomal membranes using assays based on fluorescence. The results of these studies will be used to compare the characteristics of virus fusion toward liposomes with those toward biological membranes. Our goal is to identify some relevant features of virus-liposome interactions in an attempt to build the conceptual framework for understanding the mechanism of virus-cell interactions, as a crucial step in the infectious entry of viruses into cells.

II. VIRAL ENVELOPE PROTEINS AND THEIR ROLE IN MEMBRANE FUSION

As depicted in the Introduction, viral envelope proteins mediate the ability of a viral membrane to bind to and fuse with a cellular target membrane. Hence, the functional role of viral envelope proteins is of crucial significance to virus penetration and cellular infection. In this context, it should be noted that entry *per se* is essential for viral reproduction, but not every penetrating virus necessarily leads to an "infective" reproduction.

Over the past decade, a large number of viral membrane fusion proteins have been identified, cloned and sequenced. Their properties and functions have been established and described in great detail in numerous papers and have been the topic of several recent reviews.[1,2,21-25] This section will briefly summarize some general aspects of the characteristics of these proteins and their functional role in viral membrane fusion. For details, the reader is referred to the above mentioned literature.

Two specific functions are displayed by these proteins; binding to receptors of the target cell surface and fusion of viral membrane with a cellular membrane. The binding and fusion activities can be confined to the same protein, which is often seen for viruses that enter via the endocytic pathway, while viruses entering cells by fusion at the cell surface frequently have separate binding and fusion proteins.[26,27] In this respect, HIV may be an exception to this latter role as the binding and fusion activity reside in the same protein in spite of the claim that HIV enters primarily at the plasma membrane.[3,4]

Examination of the viral fusion proteins at the molecular level has shown that some of these proteins are synthesized as larger inactive precursors. To acquire fusion activity, the proteins need to be processed, which for some but not all viral fusion proteins (see below) involves a cleavage by cellular proteases occurring late in the biosynthetic pathway. Some typical examples of fusion proteins that require such a proteolytic cleavage are the fusion protein, F, of paramyxoviruses; the hemagglutinin, HA, of Influenza virus; and the gp160 of HIV. Cleavage results in the generation of two polypeptide chains; the C-terminal product anchoring the protein in the viral envelope and the N-terminal product remaining associated through disulfide bonds or through non-covalent interactions.

A striking feature of the N-terminal regions is that they are exceptionally hydrophobic, usually containing 20–25 apolar amino acids residues, while they are also highly conserved within virus families.[21,27-29] In addition, site specific oligonucleotide-directed mutagenesis of these regions has deleterious effects on viral fusion activity.[30,31] These observations provide strong evidence that these domains are intimately involved in the overall mechanism of viral penetration and, given their hydrophobicity, particularly in triggering viral fusion. For this reason they are referred to as "fusion peptides."

The fusion protein, F, of Sendai virus, the best studied member of the paramyxoviruses, consists of two polypeptides, F_1 and F_2, which are derived from the proteolytic cleavage of the precursor, F_0, and are linked to one another by a disulfide bridge.[26,32] The N-terminal end, exclusively apolar, generated by this activation process is located in the F_1 polypeptide and exposed to aqueous medium. The initial virus-target membrane binding is mediated by a separate binding protein, HN, which exhibits hemagglutinin and neuraminidase activities. Apart from its binding capacity, it has been suggested that the HN protein may also be involved in the modulation of the viral fusion activity.[26] This potential regulatory role of HN in the fusion process most likely involves an indirect effect on the rotational and lateral mobility of the F protein.[33-36] The spatial arrangement of the F protein of Sendai virus is still unknown. However, a tetrameric structure of the F spike, possibly consisting of two identical dimers, has been suggested based on chemical cross-linking studies.[26,37]

A similar activation mechanism observed for HA of Influenza virus[38,39] results in the formation of two glycopeptides, HA_1 and HA_2, derived by proteolytic cleavage from a common precursor HA_0.[26] The hemagglutinin of Influenza virus has been the most extensively characterized viral fusion protein and the only viral envelope protein for which detailed structural information is available.[40] The two polypeptides, HA_1 and HA_2, are covalently linked by a disulfide bond, and the two chain monomers are associated noncovalently to form trimers, protruding radially from the viral membrane. The HA_1 polypeptide, which is entirely outside of the membrane, is involved in the binding of the virus to the host cell surface, while HA_2 constitutes the transmembrane subunit that is responsible for the actual fusion process.[21] Like the F_1 polypeptide of Sendai virus, the N-terminus of HA_2 contains a stretch of hydrophobic residues, which are interrupted by a few negatively charged amino acids. Therefore, maximal hydrophobicity will be achieved at acidic pH. Bromelain treatment of viral HA yields a water soluble trimeric ectodomain, BHA. This cleaved BHA fragment, which lacks the hydrophobic C-terminal membrane anchors, has been crystallized and its three-dimensional structure at neutral pH has been determined from X-ray diffraction studies to a resolution of 3Å.[41] The low pH-induced conformational change in HA, required for Influenza membrane fusion to occur, involves a partial dissociation of the trimer leading to exposure of the hydrophobic N-terminal region of the HA_2 polypeptide, which at neutral pH is buried in the interface between the monomers of the trimer.[41] Such a structural change appears to be a highly controlled process. Electron microscopy studies have shown that the acid-induced fusion active form of HA remains trimeric, rather than becoming denatured.[42,43] The contrast with BHA, which tends to dissociate at acid pH to monomers and dimers,[22,44] suggests that the transmembrane domain of HA plays an important role in confering stability to the acid form of the HA trimer.[22] It should also be noted that fusion is not induced by BHA-generated protein.

Entry of HIV into cells is mediated by the envelope glycoprotein gp160. Endoproteolytic cleavage, required for biological activity of HIV, leads to generation of a heterodimer consisting of two polypeptides gp120 and gp41.[14] gp120 and gp41 are weakly associated through noncovalent interactions and are not disulfide linked.[31,45] gp120 is responsible for the virions to attach to the CD4 cellular receptor, while the transmembrane glycoprotein gp41 is required for the fusion activity of HIV. The N-terminus of gp41 contains a conserved hydrophobic stretch of nearly 30 amino acids, which shows sequence homology to the fusion peptides of Sendai virus and Influenza virus glycoproteins. The structural arrangement of the gp160 protein is still largely obscure. Although some retroviral fusion proteins are reported to be trimers,[46,47] the gp160 spike of HIV has been suggested to consist, like the F protein of Sendai virus, of two identical dimers, thus forming a tetramer[48].

Intracellular proteolytic processing of viral proteins is not always required and putative fusion domains are not always located N-terminally. For example, the G protein of VSV and the E protein of SFV do not require a posttranslational cleavage and they do not have any obvious hydrophobic sequences. A putative internal hydrophobic domain located some 80 amino acids from the N-terminal has been found for the E protein of SFV and proposed to mediate viral fusion.[49] For VSV, the N-terminal peptide of G protein, which is highly conserved in different strains, has been suggested to play a role in fusion.[50] However, the location of the fusion domain remains unclear, since the N-terminus of the G protein is not particularly hydrophobic and specific mutations in this region did not affect the virus fusion activity.[51]

Each of the spike proteins of SFV consists of a heterotrimer of two transmembrane glycopeptides, E_1 and E_2, and a peripheral glycopeptide E_3.[52] The sequences E_2 and E_3 are initially synthesized as a precursor, p62, which is proteolytic processed during transport to the plasma membrane. However, as opposed to the fusion proteins of Influenza virus, Sendai virus and HIV, such proteolytic cleavage is not required for rendering fusion activity to SFV. The putative internal fusion peptide of SFV is thought to reside in E_1 and becomes exposed at the surface of the spike upon a major conformational change in the protein, which is triggered by a mild acidic pH.[49,53,54] Indeed, studies employing hydrophobic interaction chromatography have suggested that in response to low pH, the E protein becomes more hydrophobic.[55] It should be noted that although it has been shown that only E_1 is necessary and sufficient to mediate membrane fusion,[55] in vitro mutagenesis of the proteolytic cleavage site has revealed that p62 cleavage is required for bringing about the ultimate expression of E_1 fusion activity.[56,57]

The G protein of VSV also undergoes a conformational change at low pH, but whereas the conformational change of HA and E is irreversible, that of G seems to be unique in being reversible when the pH is returned to neutrality.[58] One of the aspects of the conformational change in the G protein appears to be the stabilization of the homotrimeric structure.[58] At low pH, the G protein spikes reversibly aggregate at the ends of the virus particles, a feature that might be relevant to the fusion mechanism of VSV.

The occurrence of these structural changes, which have best been described for Influenza virus, has been observed by employing antipeptide antibodies[59-62] and by analyzing changes in susceptibility to proteolytic digestion.[63,63a]

The analysis of the fusion-inducing conformational changes in the fusion proteins that act at neutral pH has been more difficult. For Sendai virus, it has been shown[64] that the activating proteolytic cleavage of the precursor protein F_0 involves a conformational change in the F protein with exposure of a new hydrophobic N-terminus of F_1, leading to an increase of hydrophobicity. The fusion protein conformation at neutral pH appears to differ from that at basic and acidic pH.[65] An irreversible low pH-induced conformational change in the F protein has been proposed to be responsible for the inhibition of the fusion activity of Sendai virus observed at mild acidic pH.[65] On the other hand, an optimal membrane-fusing activity of the virus has been observed at pH 9. This fusion optimum was found to correlate with an irreversible conformational change of the F protein occurring at basic pH, as demonstrated by a change in the circular dichroism spectrum of the protein.[65]

Obviously, detailed knowledge of the fusion protein structure is essential to obtain a complete picture of the mechanism of viral membrane fusion. From the foregoing and as will be further illustrated below (Figure 1A), it appears evident that hydrophobic sequences of viral proteins are intimately involved in the actual mechanism of virus fusion. Direct evidence that such domains presumably mediate fusion by penetrating into the target membrane has recently emerged from studies using hydrophobic photoactivatable probes. It has been demonstrated that the penetration of the hydrophobic N-terminal region of the F_1 peptide of Sendai virus and the HA_2 peptide of the HA glycoprotein of Influenza virus into the hydrophobic core of target membranes occurs during the early interaction stages and prior to virus fusion.[66-68] Also based on these hydrophobic photolabeling studies, clear evidence was provided that the bromelain-derived BHA ectodomain interacts somewhat differently with target membranes than the HA of intact virus.[68] While the fusion peptide of BHA is inserted into the bilayer as a shallow α-helix parallel to the plane of the membrane,[69] the fusion peptide of viral HA appears to be in a more perpendicular orientation.[68] It is possible that this different orientation is related to the fact that at low pH BHA does not trigger fusion as opposed to intact HA (see above). So far, these studies, indicating the actual penetration of viral protein segments into the target membrane, have been carried out using liposomes as target membranes for virus interaction. Hence, for biological legitimacy, it will be imperative to evaluate this approach in a pure biological system as well.

Obviously, the identification of functional domains in the viral proteins is of fundamental relevance for understanding the molecular details of virus-membrane interactions. It is conceivable that viral protein penetration into the target membrane will result in a hydrophobic dehydration, causing destabilization of the lipid bilayer, which ultimates in the merging of viral and target membranes. However, it is evident that a large number of details concerning conformational changes and exposure of hydrophobic domains still need to be clarified. As noted above, these hydrophobic regions are either already exposed to the external medium at neutral pH, such as in Sendai virus, or become exposed when the virus encounters a mild acidic environment, such as in Influenza virus, SFV, and VSV.

As mentioned in the Introduction, the threshold pH values at which fusion is triggered for viruses, that require a mild acidic pH for penetration, may vary depending on the virus involved. In fact, the different pH values at which viruses acquire fusogenic competence appear to reflect the distinct pH for the conformational change to occur such that the fusogenic peptide can be exposed to penetrate into the target membrane, thus triggering the fusion process. However, it should be noted that this mechanism wherein fusion results from the insertion of the fusion peptide, may not be universal. In the case of VSV, which does not display any identifiable fusion peptide and in the case of SFV, which shows its putative fusion domain downstream of the N-terminus, the eventual mechanism that destabilizes membranes (as a prerequisite for fusion) has not yet been resolved. For example, for VSV, a fusion mechanism has been proposed in which the G protein brings the viral and target membrane into close approach, destabilizing the (local) interface without actually penetrating into the target membrane[70] (see Figure 1B).

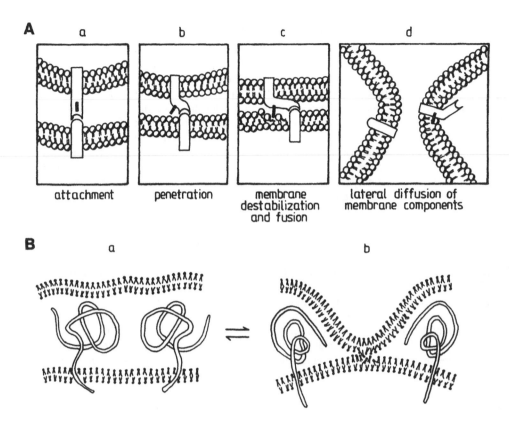

Figure 1 Hypothetical mechanisms in the fusion of the viral envelope with a cellular target membrane. (**A**) Following binding of the virus to a receptor of the target membrane (a), the fusion protein (bar) will penetrate into the hydrophobic core of the target membrane (b). The viral and target membrane will be brought into close contact causing a hydrophobic dehydration that leads to membrane destabilization (c; see also Figure 2). After fusion, the viral and target membrane components will randomize by lateral diffusion (d). As referred to in the text, this mechanism is used by viruses such as Influenza virus or Sendai virus, wherein fusion results from the penetration of the fusion peptide into the target membrane. (**B**) The fusion mechanism for VSV, which does not display any identifiable hydrophobic fusion peptide, will first involve the binding of the viral spike glycoprotein G to the target membrane (a). Upon exposure to low pH, a conformational change of the G protein will bring the viral and target membrane into close proximity, which, by an as yet unknown mechanism, may lead to interbilayer destabilization and fusion (b). It should be noted that in contrast to Influenza virus the conformational change of the G protein is reversible.

III. LIPOSOMES AS A MODEL FOR BIOLOGICAL TARGET MEMBRANES IN FUSION

Liposomes represent the lipid bilayer backbone of biological membranes. The complex topological and compositional arrangement of biological membranes renders them a difficult experimental system in which to identify the role of specific molecules and mechanism(s) involved in membrane interaction processes as they occur, e.g., during membrane fusion. In an attempt to gain an understanding of the molecular mechanism underlying the complex phenomenon of membrane fusion, liposomal systems have been developed and investigated extensively as a model of real biological membranes. The relevance of such simple models for understanding the intricate control of membrane fusion processes has been illustrated by an immense number of studies (see Chapter 7). In fact, most of our current knowledge regarding possible mechanisms of membrane interaction and fusion has been derived from investigations involving liposomal systems. These well-defined model systems have been recently introduced as appropriate targets for enveloped viruses, providing a relatively simple tool for investigating various different parameters involved in viral membrane fusion events. The phospholipid composition of the bilayer can be varied considerably from single component systems to mixture of various phospholipids, so that the head-group specificity and the influence on fusion of bilayer packing, lateral phase separations and phase transitions can be studied in great detail. Other membrane components such as cholesterol, glycolipids, and various proteins can be incorporated into the liposomal membrane, and their effects on fusion can readily be examined. Therefore, liposomes offer the possibility to investigate the role of electrostatic and steric constraints, thus permitting to identify the importance of molecular requirements during close approach of membranes to allow for the occurrence of fusion. The amenability of liposomes to variations in composition also includes the possibility to reconstitute viral glycoproteins in such artificial membranes. These reconstituted viral envelopes ("virosomes") have been shown to fuse with liposomes and to exhibit hemolytic and cell-cell fusion activity, thus allowing to investigate structural and functional properties of viral proteins. Liposomes can be easily made either as multilamellar vesicles of different sizes or as small or large unilamellar vesicles. The size of the liposomes is an important parameter, since curvature might facilitate fusion affecting thereby the relevance of the results to biological fusion, particularly with small vesicles. Indeed, small sonicated unilamellar vesicles are inherently unstable owing to packing constraints imposed by their high degree of membrane curvature. Large unilamellar vesicles constitute a more realistic model for fusion studies, as their properties more closely approach those of natural membranes. The versatility of liposomal systems has also contributed to the development and testing of fusion assays based on the incorporation of fluorescent probes in the lipid bilayer or the encapsulation of fluorescent markers in their aqueous interior, thus allowing the investigation of the kinetics of the fusion process under controlled conditions. Taken together, all these features have made liposomes particularly attractive as a model to investigate physiological fusion events such as secretion, endocytosis, and intracellular transport and also pathogenic processes like viral infectivity.

IV. INTERACTION OF ENVELOPED VIRUSES WITH LIPOSOMAL MEMBRANES

A. DETECTION OF VIRUS-LIPOSOME FUSION. ROLE OF LIPOSOME COMPOSITION AND PH

Fusion between various enveloped viruses and liposomes has been extensively investigated in the last few years. Several methodologies have been employed to study and to probe virus-liposome fusion. Electron microscopy is one of the most direct approaches and has been successfully utilized to detect virus-liposome fusion products. As early as 1974, Haywood,[71] using negative staining electron microscopy, demonstrated that Sendai virus can bind to and fuse with phospholipids bearing sialoglycolipids as viral receptors. Freeze-fracture and negative staining electron microscopy were also used by Oku et al.,[72] who showed that fusion of Sendai virus with liposomes consisting of phosphatidylcholine (PC), dicetyl phosphate (DCP), and cholesterol (Chol) depends on the presence of the virus receptor protein, glycophorin. Penetration of Semliki Forest virus into liposomes at pH 5.2 was also demonstrated by electron micros-copy using both staining and thin sectioning.[54] The images observed by negative staining showed that viral glycoprotein spikes protruded from the membrane of large liposomes composed of PC, phosphati-dylethanolamine (PE), sphingomyelin (SM), and cholesterol, reflecting the result of multiple fusion events. In samples prepared by thin sectioning, viral nucleocapsids were seen inside the large vesicles. In a very recent study on the interaction of HIV-1 with model membranes,[73] ultrastructural analysis of HIV-liposome fusion by negative staining, and freeze-fracture electron microscopy demonstrated the occurrence of fusion at pH 7.5 of HIV-1 with liposomes composed of cardiolipin (CL) or dioleoylphosphatidylcholine (DOPC). Note that fusion took place with liposomes, lacking the biological virus receptor, CD4, for the HIV envelope glycoprotein gp120. Also based on electron microscopy, Haywood and Boyer[74] reported fusion of Influenza virus X31 (H3N2) with zwitterionic liposomes containing gangliosides (G_{D1a}) at neutral pH. Morphological observation showed that after fusion, Influenza viral glycoprotein spikes were distributed in the outer lamella of multilamellar liposomes. The fusion of Influenza virus with liposomes at neutral pH has been subject to some controversy, and most authors agree that biological fusion activity of Influenza virus is mediated by the conformational change induced by low pH in the hemagglutinin glycoprotein.[53] Recently, Burger et al.[75] characterized morpho-logically the interaction of Influenza B virus with ganglioside-containing zwitterionic liposomes using sophisticated freeze-fracture electron microscopical techniques. The morphological data indicated that, although an occasional partial engulfment of the virions by the liposomes occurred at neutral pH, membrane continuity between viral and liposomal membrane was only observed upon lowering the pH.

A more quantitative method was developed using labeled virions with radioactive markers, such as ^{35}S-methionine or ^{3}H-uridine. Haywood and Boyer[74] measured the transfer of ^{35}S-methionine-labeled viral proteins into liposomes using sucrose density gradient centrifugation to allow the separation of fusion products from unfused virions. Their results indicated that fusion of Influenza virus with zwitterionic liposomes containing gangliosides could indeed occur at pH 7.5, reaching a maximum level when G_{Dla} was present at 3 mol%. Using this assay to investigate the effect of liposomal composition upon fusion of Influenza virus with liposomes at neutral pH, Haywood and Boyer[74] concluded that the presence of charged lipids was necessary for fusion with viral membranes at neutral pH. Thus, when the liposomes were composed of PC and 3 mol% G_{Dla}, fusion at neutral pH was observed. Inclusion of 5 mol% phosphatidylserine (PS) in PC resulted in some fusion, although significantly less than that occurred when G_{Dla} was also present.[74] On the other hand, fusion of liposomes, consisting solely of neutral phospholipids, was observed with Influenza virus at pH 5.2, in agreement with the low pH requirement for fusion of Influenza virus membranes that has been described by other investigators. From these studies, it appears that fusion of Influenza virus with liposomes might take place at neutral pH in the absence of gangliosides, provided they contain a net negative charge. When present, however, gangliosides will serve to promote more efficient fusion, most likely acting as virus receptors and not merely contributing a negative charge. The preferential fusion of Influenza virus observed at low pH was interpreted as a consequence of a more efficient binding at low pH than that at neutral pH.[74] Indeed, viral binding to liposomes was shown to be strongly dependent on the surface charge[74], and, therefore, it is very likely to be also affected by pH. In support of this, the binding of HA to liposomes composed of PC, PE, SM, phosphatidic acid (PA), and cholesterol was shown to increase sharply with decreasing pH (below 6.0).[22] In a subsequent study on the time dependence of Influenza virus fusion at neutral pH by using the same sucrose gradient method, Haywood and Boyer[76] showed that fusion of liposomes consisting of PC, PE, cholesterol, and G_{Dla} was significantly less efficient at neutral pH than at low pH.[77,78] Therefore, it was claimed[76] that the difference in kinetics at neutral and low pH may explain why fusion at neutral pH is often dismissed as negligible.

Fusion at neutral pH between Influenza virus and liposomes composed of PC's with different acyl chain compositions has been recently studied by monitoring the transfer of spin-labeled PC from the virus envelope to the target membrane.[79] Fusion, measured as lipid dilution, was found to occur at neutral pH, and no negative charge in the liposomes was required, thus contradicting the results of Haywood and Boyer.[74] However, at pH 5, fusion was much more efficient. Fusion with dimyristoylphosphatidylcholine (DMPC) vesicles at low pH was strongly enhanced when either glycophorin or sialylparagloboside were incorporated into the bilayer as viral receptors vesicles, the former being 50 times more effective than the latter.[79]

Very recently, a low degree of fusion at neutral pH between Influenza virus and liposomes consisting of negatively charged phospholipids, such as PS, or of PC/PE, as assessed by R_{18} fluorescence dequenching (see following section) was also reported.[80] However, fusion of Influenza virus with liposomes composed of PC required the presence of cholesterol and was observed only at pH 5. In addition, Influenza virus was able to induce the release of calcein from negatively charged calcein-loaded liposomes at both pH 5 and 7.4, but lysis of PC/Chol was dependent on the presence of gangliosides and was observed only at pH 5.[80]

Finally, it is relevant to note here that recent results on fusion of Influenza virus with PC-12 cells, using the R_{18} fluorescence assay (see next section), also suggest the occurrence of a slow virus-cell fusion at neutral pH.[81] Therefore, we may conclude that although some fusion of Influenza virus at neutral pH cannot be excluded, the most effective route for viral entry is probably extensive fusion with an internal acidic compartment following endocytosis.

Using the sucrose density gradient approach, Haywood and Boyer have also studied the fusion of Sendai virus with liposomes.[82,83] The results indicated that the presence of gangliosides as receptors was essential for fusion of Sendai virus with liposomes composed of neutral lipids, such as PC, PE, and cholesterol.[82] These studies revealed that fusion of Sendai virus with liposomes displays similar features as fusion of the virus with biological membranes, being maximal at a temperature around 37°C and at pH 7.5-9.0, and requiring the presence of virus receptors. Since liposomes contain no proteins, it was concluded that viral proteins and not host membrane proteins play the major role in the fusion of Sendai virus with liposomal membranes.[82] Further work on the effect of lipid composition upon virus-liposome fusion, using the same technique, demonstrated that similar to Influenza virus, the presence of specific receptors was not absolutely required for Sendai virus fusion.[83] Indeed, Sendai virus was able to fuse with negatively charged liposomes, such as PS, DCP, phosphatidylinositol (PI), or PA, i.e., in the absence of gangliosides. However, fusion of the virus with PC was only observed provided gangliosides were present. Cholesterol was not required and when added to liposomes containing PC and gangliosides, the degree of virus-liposome fusion decreased if gangliosides were present at low concentrations.[83]

Using virus-induced release of liposome contents as an approach to follow virus-liposome fusion, Amselem et al.,[84] observed that carboxyfluorescein was released from negatively charged liposomes lacking virus receptors upon their incubation with Sendai virus. Although lysis of such liposomes did not require the presence of receptors, proteolytic or thermal treatment of the virions inhibited the release of carboxyfluorescein, indicating that a viral protein is responsible for this lytic activity. Furthermore, the virus failed to lyse liposomes composed of PC, and neither incorporation of cholesterol nor of PE rendered the PC liposomes susceptible to the viral lytic activity.[84] These observations supporting the results of Haywood and Boyer,[83] were confirmed by Loyter and Citovsky[85] who claimed that as opposed to liposomes composed of neutral lipids, no virus receptors are needed for the lytic activity of Sendai virus toward negatively charged liposomes. The presence of cholesterol was found to reduce significantly the release of carboxyfluorescein from such liposomes upon incubation with Sendai virus.[84,85] A dependence of the fusion of Sendai virus with PC liposomes on the presence of virus receptors, has also been observed.[86] Using virus-induced lysis of calcein-loaded liposomes to follow virus-fusion, it was found[86] that in addition to glycophorin, cholesterol was absolutely required in the liposomal membrane for the lytic response of PC-containing liposomes to Sendai virus. Following the increase in the permeability of loaded liposomes upon incubation with Sendai virus, Oku et al.,[87] showed that the release of trapped methylumbelliferyl-phosphate from liposomes composed of DCP, PC, and cholesterol depended on the presence of glycophorin in the liposome membrane. The increase of permeability was shown to exhibit a temperature coefficient similar to virus-induced hemolysis. The virus-induced permeability change most likely resulted from the penetration of the viral fusion glycoprotein F into the target membrane, since trypsinization of intact virus inactivated its lytic capacity.[87] The presence of cholesterol in such liposomes containing glycophorin was not required for the lysis of the liposomes. However, there is no report of the effect of cholesterol on the viral lytic activity in liposomes composed of only PC. The results of Hsu et al.[88] appear to contradict the above results regarding the need of virus receptors in neutral liposomes for Sendai virus fusion. Like Haywood, Hsu et al.[88] used sucrose density gradient centrifugation to separate free virions from those fused with liposomes and showed that fusion of Sendai virus could occur with PC/cholesterol liposomes lacking virus receptors. Also, in contrast to the results of Haywood and Boyer,[83] where no effect of cholesterol was observed, Hsu et al.,[88] showed a cholesterol requirement in fusion, an optimum being observed at 0.3–0.4 mole fraction of cholesterol.

It is obvious that no clear picture emerges from these studies regarding the need for specific receptors, although sialic acid residues covalently linked to glycolipids and/or glycoproteins have been claimed to act as primary cell surface receptor sites for both Influenza and Sendai virus. The fact that viruses can interact efficiently with liposomes, consisting solely of negatively charged phospholipids suggest that these negatively charged phospholipid molecules can act as (high affinity) virus receptors, thus obviating the absolute need of sialic acid residues as functional specific receptors. Whether such interactions reflect the viral biological fusogenic activity required for penetration and infection of living cells is, however, questionable (see below). In this regard and of particular interest are observations that VSV strongly binds PC or PS liposomes at acidic pH values.[89] It has been suggested that this acidic pH binding to phospholipids is used by VSV to enter the cell once it has been internalized into the endosome. Another type of binding utilized by VSV appears to be receptor mediated, it predominates at physiological pH, and is used for virus attachment to the cell so that it can be internalized.[90] The receptors for VSV have not yet been identified. Possibly, the target site for VSV may not be a protein(s), a sialoglycoprotein(s), or a sialoglycolipid(s), but rather a PS-containing phospholipid domain.[89] In support of this notion is the observation that trypsin pretreatment of cells increases binding and fusion of the virus, suggesting that proteins could prevent access of the virus to phospholipid domains.[89] Using spin-labeled PC incorporated into the viral membrane, fusion of VSV was found to be negligibly small at neutral pH but was greatly activated at mild acidic pH as would be expected for enveloped viruses that, after internalization by endocytosis, fuse with the endosomal membrane.[89] Furthermore, it was observed[89] that cis-unsaturated acyl chains were required for efficient fusion of VSV, while the inclusion of cholesterol in bilayers containing these cis-unsaturated phospholipids greatly enhanced the fusion process. High fusion efficiency with cis-unsaturated phospholipids was also recently reported for Influenza virus fusing with PC liposomes, using the same spin-labeling method.[79] This stimulating effect of acyl chain unsaturation on VSV and Influenza virus fusion cannot be ascribed to a change of membrane fluidity[79,89] but may be related to higher tail-to-head volume ratios, as compared to those of trans-unsaturated and saturated phospholipids. Higher tail-to-head volume ratios have been shown to favor the hexagonal phase or inverted micelle formation.[91] Within the general concept of lipid perturbation as a requirement for fusion, lipids adopting nonbilayer structures have been proposed to play an important role as intermediates in triggering membrane fusion.[92-94]

In this context, it is interesting to note that recent work on the effect of small peptides, resembling the fusion segment of the measles and Influenza fusion proteins, on lipid polymorphism[95] has shown that these peptides can efficiently promote the conversion of bilayer to nonbilayer phases in model membranes. By contrast, peptides and proteins that stabilize the bilayer structure inhibit viral fusion. Curiously, in the case of Influenza virus, the fusogenic peptide had a marked and quite opposite effect on lipid polymorphism, when comparing such properties at neutral and at acidic pH. As demonstrated by differential scanning calorimetry,[95] addition of the peptide to DOPE vesicles at pH 5 caused a *decrease* in the bilayer to hexagonal phase transition temperature T_H, as opposed to the effect at pH 7.4 where an *increase* in T_H was observed. Hence, at neutral pH the peptide appears to have a stabilizing effect on the lipid bilayer, whereas upon acidification, penetration of the peptide will result in membrane destabilization by promoting the formation of nonbilayer phases. Thus, modulation of structural and physical properties of the lipids, leading to a bilayer destabilization or stabilization as described may play an important role in the promotion (or inhibition) of viral fusion.

As noted above, the results obtained on the role of cholesterol in virus-liposome fusion, namely for Sendai virus, are quite controversial. The discrepancy observed may depend partially on the criterion used for the interaction of the virus with liposomes, which will affect the outcome of the studies. In general, as noted, inclusion of cholesterol in the liposomal membrane decreases Sendai virus lytic activity. These observations are consistent with earlier studies showing that cholesterol increases liposome stability and reduces the leakage of their contents.[96] The range of effects of cholesterol in Sendai virus fusion may reflect the different effects of cholesterol upon membranes of variable lipid composition, thus suggesting that the lipid composition of the target membrane can modulate the effect of cholesterol on the virus interaction.

In the case of SFV, it has been clearly demonstrated that the presence of cholesterol in the liposomal membrane is an absolute requirement for the low pH-induced fusion of the virus.[54] Fusion monitored by degradation of viral RNA by RNase trapped inside the liposomes was shown[54] to be low pH-dependent, consistent with the requirement of a mild acidic pH to permit SFV fusion with the endosomal membrane of the host cell. Optimal fusion occurred with liposomes containing PC, PE, SM, and cholesterol, the extent reaching a maximal value at 33 mol% cholesterol.[54] Removal of PC had no effect on fusion, while a significant reduction in fusion was observed when PE or SM were excluded.[54]

Studies of the sterol requirement for SFV fusion with various cholesterol analogs showed that the 3-β-hydroxyl group is the critical portion of the molecule and further suggested that the action is independent of the effect cholesterol may exert on the properties of the phospholipid membranes.[97] The relevance of cholesterol as an absolute requirement for the virus infection *in vivo* can also be inferred from recent work on the interaction of SFV with sterol-enriched or sterol-depleted cells.[98] The underlying mechanism of the cholesterol dependence for SFV remains unclear. While for Sendai virus a specific interaction between the fusion peptide of F_1 and cholesterol has been proposed,[99] no direct demonstration of an interaction between the fusogenic E_1 polypeptide of SFV and cholesterol has yet been shown. However, it has been suggested[98] that E_1, once exposed by the low pH-induced conformational change of E protein, interacts in a stereo-specific manner with cholesterol in the target membrane, further changing the conformation of E_1 and leading to a disruption of the bilayer that initiates the fusion event.

B. KINETIC STUDIES OF VIRUS FUSION WITH LIPOSOMES USING FLUORESCENCE ASSAYS. A COMPARISON WITH BIOLOGICAL MEMBRANES

Recently, a number of more direct assays for lipid mixing based on the use of fluorescent lipid analogs have been developed and successfully employed to continuously monitor virus-liposome and virus-cell fusion in a very sensitive manner. Frequent use has been made of the assay based on the relief of fluorescence self-quenching of the probe, octadecylrhodamine B chloride (R_{18}).[100] Other studies have utilized resonance energy transfer (RET)[101] or the dilution of the concentration-dependent excimer-forming fluorescent probe pyrene, attached to phosphatidylcholine.[102] With these assays, the initial steps, during which viral protein penetration occurs and that are relevant to understanding the mechanism of virus fusion, can be studied. Above all, these techniques provide an attractive alternative to other methods since they are easy to apply and can be performed without removal of unfused viruses, as needed when using, for example, the assays relying on the application of radioactive markers to detect fusion (see above). In conjunction with these fluorescence assays, a mass action kinetic model has been developed and applied to viral fusion using both artificial and biological membranes. This theoretical model, which views the overall fusion reaction as a sequence of a second-order process of membrane adhesion followed by the first-order fusion reaction itself, allows to discriminate the rate constants for virus binding and of the fusion reaction itself. A quantitative characterization of the fusion process is therefore possible, thus providing insights into the mechanism of viral membrane fusion and the role of target membrane components in adhesion and fusion.

Using liposomes composed of negatively charged phospholipids as a target membrane for Influenza virus labeled with R_{18}, Hoekstra et al.[100] have shown that the pH-dependence of fusion was similar to its fusion with biological membranes, i.e., fusion occurred at pH 5.0 but not at neutral pH. Later, using RET in which two fluorescent lipid derivatives, N-(7-nitro-2, 1,3-benzoxadiazol-4-yl)-phosphatidylethanolamine (N-NBD-PE) and N-(lissamine rhodamine B sulfonyl)phosphatidylethanolamine (N-Rh-PE), were incorporated in the liposomal bilayer, Stegmann et al.[103] investigated the kinetics of pH-dependent fusion between Influenza virus and liposomes, composed of neutral and negatively charged lipids. A high degree of fusion was observed only at low pH and with liposomes composed of negatively charged phospholipids, particularly CL, while at neutral pH and with liposomes consisting of neutral lipids such as PC or SM, fusion was minimal. Using the R_{18} assay, Sendai virus was also shown to avidly fuse with CL or PS and inclusion of PC in acidic phospholipids strongly inhibited virus-liposome fusion.[104] By contrast, the incorporation of PE instead of PC sustained the fusion reaction for Influenza virus[103] and Sendai virus.[104] The inhibitory effect of PC is likely to be attributed to the strong hydration of the head group as opposed to the much less hydrated PE, precluding the establishment of close fusion-susceptible intermembrane contact between virus and target membrane. A kinetic analysis of the fusion process for Sendai virus, as monitored by the R_{18} assay, revealed that the inhibitory effect of PC in the target liposomal membrane occurs at the level of the fusion reaction *per se* rather than at that of the binding step.[105] Inclusion of PC in CL vesicles leads to almost a twofold decrease in the fusion rate constant. It is interesting to note that reconstituted VSV virions exhibit a greater fusion activity toward PA over PS and PI, as monitored by RET, consistent with a decrease in the relative hydration energies of the head groups.[106] These results indicate that repulsive hydration forces may modulate the ability of a virus to fuse, and therefore membrane dehydration may be essential for virus-liposome fusion to occur. In this regard, it was observed that the presence of small amounts of the dehydrating agent poly(ethylene glycol) greatly enhances the kinetics of fusion of Sendai virus[105,106a] and Influenza virus[106a] with liposomes. The significance of this observation in virus-liposome fusion has been evaluated in a biological system by examining the effect of poly(ethylene glycol) on the kinetics and extent of fusion between Sendai virus and erythrocyte ghosts. It was found[35] that in the presence of poly(ethylene glycol), the adhesion rate constant increased slightly, while the fusion rate constant increased by almost an order of magnitude, implying that hydrophobic interaction forces govern to a major extent the actual fusion reaction.

The lipid packing density may also govern the ability of a virus to fuse with liposomal systems, as revealed by the observation that the susceptibility of Sendai virus to fuse with CL liposomes is higher than that with PS and zwitterionic G_{D1a}-containing vesicles,[104] i.e., with increasing packing density, a decreasing fusion activity is seen. These results would be consistent with a fusion mechanism involving an hydrophobic penetration of the viral fusion protein followed by local membrane dehydration as a key element in triggering fusion activity. As pointed out earlier, direct evidence for the hydrophobic penetration concept has been obtained from experiments in which such an event was probed during virus interaction with liposomes, using hydrophobic photoafinity labels.[66,67]

The observation that both Influenza and Sendai virus fuse with liposomes of various lipid compositions suggests that the virus-liposome interaction is rather non-specific. However, the initial rate and final extent of fusion depend strongly on the lipid composition of the liposomal membrane, fusion being as already mentioned, most efficient with liposomes consisting of negatively charged lipids.[103,104] A kinetic analysis of the overall fusion process showed that preferential fusion of Influenza virus with CL liposomes at pH 5 arises at the level of the fusion reaction itself and not at the level of initial binding to the liposomes.[107] Indeed, the binding rate constant is the same at both pH 5 and 6, indicating that the affinity of Influenza virus for CL liposomes is relatively independent of the pH, whereas the fusion rate constant is about fivefold higher at pH 5 than that at pH 6.

As mentioned before, the fact that fusion of Influenza virus and Sendai virus occurs with vesicles solely consisting of negatively charged lipids would raise the possibility that these negatively charged molecules can substitute for the sialic acid residues as efficient functional virus receptors. However, the characteristics of fusion of Influenza virus with CL liposomes deviate from those observed with PC/PE/G_{Dla} liposomes or erythrocyte ghosts.[108,109] This is reflected by several observations of the effect of the nature and composition of target membranes on the kinetics of the low pH-dependence of Influenza virus fusion as monitored by the R_{18} and RET assays. For example, fusion with CL liposomes, although protein-mediated as revealed by an inhibition of fusion after proteolytic treatment of the virus,[103] does not exhibit a strict dependence on the acid-induced conformational change in the HA as observed with PC/PE/G_{Dla} or ghosts, thus indicating that part of the fusion between Influenza virus and CL liposomes occurs in a non-specific electrostatic manner.[108] Furthermore, in the CL liposome system, the lag phase as seen with PC/PE/G_{Dla} or ghosts is absent at any pH value.[108] The aberrant behavior of pure CL liposomes is also reflected by a significant larger fusion rate constant than for most other liposome compositions and erythrocyte ghosts or cells.[110-112] Another prominent difference between CL liposomes and ghosts or PC/PE/G_{Dla} as targets for Influenza virus is that fusion with the latter systems is strongly inhibited by a low pH preincubation of the virus alone whereas no significant inactivation is observed with CL liposomes. In this regard, it is interesting to note that "low pH inactivation" of Influenza virus is not as pronounced toward certain types of living cells as toward ghosts,[111] suggesting that viral inactivation may not be fully irreversible. This would imply that fusion activity can be reverted upon interaction of viral glycoproteins with cellular components. The artificial nature of fusion between Influenza virus and acidic phospholipids does not appear to be restricted to only pure CL liposomes. Although fusion of Influenza virus with pure PS- liposomes and with PS- or CL- containing zwitterionic liposomes displays characteristics that are similar to its fusion with biological target membranes,[110] fusion of Influenza virus with PS liposomes was recently claimed not to reflect the fusogenic activity utilized by the virus for penetration *in vivo*.[80] This consideration was based on the finding that incubation of non-fusogenic virions with liposomes consisting solely of PS resulted in a high degree of R_{18} dequenching. In agreement with previous suggestions regarding fusion of Influenza virus with CL liposomes,[108] Nussbaum et al.[80] proposed that fusion between Influenza virus and negatively charged liposomes rely on an electrostatic interaction between the negatively charged lipids and positively charged groups on the viral spike glycoproteins. On the other hand, fusion of Influenza virus with PC/cholesterol vesicles, lacking or bearing gangliosides, could be attributed to the biological activity of the viral HA glycoprotein.[80]

Interestingly, fusion between Influenza virus and both CL liposomes and ghosts at 37°C can be reversibly switched on and off by readjustments of the pH from acidic to neutral and vice versa.[103,108] These results suggest that the low pH-induced conformational change of HA itself is not sufficient to ensue the viral fusion activity, thus pointing to the need for continuous acidic conditions during the fusion process. In contrast to these results, fusion at 4°C with zwitterionic liposomes consisting of PC/PE/G_{Dla} as followed by RET is not arrested upon neutralization.[113] Based on these results it was suggested[113] that, while the reactions that lead up to the fusion-competent complex are dependent on low pH, the final fusion event itself is not. However, recent results have demonstrated that fusion of Influenza virus with PC-12 cells at 37°C and even at 20°C is essentially arrested upon neutralization.[81] It was proposed[81] that at 37°C or 20°C, neutralization leads to rapid dissociation of the oligomeric fusion complex, most likely due to repulsion between adjacent charged amino acids at neutral pH. At low temperatures, such as 0°C or 4°C, the fusion complex, whose rates of dissociation and formation are very low, remains in an associated form/state for very long periods[81] (see Figure 2).

As for Influenza virus, fusion of Sendai virus with pure acidic phospholipids appears to display some artificial features. Indeed, fusion of Sendai virus with CL or PS liposomes is immediately triggered at neutral pH as opposed to fusion of the virus with erythrocyte ghosts or G_{Dla}-containing vesicles, where a lag-phase is consistently apparent.[34,104,114] This lag-phase most likely reflects a spatial and/or lateral reorganization of the viral proteins as required to establish a close interbilayer contact, possibly forming the appropriate fusion complex, and triggering the subsequent fusion reaction.[104] As a function of pH, the rate and extent of fusion of Sendai virus with CL and PS liposomes dramatically increase with decreasing pH, an optimal fusion activity being observed around pH 4.0.[104] Similar results have been obtained for Sendai virus fusing with PS and phosphatidylglycerol (PG), using pyrene-labeled phospholipid liposomes.[115] Detailed kinetic studies of the fusion of Sendai virus with CL, PS, and CL/DOPC liposomes, as monitored by the R_{18} assay, showed that the fusion ability of the virus toward acidic liposomes as a

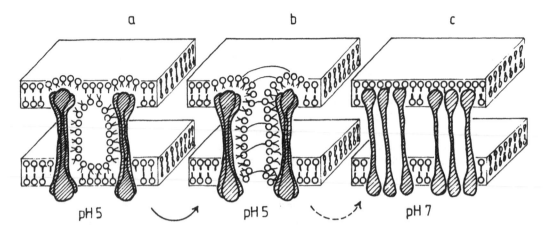

Figure 2 Schematic representation of the postulated stages during Influenza virus HA-mediated fusion. **(a)** At pH5, several unfolded HA trimers associate and interfacial water is expelled, thus leading to the formation of a fusogenic intermediate. **(b)** Following the mixing of the outer leaflets of the fusing bilayers, the inner leaflets will mix with a subsequent formation of a small pore causing membrane fusion. **(c)** Upon neutralization, the dissociation of the oligomeric fusion complex would arrest the bilayer fusion. As referred to in the text, this latter stage will be rapid at 20°C or 37°C but it will be very slow at 0°C or 4°C.

function of pH is primarily determined by the fusion rate *per se*, rather than the adhesion rate constant.[105] With liposomes composed of DOPC/DOPE/Chol/G_{Dla}, the rate of fusion was shown to be considerably slower than that in the case of CL or PS liposomes and had a local optimum at pH 7; again the rate of fusion increased dramatically below pH 5.[104] The effect of lowering the pH on the fusion capacity of Sendai virus toward acidic liposomes is in sharp contrast with that observed using erythrocytes or murine cell lines as target membranes, where an optimal fusion activity is seen around pH 7.5.[26,34] At mild acidic pH (5.0) the viral fusion activity toward the latter systems is almost an order of magnitude lower. Also in contrast to the reversible low pH-induced fusion activity of Sendai virus toward CL liposomes, mild acidic pH (5–6) drastically inhibits the fusion activity of the virus with erythrocyte ghosts and mouse spleen cells.[26,34] As mentioned before, this low pH-induced decrease of Sendai virus fusion activity has been related to an irreversible pH-induced conformational change in the F protein.[65]

Recently, it was observed[116] that fusion of Sendai virus with PC-12 cells, as monitored by the R_{18} assay seems essentially pH-independent, unlike the case of liposomes and erythrocyte ghosts. Viral preincubation at 37°C resulted in a marked, though slow, inhibition in fusion. With time the fusion rate constant eventually reduced eightfold.[116] Curiously, but quite unexpectedly, an isolate of Sendai virus displayed enhanced fusion activity with decreasing pH toward human cell lines, while for another isolate of the same strain, fusion was inhibited.[117] However, reducing the pH after a short preincubation at neutral pH enhanced fusion for both of the isolates. Furthermore, similar to fusion with CL liposomes, a low pH preincubation of Sendai virus had no significant effect on the fusion activity toward the human cells. The enhancement effect observed for the fusion activity of Sendai virus toward the human cell lines upon lowering the pH is quite intriguing and remains to be clarified in molecular terms. *A priori*, this observation might suggest that negatively charged liposomes, although somewhat artificial, could provide valuable insights that might be missed using erythrocyte ghosts as target membrane. However, evidence has been obtained[104] on the nonspecific fusion process at low pH between Sendai virus and negatively charged liposomes, particularly CL. Indeed, a considerable part of the fusion reaction with CL at low pH, as monitored by the R_{18} assay, is accomplished by an F protein-independent mechanism as opposed to fusion at neutral pH where fusion of trypsinized virus is significantly reduced. Furthermore, both the rate and extent of Sendai virus fusion with CL and PS liposomes are essentially unaffected at low pH (5.0) upon altering the protein conformation by treatment of the virus with the reducing agent dithiothreitol (DTT).[104] Clearly, these results contrast markedly with those on the interaction of Sendai virus with PC/PE/Chol/G_{Dla} vesicles or erythrocyte membranes, where fusion is strongly inhibited upon treatment of the virus with trypsin or DTT.[104,34] The nonspecific fusion process between Sendai virus and negatively charged liposomes can also be inferred from fluorescence dequenching studies showing that DTT- or phenylmethanesulfonyl fluoride (PMSF)-treated virions and non-hemolytic reconstituted Sendai virus envelopes readily fuse with liposomes composed of PS or PC/Chol/DCP but not of PC/Chol.[118] The view that fusion of Sendai virus with negatively charged liposomes does not reflect the biological fusion activity is further strengthened by results showing that reconstituted vesicles containing only the HN glycoprotein are able to fuse with PS liposomes but not with PC liposomes.[119] Interestingly, fusion between HN vesicles and PS liposomes was highly pH-dependent, being maximal at low pH.[119,85] Neither treatment with PMSF nor with DTT caused inhibition of the lytic and fusogenic activities of the HN vesicles toward PS liposomes. On the other hand, neither fusion nor induction of lysis was observed upon incubation of F vesicles with PS liposomes, indicating that fusion of Sendai virions with these liposomes, unlike the fusion of the virus with biological membranes where both F and HN are absolutely required,[120,121] is due to the activity of the viral HN glycoprotein.[119,85]

As pointed out earlier, electron microscopy has provided visual evidence that HIV fusion does not require CD4 in the target liposome.[73] The kinetics and extent of fusion of SIV_{mac} and HIV with liposomes of several lipid compositions and with erythrocyte ghosts have recently been examined, using the R_{18} assay.[73,122] Viral fusion activity was shown to be qualitatively similar for both SIV_{mac} and HIV and, as with Influenza and Sendai viruses, was strongly dependent on the liposomal membrane composition. Indeed, fusion was most efficient with liposomes containing anionic lipids. The presence of cholesterol in the target membrane had no effect on HIV-1 fusogenic activity.[73] Interestingly, physiological levels of Ca^{2+} enhanced HIV-1 fusion activity *per se* and the extent of fusion, using acidic liposomes as target membranes.[73] Similar levels of Ca^{2+} have no effect on the fusion activity of Influenza or Sendai virus toward liposomes. Fusion of HIV-1 with cardiolipin liposomes was shown to be greatly inhibited by prior trypsinization of the intact virus at pH 7.5 in the presence of Ca^{2+}, which indicates that the envelope glycoproteins are required for virus fusion with this liposomal system.[73] The initial rate and extent of fusion of both SIV_{mac} and HIV with liposomes composed of negatively charged lipids are enhanced by a reduced pH value,[73,122] most likely due to viral glycoprotein protonation. The physiological significance of the enhanced fusion activity of SIV_{mac} and HIV at mildly acidic pH remains unclear. Studies of HIV, fusing with CD4-positive cells as monitored by the R_{18} assay, have indicated that although fusion can occur at pH 5, it does not require low pH, since fusion was most efficient at pH 7.4.[5] It should be noted that the fusion event at lower pH was only observed with acidic liposomes; with erythrocyte ghosts no significant effect at mild acidic pH was observed.[73] Indeed, as pointed out earlier, the prevailing view is that HIV entry occurs at the plasma membrane and is not dependent on acidification of the endosome lumen.

The observations that HIV and SIV are capable of fusing with target membranes lacking the CD4 receptor, such as liposomes composed of negatively charged lipids and erythrocyte ghosts, would be consistent with the finding that CD4-independent cellular entry and infection of SIV_{mac} and HIV occur. It should be noted that galactosyl ceramide has also been proposed as a receptor for HIV-1 in certain CD4-negative cells.[123,124] Finally, it is relevant to note here that preliminary observations on fusion of Influenza virus, with CEM and MDCK cells using the R_{18} assay, have indicated that removal of sialic acid from the cell surface by neuraminidase treatment drastically reduces viral binding at neutral pH while enhancing viral association at acidic pH.[125] However, it should be noted that viral fusion is strongly inhibited, suggesting that the virus bound to non-sialic acid sites at low pH may not engage in fusion. It will be interesting to investigate which target membrane components, apart from the identified cellular receptors, are playing a role in the overall viral fusion events. Evidently, such insight will be needed to further elucidate the molecular details of the mechanism involved in virus infectious cell entry.

V. BIOLOGICAL RELEVANCE OF VIRUS-LIPOSOME INTERACTIONS: CONCLUDING REMARKS

To investigate the types of molecules with which the membrane of enveloped viruses can interact and undergo fusion, extensive work has been carried out using liposomes as target membranes. From the brief summary of the results on virus-liposome interactions presented in this review, it is clear that enveloped viruses are able to fuse with liposomes lacking functional receptors. Even HIV, which seems to have a relatively restricted host cells range and preferentially infects CD4-expressing peripheral blood cells,[126] does not require a specific protein or lipid molecules in the target membrane for fusion activity. Although the presence of a receptor enhances the rate of adhesion between virus and target membranes, it does not necessarily affect the fusion rate, as revealed for Influenza and Sendai viruses. It would be interesting to see how reconstituted CD4 and/or galactosyl ceramide into liposomes will affect the kinetics of fusion for HIV. The above observations suggest that the presence of specific receptors may not be a prerequisite for viral fusion, although the lipid composition plays a major role in determining the efficiency of fusion. Most important is the fact that the target membrane itself may modulate the fusogenic properties of a virus and so the choice of liposomal composition must be carefully considered in order to represent a biologically relevant target membrane. As has been shown, pure negatively charged bilayers, consisting of PS or CL, are particularly amenable to fuse with various viruses. However, the outcome of the fusion characteristics suggests that such fusion may not represent a physiological model for viruses, fusing with host cell membranes. In addition, it should be noted that upon infection, enveloped viruses will interact first with the outer leaflet of the membrane, which is commonly poorly endowed with negatively charged lipids. It is thus reasonable to assume that fusion with liposomes consisting of negatively charged lipids is too simple to resemble the complex protein-mediated biological fusion process. However, the acidic phospholipid PS should not be dismissed completely as a potentially irrelevant cellular target. A PS-dependent binding of VSV to a target membrane has been suggested[89] and infectious entry of Sindbis virus,[127] a togavirus like SFV, was reported to require PS and/or PE. In this context, it remains to be determined whether lipid asymmetry will be maintained during virus entry. In fact, during fusion of hamster fibroblasts induced by Sendai virus, a transmembrane flip-flop of a fluorescent phospholipid analogue inserted into the outer leaflet of cell plasma membrane was observed.[128] If membrane lipid asymmetry is indeed perturbed upon viral interaction, it would be particularly interesting to see whether the kinetics of exposure of lipids, such as PS, could correlate with the kinetics of virus fusion.

As already referred to in the previous section, fusion of Sendai and Influenza viruses with liposomes consisting of neutral lipids such as PC and cholesterol with added gangliosides, unlike fusion with negatively charged liposomes, displays the same features that characterize fusion with biological membranes, thus reflecting the viral biological activities necessary for penetration and infectivity. Also, as opposed to liposomes consisting of negatively charged lipids, it turns out that fusion of Sendai and Influenza viruses with liposomes composed of PC requires either the presence of gangliosides and/or cholesterol. As already pointed out, this observation, among others, raises the possibility that cholesterol may constitute a putative membrane receptor, particularly for Sendai virus. However, a number of reports seem to disclaim this possibility. Unfortunately, knowledge as to the effect of cholesterol upon fusion of viruses with biological membranes is extremely scanty. In a recent study[129] it was reported that the fusion of Sendai virus and Influenza virus with prokaryotic mycoplasma cells required the presence of cholesterol. Little fusion was observed to occur with cells that had not been supplemented with exogenous cholesterol. By contrast, Yoshimura et al.[130] have shown that the amphotericin B-resistant mutant of Chinese hamster V79 cells, which is defective of sterol synthesis, is much more susceptible to Sendai virus-induced cytolysis than the parent V79 cell. Moreover, the kinetics and the extent of fusion of Sendai virus with the mutant cells, as measured by electron spin resonance spectroscopy, are much higher than in the parent cells. The increase in fusion and cytolysis susceptibility with decreasing cholesterol was interpreted in terms of an increased membrane fluidity and decreased stability of the cell plasma membrane.[130] Clearly, the relevance of liposomes as target membranes to identify a critical role of cholesterol for virus infectious entry has been best illustrated for SFV, where, as already mentioned, cholesterol is indispensable for viral fusion activity.

The possibility of identifying the fusion-initiating proteins by means of photoactivatable probes inserted into a liposomal membrane highlights the relevance of liposomes as valuable tools for understanding the molecular basis of protein-mediated virus-cell interaction. In addition, a detailed picture of how the viral hydrophobic peptides trigger fusion has also emerged from the extensive work on fusion of viruses with liposomes of variable composition. Particularly, these studies, as has been illustrated, revealed that various physical parameters, such as bilayer packing and state of surface hydration, may modulate the ability of a virus to fuse. The biological relevance of these studies can be inferred from recent work using erythrocytes with a symmetric or asymmetric lipid distribution, where membrane packing density is shown to be an important parameter in determining fusion susceptibility of VSV with these cells.[131] As pointed out, the importance of repulsive hydration forces as the ultimate fusion barrier in a biological system has been illustrated by the increase in the kinetics of fusion between Sendai virus and erythrocyte membranes observed in the presence of poly(ethylene glycol).[35] Virus fusion thus appears to be governed by hydrophobic interactions and impeded by repulsive hydration forces. It is reasonable to assume, therefore, that for fusion to occur, the viral fusion glycoprotein has to bring viral and target membrane into close proximity, overcoming the repulsive hydration shells and perturbing the target bilayer via hydrophobic interactions. It should be emphasized, however, that the function of liposomes as a model for a biological target membrane should not be seen as a system imitating all the requirements of a biological virus fusion event. Therefore, extrapolation of the mechanistic features observed in virus-liposome interaction should be done with caution. However, from the studies presented in this review one may conclude that the information gained from investigation on the interaction between viruses and liposomes might be valuable for helping to elucidate the molecular mechanisms relevant to virus-membrane fusion and infectious entry into cells.

ACKNOWLEDGMENT

Parts of this work were supported by JNICT, Portugal, and by a NATO Collaborative Research Grant, CRG 900333 (M.C.P. Lima and N. Düzgüneş)

REFERENCES

1. White, J. M., Viral and cellular membrane fusion proteins, *Annu. Rev. Physiol.,* 52, 675, 1990.
2. Hoekstra, D. and de Lima, M. C. P., Molecular mechanisms of enveloped viruses entry into host cells. Protein dynamics and membrane fusion, in *Membrane interactions of HIV: implications for pathogenesis and therapy in Aids,* Aloia, R. C. and Curtain, C. C., Eds., Wiley-Liss, Inc., New York, 1992, 71.
3. Stein, B. S., Gowda, S. D., Lifson, J. D., Penhallow, R. C., Bensch, K. G., and Engleman, E. G., pH-independent HIV entry into CD4-positive T cells via virus envelope fusion to the plasma membrane, *Cell,* 49, 659, 1987.
4. McClure, M. O., Marsh, M., and Weiss, R.A., Human immunodeficiency virus infection of CD4-bearing cells occurs by a pH-independent mechanism, *EMBO J.,* 7, 513, 1988.
5. Sinangil, F., Loyter, A., and Volsky, D. J., Quantitative measurement of fusion between human immunodeficiency virus and cultured cells using membrane fluorescence dequenching, *FEBS Lett.,* 239, 88, 1988.
6. Pauza, C. D. and Price, T. M., Human immunodeficiency virus infection of T cells and monocytes proceeds via receptor-mediated endocytosis, *J. Cell Biol.,* 107, 959, 1988.
7. Maddon, P. J., Dalgleish, A. G., McDougal, J. S., Clapham, P. R., Weiss, R. A., and Axel, R., The T4 gene encodes the AIDS virus receptor and is expressed in the immune system and the brain, *Cell,* 47, 333, 1986.
8. Fields, A. P., Bednarik, D. P., Hess, A., and May, W. S., Human immunodeficiency virus induces phosphorylation of its cell surface receptor, *Nature,* 333, 278, 1988.
9. Paterson, R. G., Hiebert, S. W., and Lamb, R. A., Expression at the cell surface of biologically active fusion and hemagglutinin/neuraminidase proteins of the paramyxovirus simian virus 5 from cloned cDNA., *Proc. Natl. Acad. Sci., USA,* 82, 7520, 1985.
10. Wertz, G. W., Stott, E. J., Young, K. K. Y., Anderson, K., and Ball, L. A., Expression of the fusion protein of human respiratory syncytial virus from recombinant vaccinia virus vectors and protection of vaccinated mice, *J. Virol.,* 61, 293, 1987.
11. White, J., Helenius, A., and Gething, M.-J., Haemagglutinin of influenza virus expressed from a cloned gene promotes membrane fusion, *Nature,* 300, 658, 1982.
12. Kondor-Kock, C., Burke, B., and Garoff, H., Expression of Semliki Forest virus proteins from cloned complementary DNA. The fusion activity of the spike glycoprotein, *J. Cell Biol.,* 97, 644, 1983.
13. Florkiewicz, R. Z. and Rose, J. K., A cell line expressing the vesicular stomatitis virus glycoprotein fuses at low pH, *Science,* 225, 721, 1984.
14. McCune, J. M., Rabin, L. B., Feinberg, M. B., Lieberman, M., Kosek, J. C., Reyes, C. R., and Weissman, J. L., Endoproteolytic cleavage of gp160 is required for the activation of human immunodeficiency virus, *Cell,* 53, 55, 1988.
15. Boulay, F., Doms, R. W., Webster, R. G., and Helenius, A., Posttranslational oligomerization and cooperative acid activation of mixed influenza hemagglutinin trimers, *J. Cell Biol.,* 106, 629, 1988.
16. Sambrook, J., Rodgers, L., White, J., and Gething, M. J., Lines of BPV-transformed murine cells that constitutively express influenza virus hemagglutinin, *EMBO J.,* 4, 91, 1985.
17. Sechoy, O., Philippot, J. R., and Bienvenue, A., Preparation and characterization of F-protein vesicles isolated from Sendai virus by means of octyl glucoside, *Biochim. Biophys. Acta,* 857, 1, 1986.

18. Stegmann, T., Morselt, H. W. M., Booy, F. P., van Breemen, J. F. L., and Scherphof, G., Functional reconstitution of influenza virus envelopes, *EMBO J.*, 6, 2651, 1987.

19. Metsikko, K., van Meer, G., and Simons, K., Reconstitution of the fusogenic activity of vesicular stomatitis virus, *EMBO J.*, 5, 3429, 1986.

20. Marsh, M., Bolzau, E., White, J., and Helenius, A., Interactions of Semliki Forest virus spike glycoprotein rosettes and vesicles with cultured cells, *J. Cell Biol.*, 96, 455, 1983.

21. Wiley, D. C. and Skehel, J. J., The structure and function of the hemagglutinin membrane glycoprotein of influenza virus, *Annu. Rev. Biochem.*, 56, 365, 1987.

22. Doms, R. W. and Helenius, A., Properties of a viral fusion protein, in *Molecular Mechanisms of Membrane Fusion*, Ohki, S., Doyle, D., Flanagan, T. D., Hui, S. W., and Mayhew, E., Eds., Plenum Press, New York, 1988, 385.

23. Kielian, M. and Helenius, A., Entry of alphaviruses, in *The Togaviridae and Flaviridae*, Schlesinger, S. S. and Schlesinger, M. J., Eds., Plenum Press, New York, 1986, 91.

24. Kreis, T. E. and Lodish, H. F., Oligomerization is essential for transport of vesicular stomatitis viral glycoprotein to the cell surface, *Cell*, 46, 929, 1986.

25. Blumenthal, R., Membrane fusion, *Curr. Top. Membr. Transp.*, 29, 203, 1987.

26. Hoekstra, D. and Kok, J. W., Entry mechanisms of enveloped viruses. Implications for fusion of intracellular membranes, *Bioscience Rep.*, 9, 273, 1989.

27. Hoekstra, D., Membrane fusion of enveloped viruses: especially a matter of proteins, *J. Bioenerg. Biomembr.*, 22, 121, 1990.

28. Gallaher, W. R., Detection of a fusion peptide sequence in the transmembrane protein of human immunodeficiency virus, *Cell*, 50, 327, 1987.

29. Ohnishi, S.-I., Fusion of viral envelopes with cellular membranes, *Curr. Top. Membr. Transp.*, 32, 257, 1988.

30. Gething, M.-J., Doms, R. W., York, D., and White, J., Studies on the mechanism of membrane fusion. Site-specific mutagenesis of the hemagglutinin of influenza virus, *J. Cell Biol.*, 102, 11, 1986.

31. Kowalski, M., Potz, J., Basiripour, L., Dorfman, T., Goh, W. C., Terwilliger, E., Dayton, A, Rosen, C., Haseltine, W., and Sodroski, J., Functional regions of the envelope glycoprotein of human immunodeficiency virus type 1, *Science*, 237, 1351, 1987.

32. Morrison, T., Structure, function, and intracellular processing of paramyxovirus membrane proteins, *Vir. Res.*, 10, 113, 1988.

33. Lee, P. M., Cherry, R. J., and Bachi, T., Correlation of rotational mobility and flexibility of Sendai virus spike glycoproteins with fusion activity, *Virology*, 128, 65, 1983.

34. Hoekstra, D., Klappe, K., de Boer, T., and Wilschut, J., Characterization of the fusogenic properties of Sendai virus: kinetics of fusion with erythrocyte membranes, *Biochemistry*, 24, 4739, 1985.

35. Hoekstra, D., Klappe K., Hoff, H., and Nir, S., Mechanism of fusion of Sendai virus: role of hydrophobic interactions and mobility constraints of viral membrane proteins. Effects of poly(ethylene glycol), *J. Biol. Chem.*, 264, 6786, 1989.

36. Henis, Y. I., Herman-Barhom, Y., Aroeti, B., and Gutman, O., Lateral mobility of both envelope proteins (F and HN) of Sendai virus in the cell membrane is essential for cell-cell fusion, *J. Biol. Chem.*, 264, 17119, 1989.

37. Sechoy, O., Philippot, J. R., and Bienvenue, A., F-protein-F-protein interaction within the Sendai virus identified by native binding or chemical cross-linking, *J. Biol. Chem.*, 262, 11519, 1987.

38. Klenk, H.-D., Rott, R., Orlich, M., and Blödorn, J., Activation of influenza A viruses by trypsin treatment, *Virology*, 68, 426, 1975.

39. Gething, M.-J. and Sambrook, J., Cell-surface expression of influenza haemagglutinin from a cloned copy of the RNA gene, *Nature*, 293, 620, 1981.

40. Skehel, J. J., Bayley, P. M., Brown, E. B., Martin, S. R., Waterfield, M. D., White, J. M., Wilson, I. A., and Wiley, D. C., Changes in the conformation of influenza hemagglutinin at the pH optimum of virus-mediated membrane fusion, *Proc. Natl., Acad. Sci., USA*, 79, 968, 1982.

41. Wilson, I. A., Skehel, J. J., and Wiley, D. C., Structure of the haemagglutinin membrane glycoprotein of influenza virus at 3Å resolution, *Nature*, 289, 366, 1981.

42. Ruigrok, R. W. H., Wrigley, N. G., Calder, L. J., Cusack, S., Wharton, S. A., Brown, E. B., and Skehel, J. J., Electron microscopy of the low pH stucture of the influenza virus hemagglutinin, *EMBO J.*, 5, 41, 1986.

43. Doms, R. W. and Helenius, A., Quaternary structure of influenza virus hemagglutinin after acid treatment, *J. Virol.* 60, 833, 1986.

44. Nestorowicz, A., Laver, G., and Jackson, D. C., Antigenic determinants of influenza virus hemagglutinin. A comparison of the physical and antigenic properties of monomeric and trimeric forms, *J. Gen. Virol.*, 66, 1687, 1985.

45. McDougal, J. S., Nicholson, J. K. A., Cross, G. D., Cort, S. P., Kennedy, M. S., and Mawle, A. C., Binding of the human retrovirus HTLV-III/LAV/ARV/HIV to the CD4 (T4) molecule: conformation dependence, epitope mapping, antibody inhibition and potential for idiotypic mimicry, *J. Immunol.*, 137, 2937, 1986.

46. Racevskis, J. and Sarkar, N., Murine mammary tumor virus structural protein interactions: formation of oligomeric complexes with cleavable cross-linking agents, *J. Virol.*, 35, 937, 1980.

47. Einfeld, D. and Hunter, E., Oligomeric structure of a prototypic retrovirus glycoprotein, *Proc. Natl. Acad. Sci., USA*, 85, 8688, 1988.

48. Schawaller, M., Smith, G. E., Skehel, J. J., and Wiley, D. C., Studies with crosslinking reagents on the oligomeric structure of the env glycoprotein of HIV, *Virology*, 172, 367, 1989.

49. Garoff, H., Frischauf, A. M., Simons, K., Lerach, H., and Delius, H., Nucleotide sequence of cDNA coding for Semliki Forest virus membrane glycoproteins, *Nature*, 288, 236, 1980.

50. Schlegel, R. and Wade, M., Biologically active peptides of the vesicular stomatitis virus glycoprotein, *J. Virol.*, 53, 319, 1985.

51. Woodgett, C. and Rose, J. K., Amino-terminal mutation of the vesicular stomatitis virus glycoprotein does not affect its fusion activity, *J. Virol.*, 59, 486, 1986.

52. Simons, K. and Warren, G., Semliki Forest virus: a probe for membrane traffic in the animal cells, *Adv. Protein Chem.*, 36, 79, 1984.

53. White, J., Kielian, M., and Helenius, A., Membrane fusion proteins of enveloped animal viruses, *Q. Rev. Biophys.*, 16, 151, 1983.

54. White, J. and Helenius, A., pH-dependent fusion between the Semliki Forest virus membrane and liposomes, *Proc. Natl. Acad. Sci., USA*, 77, 3273, 1980.

55. Omar, A. and Koblet, H., Semliki Forest virus particles containing only the E1 envelope glycoprotein are infectious and can induce cell-cell fusion. *Virology*, 166, 17, 1988.

56. Lobigs, M., Hongxing, Z., and Garoff, H., Function of Semliki Forest virus E3 peptide in virus assembly: Replacement of E3 with an artificial signal peptide abolishes spike heterodimerization and surface expression of E1, *J. Virol.*, 64, 4346, 1990.

57. Lobigs, M., Wahlberg, J. M., and Garoff, H., Spike protein oligomerization control of Semliki Forest virus fusion, *J. Virol.*, 64, 5214, 1990.

58. Doms, R. W., Keller, D. S., Helenius, A., and Balch, W. E., Role for adenosine triphosphate in regulating the assembly and transport of vesicular stomatitis virus G protein trimers, *J. Cell Biol.*, 105, 1957, 1987.

59. Daniels, R. S., Douglas, A. R., Skehel, J. J., and Wiley, D. C., Analysis of the antigenicity of influenza hemagglutinin at the pH optimum for virus-mediated membrane fusion, *J. Gen. Virol.*, 64, 1657, 1983.

60. Daniels, R. S., Jefferies, S., Yates, P., Schild, G. C., Rogers, G. N., Paulson, J. C., Wharton, S. A., Douglas, A. R., Skehel, J. J., and Wiley, D. C., The receptor-binding and membrane-fusion properties of influenza virus variants selected using antihemagglutinin monoclonal antibodies, *EMBO J.*, 6, 1459, 1987.

61. Copeland, C. S., Doms, R. W., Bolzau, E. M., Webster, R. G., and Helenius, A., Assembly of influenza hemagglutinin trimers and its role in intracellular transport, *J. Cell Biol.*, 103, 1179, 1986.

62. Kielian, M. and Jungerwirth, S., Mechanisms of enveloped virus entry into cells, *Mol. Biol. Med.*, 7, 17, 1990.

63. Doms, R. W., Helenius, A. H., and White, J., Membrane fusion activity of the influenza virus hemagglutinin. The low pH-induced conformational change, *J. Biol. Chem.*, 260, 2973, 1985.

63a. Kielian, M. and Helenius, A., pH-induced alterations in the fusogenic spike protein of Semliki Forest virus, *J. Cell Biol.*, 101, 2284, 1985.

64. Hsu, M.-C., Scheid, A., and Choppin, P., Activation of the Sendai virus fusion protein (F) involves a conformational change with exposure of a new hydrophobic region, *J. Biol. Chem.*, 256, 3557, 1981.

65. Hsu, M.-C., Scheid, A., and Choppin, P. W., Enhancement of membrane fusion activity of Sendai virus by exposure of the virus to basic pH is correlated with a conformational change in the fusion protein, *Proc. Natl. Acad. Sci., USA*, 79, 5862, 1982.

66. Novick, S. L. and Hoekstra, D., Membrane penetration of Sendai virus glycoproteins during the early stages of fusion with liposomes as determined by hydrophobic photoaffinity labeling, *Proc. Natl. Acad. Sci., USA*, 85, 7433, 1988.

67. Brunner, J., Zugliani, C., and Mischler, R., Fusion activity of influenza virus PR8/34 correlates with a temperature-induced conformational change within the hemaggutinin ectodomain detected by photochemical labeling, *Biochemistry*, 30, 2432, 1991.

68. Stegmann, T., Delfino, J. M., Richards, F. M., and Helenius, A., The HA_2 subunit of influenza hemagglutinin inserts into the target membrane prior to fusion, *J. Biol. Chem.*, 266, 18404, 1991.

69. Brunner, J., Testing topological models for the membrane penetration of the fusion peptide of influenza virus hemagglutinin, *FEBS Lett.*, 257, 369, 1989.

70. Blumenthal, R., Puri, A., Walter, A., and Eidelman, O., pH-dependent fusion of vesicular stomatitis virus with cells: studies of mechanism based on an allosteric model, in *Molecular Mechanisms of Membrane Fusion*, Ohki, S., Doyle, D., Flanagan, T.D., Hui, S.W., and Mayhew, E., Eds., Plenum Press, New York, 1988, 367.

71. Haywood, A. M., Fusion of Sendai viruses with model membranes, *J. Mol. Biol.*, 87, 625, 1974.

72. Oku, N., Inoue, K., Nojima, S., Sekiya, T., and Nosawa, Y., Electron microscopic studies on the interaction of Sendai virus with liposomes containing glycophorin, *Biochim. Biophys. Acta*, 691, 91, 1982.

73. Larsen, C. E., Nir, S., Alford, D. R., Jennings, M., Lee, K.-D., and Düzgünes, N., Human immunodeficiency virus type 1 (HIV-1) fusion with model membranes: kinetic analysis and the role of lipid composition, pH and divalent cations, *Biochim Biophys. Acta*, in press, 1993.

74. Haywood, A. M. and Boyer, B. P., Fusion of influenza virus membranes with liposomes at pH 7.5, *Proc. Natl. Acad. Sci., USA*, 82, 4611, 1985.

75. Burger, K. N. J., Knoll, G., and Verkleij, A. J., Influenza virus-model membrane interaction. A morphological appproach using modern cryotechiques, *Biochim. Biophys. Acta*, 939, 89, 1988.

76. Haywood, A. M. and Boyer, B. P., Time and temperature dependence of influenza virus membrane fusion at neutral pH, *J. Gen. Virol.*, 67, 2813, 1986.

77. Yoshimura, A., Kuroda, K., Kawasaki, K., Yamashina, S., Maeda, T., and Ohnishi, S.-I., Infectious cell entry mechanism of influenza virus, *J. Virol.*, 43, 284, 1982.

78. White, J., Kartenbeck, J., and Helenius, A., Membrane fusion activity of influenza virus, *EMBO J.*, 1, 217, 1982.

79. Kawasaki, K. and Ohnishi, S.-I., Membrane fusion of influenza virus with phosphatidylcholine liposomes containing viral receptors, *Biochem. Biophys. Res. Commun.*, 166, 378, 1992.

80. Nussbaum, O., Rott, R., and Loyter, A., Fusion of influenza virus particles with liposomes: requirement for cholesterol and virus receptors to allow fusion with and lysis of neutral but not of negatively charged liposomes, *J. Gen. Virol.*, 73, 2831, 1992.

81. Ramalho-Santos, J., Nir, S., Düzgünes, N., Carvalho, A. P., and Pedroso de Lima, M. C., A common mechanism for Influenza virus fusion activity and inactivation, *Biochemistry*, 32, 2771, 1993.

82. Haywood, A. M. and Boyer, B. P., Sendai virus membrane fusion: time course and efect of temperature, pH, calcium and receptor concentration, *Biochemistry*, 21, 6041, 1982.

83. Haywood, A. M. and Boyer, B. P., Effect of lipid composition upon fusion of liposomes with Sendai virus membranes, *Biochemistry*, 23, 4161, 1984.

84. Amselem, S., Loyter, A., Lichtenberg, D., and Barenholtz, Y., The interaction of Sendai virus with negatively charged liposomes: virus-induced lysis of carboxyfluorescein-loaded small unilamellar vesicles, *Biochim. Biophys. Acta*, 820, 1, 1985.

85. Loyter, A. and Citovsky, V., The role of envelope glycoproteins in the fusion of Sendai virus with liposomes, in *Membrane Fusion*, Wilschut J. and Hoekstra, D., Eds., Marcel Dekker, New York, 1991, 375.

86. Kundrot, C. E., Spangler, E. A., Kendall, D. A., MacDonald, R. C., and MacDonald, R.I., Sendai virus-mediated lysis of liposomes requires cholesterol, *Proc. Natl. Acad. Sci., USA*, 80, 1608,1983.

87. Oku, N., Nojima, S., and Inoue, K., Studies on the interaction of HVJ (Sendai virus) with liposomal membranes induced permeability increase of liposomes containing glycophorin, *Virology*, 116, 419, 1982.

88. Hsu, M.-C., Scheid, A., and Choppin, P. W., Fusion of Sendai virus with liposomes: dependence on the viral fusion protein (F) and the lipid composition of liposomes, *Virology*, 126, 361, 1983.

89. Yamada, S. and Ohnishi, S.-I., Vesicular stomatitis virus binds and fuses with phospholipid domain in target cell membranes, *Biochemistry*, 25, 3703, 1986.

90. Wickham, T. J., Granados, R. R., Wood, H. A., Hamxner, D. A., and Schuler, M. L., General analysis of receptor-mediated viral attachment to cell surfaces, *Biophys. J.*, 58, 1501,1990.

91. Israelachvili, J. N., Marcelja, S., and Horn, R. G., Physical principles of membrane organization, *Q. Rev. Biophys.*, 13, 121, 1980.

92. Cullis, P. R. and De Kruijff, B., Lipid polymorphism and the functional roles of lipids in biological membranes, *Biochim. Biophys. Acta*, 559, 399, 1979.

93. Cullis, P. R., De Kruijff, B., Hope, M. J., Verkleij, A. J., Nayar, R., Farren, S. B., Tilcock, C. P. S., Madden, T. D., and Bally, M. B., Structural properties of lipids and their functional roles in biological membranes, in *Membrane Fluidity*, Aloia, R. C., Ed., Academic Press, New York, 1983, 39.

94. Verkleij, A. J., Lipid intramembranous particles, *Biochim. Biophys. Acta*, 779, 43, 1984.

95. Epand, R. H., Cheetham, J. J., Epand, R. F., Yeagle, P. L., Richardson, C. D., Rockweel, A., and DeGrado, W. F., Peptide models for the membrane destabilizing actions of viral fusion proteins, *Biopolymers*, 32, 309, 1992.

96. Papahadjopoulos, D., Nir, S., and Ohki, S., Permeability properties of phospholipid membranes: effect of cholesterol and temperature, *Biochim. Biophys. Acta*, 266, 571, 1971.

97. Kielian, M. and Helenius A., The role of cholesterol in the fusion of Semliki Forest virus with membranes, *J. Virol.*, 52, 281, 1984.

98. Phalen, T. and Kielian, M., Cholesterol is required for infection by Semliki Forest virus, *J. Cell Biol.*, 112, 615, 1991.

99. Asano, K. and Asano, A., Binding of cholesterol and inhibitory peptide derivatives with the fusogenic hydrophobic sequence of F-glycoprotein of HVJ (Sendai virus): possible implication in the fusion reaction, *Biochemistry*, 27, 1321, 1988.

100. Hoekstra, D., de Boer, T., Klappe, K., and Wilschut, J., Fluorescence method for measuring the kinetics of fusion between biological membranes, *Biochemistry*, 23, 5675, 1984.

101. Struck, D. K., Hoekstra, D., and Pagano, R. E., Use of resonance energy transfer to monitor membrane fusion, *Biochemistry*, 20, 4093, 1981.

102. Stephen, J. M., Bradley, D., Gibson, C. C., Smith, P. D., and Blumenthal, R., Use of membrane-associated fluorescence probes to monitor fusion of bilayer vesicles: application to rapid kinetics using pyrene excimer/monomer fluorescence, in *Spectroscopic Membrane Probes*, Loew, L. M., Ed., CRC Press Inc., Boca Raton, FL, 1988, 161.

103. Stegmann, T., Hoekstra, D., Scherphof, G., and Wilschut, J., Kinetics of pH-dependent fusion between Influenza virus and liposomes, *Biochemistry*, 24, 3107, 1985.

104. Klappe, K., Wilschut, J., Nir, S., and Hoekstra, D., Parameters affecting fusion between Sendai virus and liposomes. Role of viral proteins, liposome composition, and pH, *Biochemistry*, 25, 8252, 1986.

105. Nir, S., Klappe, K., and Hoekstra D., Mass action analysis of kinetics and extent of fusion between Sendai virus and phospholipid vesicles, *Biochemistry*, 25, 8261, 1986.

106. Eidelman, O., Schlegel, R., Tralka, T. S., and Blumenthal, R., pH-dependent fusion induced by vesicular stomatitis virus glycoprotein reconstituted into phospholipid vesicles, *J. Biol. Chem.*, 259, 4622, 1984.

106a. Ramalho-Santos, J., Negrão, R., and Pedroso de Lima, M. C., Role of hydrophobic interactions in the fusion activity of Influenza and Sendai viruses towards model membranes, *Bioscience Reports*, 14, 15, 1994.

107. Nir, S., Stegmann, T., and Wilschut, J., Fusion of influenza virus with cardiolipin liposomes at low pH: mass action analysis of kinetics and extent, *Biochemistry*, 25, 257, 1986.

108. Stegmann, T., Hoekstra D., Scherphof G., and Wilschut, J., Fusion activity of Influenza virus: a comparison between biological and artificial target membrane vesicles, *J. Biol. Chem.*, 261, 10966, 1986.

109. Stegmann, T., Booy, F. P., and Wilschut, J., Effects of low pH on influenza virus: activation and inactivation of the membrane fusion capacity of the hemagglutinin, *J. Biol. Chem.*, 262, 17744, 1987.

110. Stegmann, T., Nir, S., and Wilschut, J., Membrane fusion activity of influenza virus. Effects of gangliosides and negatively charged phospholipids in target liposomes, *Biochemistry*, 28, 1698, 1989.

111. Düzgüneş, N., Pedroso de Lima, M. C., Stamatatos, L., Flasher, D., Alford, D., Friend, D. S., and Nir, S., Fusion activity and inactivation of influenza virus: kinetics of low pH-induced fusion with cultured cells, *J. Gen. Virol.*, 73, 27, 1992.

112. Nir, S., Düzgüneş, N., Pedroso de Lima, M. C., and Hoekstra, D., Fusion of enveloped viruses with cells and liposomes, *Cell Biophys.*, 17, 181, 1990.

113. Stegmann, T., White, J., and Helenius, A., Intermediates in influenza induced membrane fusion, *EMBO J.*, 9, 4231, 1990.

114. Hoekstra, D. and Klappe, K., Sendai virus-erythrocyte membrane interaction: quantitative and kinetic analysis of viral binding, dissociation, and fusion, *J. Virol.,* 58, 87, 1986.

115. Amselem, S., Barenholtz, Y., Loyter, A., Nir, S., and Lichtenberg, D., Fusion of Sendai virus with negatively charged liposomes as studied by pyrene-labeled phospholipid liposomes, *Biochim. Biophys. Acta,* 860, 301, 1986.

116. Pedroso de Lima, M. C., Ramalho-Santos, J., Martins, M. F., Pato de Carvalho, A., Bairos, V., and Nir, S., Kinetic modeling of Sendai virus fusion with PC-12 cells. Effect of pH and temperature on fusion and viral inactivation, *Eur. J. Biochem.,* 205, 181, 1992.

117. Pedroso de Lima, M. C., Nir, S., Flasher D., Klappe, K., Hoekstra, D., and Düzgüneş, N., Fusion of Sendai virus with human HL-60 and CEM cells: different kinetics of fusion for two isolates, *Biochim. Biophys. Acta,* 1070, 446, 1991.

118. Citovsky, V., Blumenthal, R., and Loyter, A., Fusion of Sendai virions with phosphatidylcholine-cholesterol liposomes reflects the viral activity required for fusion with biological membranes, *FEBS Lett.,* 193, 135, 1985.

119. Chejanovsky, N., Zakai, N., Amselem, S., Barenholz, Y., and Loyter, A., Membrane vesicles containing the Sendai virus binding glycoprotein, but not the viral fusion protein, fuse with phosphatidylserine liposomes at low pH, *Biochemistry,* 25, 4810, 1986.

120. Loyter, A. and Volsky, D. J., Reconstituted virus envelopes as carriers for the introduction of biological material into animal cells, in *Membrane Reconstitution,* Poste, G. and Nicolson, G. L., Eds., Elsevier, Amsterdam, 1982, 215.

121. Poste, G. and Pasternak, C. A., Virus-induced cell fusion, in *Cell Surface Reviews,* Poste, G. and Nicolson, G. L., Eds., Elsevier, Amsterdam, 1978, 305.

122. Larsen, C. E., Alford, D. R., Young, L. J. T., McGraw, T. P., and Düzgüneş, N., Fusion of simian immunodeficiency virus with liposomes and erythrocyte ghost membranes: effects of lipid composition, pH and calcium, *J. Gen. Virol.,* 71, 1947, 1990.

123. Harouse, J. M., Bhat, S., Spitalnik, S. L., Laughlin, M., Stefano, K., Silberberg, D. H., and Gonzalez-Scarano, F., Inhibition of entry of HIV-1 in neural cell lines by antibodies against galactosyl ceramide, *Science,* 253, 320, 1991.

124. Yahi, N., Baghdiguian, S., Moreau, H., and Fantini, J., Galactosyl ceramide (or a closely related molecule) is the receptor for human immunodeficiency virus type 1 on human colon epithelial HT29 cells, *J. Virol.* 66, 4848, 1992.

125. Pedroso de Lima, M. C. and Düzgüneş, N., unpublished data, 1992.

126. Sattentau, Q. J. and Weiss, R. A., The CD4 antigen: physiological ligand and HIV receptor, *Cell,* 52, 631, 1988.

127. Kuge, O., Akamatsu, Y., and Nishijima, M., Abortive infection with Sindbis virus of a Chinese hamster ovary cell mutant defective in phosphatidylserine and phosphatidylethanolamine biosynthesis, *Biochim. Biophys. Acta,* 986, 61, 1989.

128. Hoekstra, D., Topographical distribution of a membrane-inserted fluorescent phospholipid analogue during cell fusion, *Exp. Cell. Res.,* 144, 482, 1983.

129. Citovsky, V., Rottem, S., Nussbaum, O., Laster, Y., Rott, R., and Loyter, A., Animal viruses are able to fuse with prokaryotic cells. Fusion between Sendai or influenza virions with mycoplasma, *J. Biol. Chem.,* 263, 461, 1988.

130. Yoshimura, A., Kobayashi, T., Hidaka, K., Kuwano, M., and Ohnishi, S.-I., Altered interaction between Sendai virus and a Chinese hamster cell mutant with defective cholesterol synthesis, *Biochim. Biophys.* Acta, 904, 159, 1987.

131. Herrmann, A., Claque, M. J., Puri, A., Morris, S. J., Blumenthal, R., and Grimaldi, S., Effect of erythrocyte transbilayer phospholipid distribution on fusion with vesicular stomatitis virus, *Biochemistry,* 29, 4054, 1990.

PART III
New Developments
of
Liposomes

Chapter 9

Gene Transfer with Cationic Amphiphiles

Jean-Serge Remy, Claude Sirlin and Jean-Paul Behr

CONTENTS

I. INTRODUCTION

The introduction of genes into cells of various origins has been a major technique of cell biology research for over a decade. Gene transfer is indeed the most straightforward way to study gene and protein function and regulation, from a single cell context up to complex multicellular processes such as embryogenesis, or to the creation of animal models of human diseases. Besides being a powerful research tool, transfection also has important economic implications with the genetic engineering of microorganisms, plants, and animals, for the production of proteins, as well as for crop and livestock improvement. Finally, gene transfer is the prerequisite for a conceptually new therapeutic approach for curing inherited and many other diseases, namely, gene therapy.

While most current gene therapeutical protocols rely on recombinant viral vectors, the diversity of gene transfer uses has resulted in the development of a variety of artificial techniques, such as direct DNA microinjection, DNA coprecipitation or compaction with polycations, DNA encapsulation into liposomes, and cell membrane perturbation by chemicals (organic solvents, detergents, polymers, enzymes), or by physical means (mechanic, osmotic, thermic, electric shocks). Within the context of this book, however, only liposomes and cationic lipid-mediated DNA delivery techniques are of interest.

Liposome encapsulation of a gene is done most efficiently by reverse phase evaporation,[1] especially when performed in the presence of DNA compacting molecules.[2,3] Over the past fifteen years, many liposome-based DNA delivery systems have been described and reviewed,[2-7] including molecular components for targeting to given cell surface receptors or to the nucleus, or for escape from the lysosomal compartment. However, none compared favorably with commonly used recent techniques, unless viral fusion proteins were included.[8] Therefore, we shall focus here on the more recent and presumably much more efficient[9] transfection technique using cationic lipids. These transfection vehicles are often erroneously called "cationic liposomes" (hence the editor's request for this chapter). The latter denomination is confusing, since it implicitly suggests both the occurrence during transfection of spherical lipidic particles with an aqueous interior[10] and a DNA encapsulation step, none of which is needed as will be seen below.

II. CATIONIC MOLECULES USED FOR TRANSFECTION

The rational design of synthetic gene carriers is recent,[11-13] and within a few years several groups reported on the synthesis or the use of already available commercial cationic DNA binding amphiphiles for gene transfer purposes.

Felgner and co-workers synthesized the quaternary ammonium amphiphile DOTMA (dioleoyloxypropyl-trimethylammonium bromide, see below), and were the first to describe DNA[14] as well as RNA[15] transfer into eukaryotic cell lines following *mixing* with a cationic lipid (as opposed to *encapsulation* into a liposome). This new technique was shown to be up to a hundredfold more efficient than calcium phosphate or DEAE-dextran coprecipitation. DOTMA was commercialized (Lipofectin®,

160

GIBCO-BRL) as a one-to-one mixture with dioleoyl-phosphatidylethanolamine (DOPE) and has been widely used since to transfect a large variety of animal and plant eukaryotic cells.[14-31]

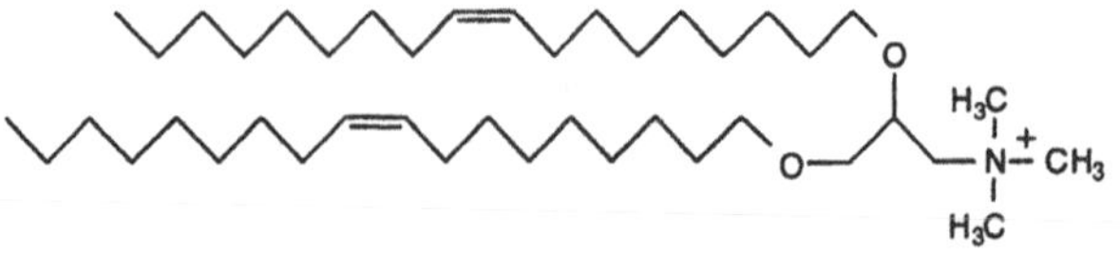

DOTMA

Our own efforts were directed towards the design of cationic amphiphiles able to *compact* genomic DNA, namely, lipopolyamines.[32] The metabolizable parent lipids, DOGS and DPPES, were shown to be able to transfect efficiently (two orders of magnitude better than Ca phosphate) established cell lines as well as primary neuronal cultures. Most importantly, *DNA coating with excess cationic lipid* (rather than liposome binding to the nucleic acid) and subsequent binding of the resulting particle to the negatively charged cell surface via ionic forces was inferred.[33] Such an excess of lipid cationic charges over the DNA anionic phosphates proved to be a general requirement for optimal in vitro transfection with cationic lipids (see below). DOGS also has been commercialized (Transfectam®, Promega) and has been shown to very efficiently transfect many animal cells.[33-53]

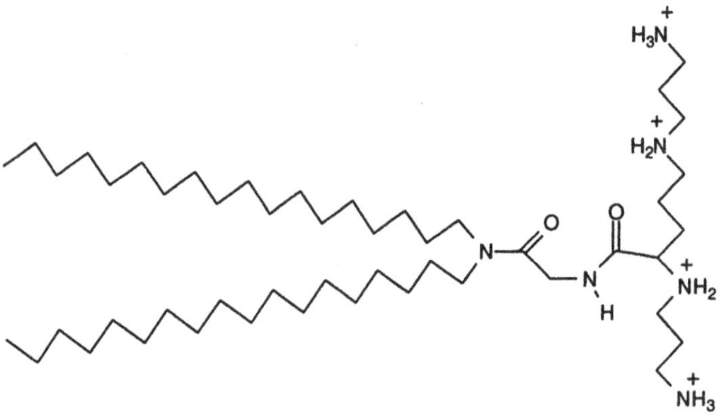

DOGS

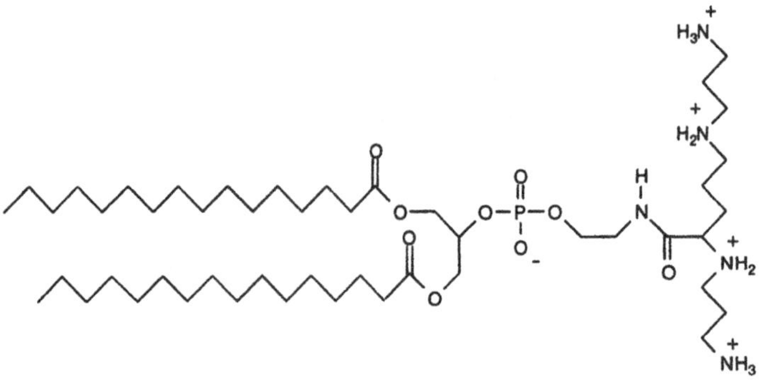

DPPES

Kunitake et al.[54] synthesized a whole series of lipophilic glutamate diesters with pendent trimethylammonium heads. Whereas transfection properties were similar for dodecyl- and tetradecylesters, they decreased drastically as the link between the lipid and cationic parts increased. $C_{12}GluPhC_2N^+$ was as efficient as Lipofectin.

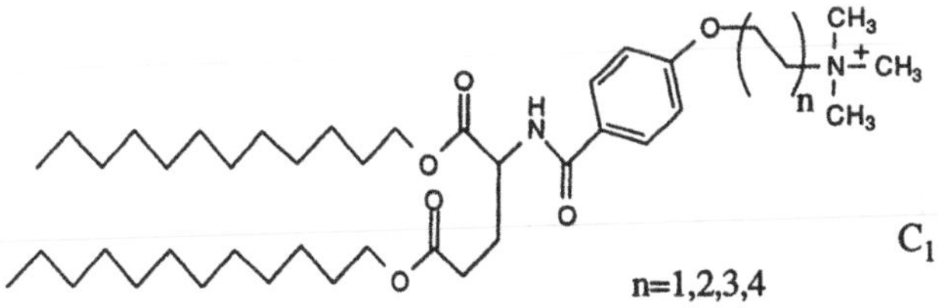

$$C_{12}GluPhC_nN^+$$

n=1,2,3,4

$$C_{14}GluC_nN^+$$

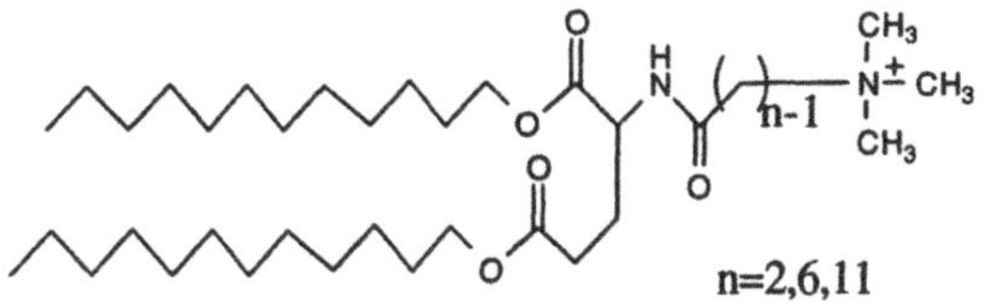

n=2,6,11

In an effort to reduce the cytotoxicity of DOTMA, Silvius et al.[55] synthesized a series of metabolizable quaternary ammonium salts, some of which (DOTB, DOTAP dioleoyl esters) had efficiencies comparable to that of Lipofectin when dispersed with DOPE. Mixtures with cholesterol-derived cations (ChoTB, ChoSC), and most surprisingly, the dioleoyl ester with a slightly longer head group spacer (DOSC), were much less efficient. DOTAP also is available (Boeringer, Mannheim).

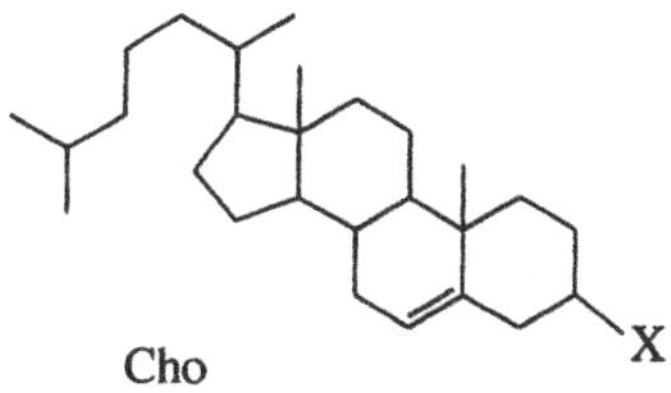

Cho

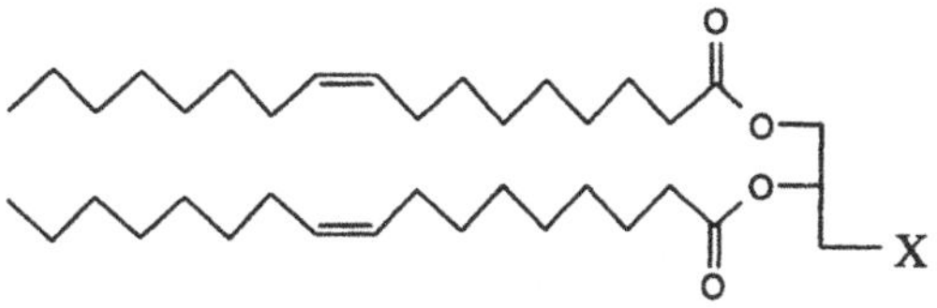

DO

X=

N^+Me_3	DOTAP
$OOC(CH_2)_3N^+Me_3$	DOTB ChoTB
$OOCH_2CH_2COOCH_2CH_2N^+Me_3$	DOSC ChoSC

Huang et al.[56] thought along the same lines and synthesized DC-Chol having a hydrolyzable dimethylethylenediamine headgroup. DOPE/DC-Chol one-to-one mixtures were able to transfect several cell lines, with efficiencies slightly better than with Lipofectin. The same group also reported about lipophilic polylysines (LPLL),[57] which were up to threefold more efficient than Lipofectin, but only when the fibroblast cells were mechanically scraped following the transfection period.

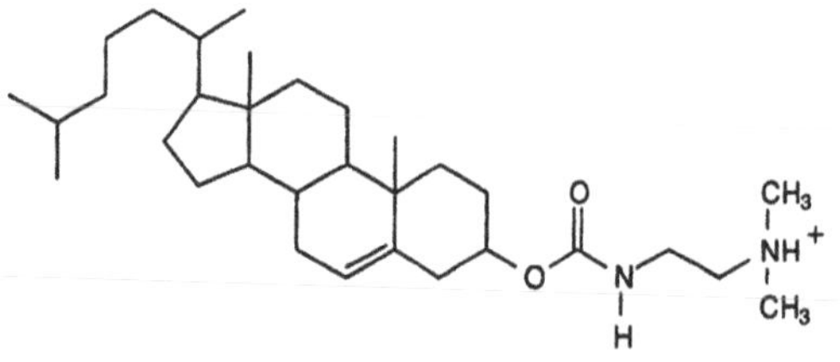

DC-Chol

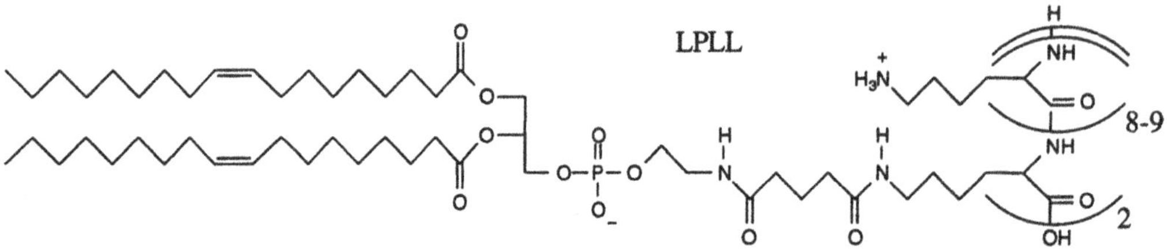

LPLL

Meanwhile, biologists had also discovered that some commercial cationic detergents, when mixed with true lipids, could function as transfection agents. Loyter et al.[58] showed that DEBDA hydroxide added to excess phosphatidylcholine/cholesterol is able to introduce tobacco mosaic virus-RNA into tobacco and petunia protoplasts. Unfortunately, no comparison was made with the classical calcium/ polyethyleneglycol technique for transfection of plant protoplasts.

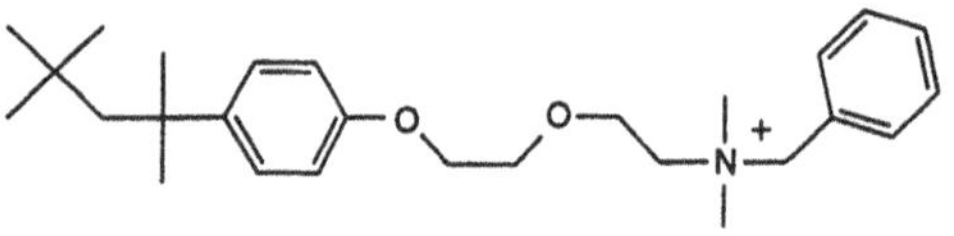

DEBDA

Huang et al.[59] compared the transfection properties of several quaternary ammonium detergents on fibroblasts, and found that CTAB/DOPE mixtures were the most efficient, although somewhat below Lipofectin.

R=

H	DTAB
CH_3CH_2	TTAB
$CH_3(CH_2)_3$	CTAB

Yagi et al.[60] used a lipophilic diester of glutamic acid (TMAG) with DOPE, and found it as efficient as calcium phosphate for fibroblasts transfection.

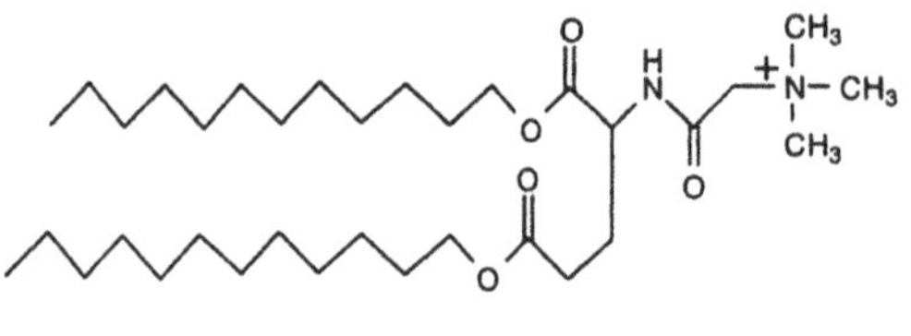

TMAG

Rose et al.[61] compared common detergents of diverse structures: CTAB, DEBDA (see structures above), DDAB, and stearylamine in admixture with phosphatidylethanolamine. Comparative transfection efficiencies were highly variable, depending on the cell line; DDAB (which had been shown earlier to be efficient,[33] albeit very toxic) seemed to be the most promising and the DDAB/DOPE formulation was patented (TransfectACE®, GIBCO BRL).

DDAB

III. COMPARATIVE EFFICIENCIES

The level of transfection is best assessed by using a gene which is absent from the cells to be transduced (reporter gene), driven by a strong viral promoter/enhancer element. Normally this exo-gene must reach the eukaryotic cell nucleus where it is transiently transcribed, or an alternatively stable expression may be seeked in a sub-population of daughter cells. In any case the net result is the synthesis of a foreign protein, which may be detected immunologically or enzymatically. Widely used reporter genes include those for the following enzymes:

- The bacterial chloramphenicol-acetyltransferase (CAT) with either radioactive or fluorescent detection of acetylchloramphenicol, or immunochemical ELISA quantitation of the CAT protein;
- The bacterial beta-galactosidase (β-Gal) sometimes bearing a nuclear localization signal, which is mostly used for in vivo histochemical visualization of transfected cells: a synthetic substrate (X-Gal) is diffused into the fixed tissue, leading to a deep blue product in cells expressing high levels of β-Gal;
- The firefly luciferase (Luc) which is quantitated in the cell extract by photon counting of a phosphorescent product it generates catalytically; this technique has by far the highest dynamic range (circa seven orders of magnitude).

Besides the vehicle and nucleic acid, transfection also depends to a large extent on cell type as well as on many other (often unpredictable) factors.[37,57] Efforts to draw up a hit parade of gene transfer vectors therefore seems hopeless. Nonetheless, the best transfection experiments performed with each of the cationic amphiphiles mentioned above have been summarized in a table, for several pertinent conclusions emerge from such an overview.

The extent of electrical neutralization of DNA anionic charges by the cationic amphiphile has been calculated from the concentration data given for each system (+/– ratio in the table). It shows that regardless of the structure and number of charges borne by the molecule, transfection becomes efficient only when the nucleolipidic particles have a net mean positive charge.[33] Transfection efficiency increases with the cationic lipid/DNA ratio up to the onset of cytotoxicity. Bilayer-disrupting detergents and DDAB are among the most toxic, which limits their performances.

Genuine double chain cationic lipids (upper part of the table) are active by themselves, whereas cationic detergent-like molecules require an additional excess of neutral lipid; the detergent/lipid ratio is such that non-micellar structures prevail. Such a behavior suggests that cationic amphiphiles act as gene transfer vectors by providing a cationic multimolecular surface which glues the anionic DNA and anionic cell surface together (see next paragraph).

A dangling cationic head seems to be a penalty for a vector, as seen in the series $C_{12}GluPhC_nN^+$, $C_{14}GluC_nN^+$, and DOTB/DOSC, where the shortest spacers lead to the highest surface charge density, hence the best transfection efficacy.

The new vectors have been compared to popular techniques such as calcium phosphate or DEAE-dextran coprecipitation, or to the widely distributed DOTMA (Lipofectin). Efficiencies are at least those of classical coprecipitation techniques, and most often are 1–2 orders of magnitude over that of calcium phosphate. Surprisingly, several compounds of widely different stuctures, including DOTMA, seem roughly equivalent to each other. In order to check whether this truly reflects an upper limit to the transfection level obtainable with the cationic lipid method (rather than saturation of the detection

Table 1 Gene transfer efficiencies of various cationic amphiphiles

Cationic Amphiphile	Reporter Gene[a]	Cell Line	+/– Ratio[b]	Efficiency	Ref.
Lipid:					
DOTMA	CAT	CV1, COS7...	>2	5–100 × DEAE Dextran	14
	Neo	LMTK, Ψ2...		10–30 × Calcium phosphate	
DOGS	CAT	Prim. neurons, CHO, S49...	4	>10 × Calcium phosphate	33
	DHFR	CHO⁻...		10–100 × Calcium phosphate	
DDAB	CAT	Prim. neurons	3	< DOGS	33
$C_{12}GluC_2N^+$	Cyt c_5	COS1	>2	~ DOTMA	54
DOTAP	CAT	CV1, 3T3	2.5	20–200 × DEAE Dextran ~ DOTMA	55
LPLL	CAT	L929, HeLa...	6	3 × DOTMA	57
Detergent/neutral lipid:					
DEBDA 1/ Lecithin 1.3/ Chol. 0.7	TMV-RNA	Tobacco, petunia protoplasts	>10	?	58
DC-Chol 1/DOPE 1	CAT	A431, L929...	~ 1	2 × DOTMA	56
CTAB 1/ DOPE 4	CAT	L929	~ 1	< DOTMA	59
TMAG 1/ Lecithin 4	β-Gal	COS1	3	~ Calcium phosphate	60
DDAB 1/ PE 2	VSV-G protein	VTF 7-3 infected HeLa, BHK[c]	1.5	~ DOTMA	61

[a]CAT (chloramphenicol-acetyl-transferase), Neo (neomycin), DHFR (dihydrofolate reductase), Cyt c_5 (cytochrome c_5), TMV (tobacco mosaic virus), β-Gal (beta-galactosidase), VSV (vesicular stomatitis virus). [b]The molar ratio of cationic amphiphile to anionic DNA phosphate, as estimated from concentration data given in the Ref. [c]These cells express bacteriophage T7 RNA polymerase in their cytoplasm.

method), several available compounds were compared using the luciferase reporter gene which has the most extended measurement range. Results obtained with various cell types (Figure 1) favor the second hypothesis: relative efficiencies now are indeed split over more than an order of magnitude. The order of efficiencies varies with the cell type, emphasizing the precautions already mentioned above, yet invariably the DNA-compacting lipopolyamine is found at least as efficient as quaternary ammonium lipids. These remarks shed some light on the way cationic lipids help a gene to enter a cell.

IV. TRANSFECTION MECHANISM

Broadly speaking, a gene is a huge piece of DNA (typically of micron size and 10^4 anionic charges) and it therefore has two good reasons not to enter a cell of comparable size, itself surrounded by a negatively charged lipid bilayer. Consequently, a synthetic gene transfer vector should be designed to compact DNA and to bind it to the cell surface in such a way as to favor membrane destabilization or endocytosis. As depicted in Figure 2, bilayer-forming cationic lipids schematically could provide a kind of "double-faced sticky tape" for linking polyanionic DNA and cell surface together.

Our own efforts toward the development of synthetic gene carriers were directed along the following lines.[33] The simplest natural organic cations able to condense DNA are the polyamines — spermidine and spermine. Regardless of its size, DNA is condensed into toroïds and rods of circa 70–100 nm with stoichiometric amounts of spermine in low salt.[62] In physiological conditions, however, this interaction is quickly reversed. We therefore designed *lipo*polyamines, i.e., amphiphiles with a self-aggregating hydrocarbon tail linked to a polycationic DNA-compacting headgroup, and showed them to be able to *stably* condense DNA into discrete particles.[32] In particular, when DNA is mixed with excess bilayer-forming lipopolyamines such as DOGS and DPPES, the nucleic acid is condensed into circa 100-nm nucleolipidic particles surrounded by a lipid bilayer, as confirmed by electron microscopy[63] and quasi-elastic light scattering[65] (Figure 3). Such particles are clearly not liposomes since their interiors are composed of several compacted plasmid molecules glued together with the lipopolyamine.

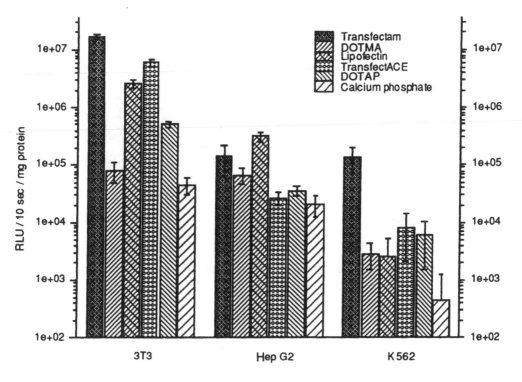

Figure 1 Comparison of transfection efficiencies of various commercially available cationic lipids to that of the calcium phosphate method. 3T3 (murine fibroblasts), HepG2 (human hepatoma), and K562 (human leukemia) derived cell lines (ca. 10^5 cells/15 mm well) were transfected with 2 µg pGL2-Luc plasmid (Promega) and a fourfold charge excess of cationic lipid, essentially following the manufacturer's protocol. Control values (250 RLU for DNA alone) were substracted; the bars are the mean of 3 experiments and S.D. is given.

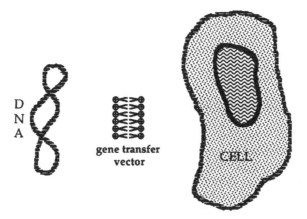

Figure 2 The "double-faced sticky tape" hypothesis for binding similarly charged DNA and cell surface together.

The lipid composition of cytoplasmic membranes is highly variable, yet natural lipids are either zwitterionic or anionic, *never cationic*. It is therefore conceivable that sub-micron size particles with a high cationic surface charge density bind cooperatively to anionic residues on the cell surface (Figure 4). This hypothesis[33,38] would explain the common transfection property shared by vehicles as different as high molecular weight cationic polymers (DEAE-dextran, polybrene), a mineral cationic particle (asbes-

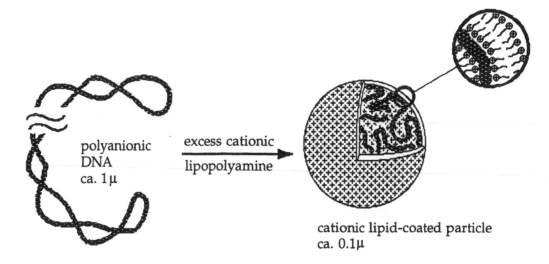

Figure 3 Lipopolyamines condense extended plasmids into a small nucleolipidic particle.

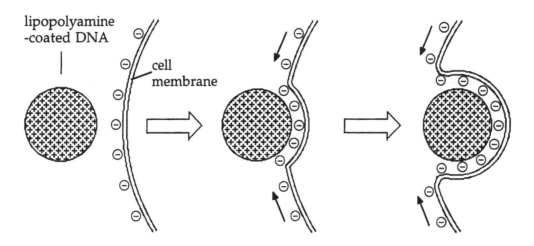

Figure 4 Spontaneous "zipper" engulfment of a rigid cationic particle.

tos fiber), alkaline earth insoluble salts (e.g., calcium phosphate), and aggregated cationic amphiphiles: all may provide a cationic surface for binding DNA to the cell membrane. There, lateral diffusion of cellular lipids together with cytoplasmic membrane deformability (as opposed to the hard cationic particle) result in engulfment of the latter[13] (Figure 4). Subsequent cell entry may then depend on many factors such as cell type or the particle nature and size. Cationic lipid-coated particles may take either of two routes reminiscent of viral infection:[64] membrane fusion at the most curved edge of the cell surface or spontaneous endocytosis. The intracellular fate of the DNA-containing particles is even more obscure, but some of the nucleic acid must reach the nucleus and become uncoated at some stage, since active exogene transcription is observed.

This rather putative view of the mechanism by which the nucleolipidic particles enter a cell is consistent with the aforementioned general observations (i.e., the requirement for positively charged particles, and the inability of cationic detergents to promote transfection by themselves). Further, lipids with other strong DNA-binding heads (see structures below) which either intercalate between base pairs (1) or fill the minor groove (2) are unable to transfect cells.[65]

More speculative arguments go along the same lines. DOGS, DDAB, and oleoyl containing lipids (DOTMA, DOPE, DOTAP, LPLL) do not form stable bilayer liposomes in high salt physiological media; a similar behavior probably holds for cationic detergents/DOPE mixtures. This property favors several lipidic structures initially bound to a plasmid (as dewdrops on a spider's thread) to merge into a lipid coated DNA particle. It may also destabilize cytoplasmic and internal endosomal and nuclear membranes, thus helping the particles to sequentially reach the cytoplasm and the nucleus. There, uncoating may simply be the result of competitive distribution of the cationic lipid between DNA from exogenous plasmids and that from a large pool of chromatin.

The much higher transfection efficiency observed with lipopolyamines, as compared to quaternary ammonium-bearing lipids (Figure 1) may be a consequence of the high charge density of N–H$^+$ as compared to the more diffuse $N(CH_3)_3{}^+$ cation, where the charge is spread over some ten aliphatic hydrogens (hence a weaker interaction with the cell membrane phosphate or carboxylate groups). Furthermore, the weakest secondary amine of DOGS (pK$_4$ circa 5.4) would be able to buffer acidic lysosomes, thus protecting DNA from degradation. Finally, thanks to its headgroup, DOGS possesses the unique property of DNA *compaction*, whereas DOTMA probably coats DNA without significant size reduction.

V. CONCLUSION

Cationic lipid-mediated gene transfer has become a very attractive alternative to other in vitro techniques: *it is* as straightforward as calcium phosphate coprecipitation (requiring only mixing of DNA with the lipid); it is of general use with respect to cell type and DNA size, since it is entirely driven by nonspecific ionic interactions; and it may be of low toxicity if the carrier was designed to be biodegradable. As such, cationic lipids are being widely used in cell biology research as well as for gene therapy trials where ex vivo transfection is possible.

In vivo gene transfer with synthetic vectors is less obvious (see next chapter): except for endothelial cells, the nucleolipidic particles must penetrate a tissue, and in doing so may interact with circulating serum proteins (heparin, albumin...) as well as with the tissue matrix. Advances in this field will probably arise from other formulations of molecules already existing, or from vectors based on a different principle.

REFERENCES

1. Szoka, F., Papahadjopoulos, D. (1978) Procedure for preparation of liposomes with large internal aqueous space and high capture by reverse-phase evaporation, *Proc. Natl. Acad. Sci. U.S.A.*, 75, 4194-4198.
2. Tikchonenko, T.I., Glushakova, S.E., Kislina, O.S., Grodnitskaya, N.A., Manykina, A., Naroditsky, B.S. (1988) Transfer of condensed viral DNA into eukaryotic cells using proteoliposomes, *Gene, 63*, 321-330.

3. Kaneda, Y., Iwai, K., Uchida, T. (1989) Increased expression of DNA cointroduced with nuclear protein in adult rat liver, *Science, 243*, 375-378.

4. Mukherjee, A.B., Orloff, S., Butler, J.D., Triche, T., Lalley, P., Schulman, J.D. (1978) Entrapment of metaphase chromosomes into phospholipid vesicles (lipochromosomes): carrier potential in gene transfer, *Proc. Natl. Acad. Sci. U.S.A., 75*, 1361-1365.

5. Fraley, R., Straubinger, R.M., Rule, G., Springer, E.L., Papahadjopoulos, D. (1981) Liposome-mediated delivery of deoxyribonucleic acid to cells: enhanced efficiency of delivery related to lipid composition and incubation conditions, *Biochemistry, 20*, 6978-6987.

6. Nicolau, C., Legrand, A., Grosse, E. (1987) Liposomes as carriers for in vivo gene transfer and expression, *Methods Enzymol., 149*, 157-176.

7. Mannino, R.J., Gould-Fogerite, S., (1988) Liposome-mediated gene transfer, *Biotechniques, 6*, 682-690.

8. Gould-Fogerite, S., Mazurkiewicz, J.E., Raska, K., Voelkerding, K., Lehman, J.M., Mannino, R.J. (1989) Chimerasome-mediated gene transfer in vitro and in vivo, *Gene, 84*, 429-438.

9. Legendre, J.Y., Szoka, F.C. (1992) Delivery of plasmid DNA into mammalian cell lines using pH-sensitive liposomes: comparison with cationic liposomes, *Pharm. Res., 9*, 1235-1242.

10. Felgner, P.L., Ringold, G.M. (1989) Cationic liposome-mediated transfection, *Nature, 337*, 387-388.

11. Felgner, P.L. (1990) Particulate systems and polymers for in vitro and in vivo delivery of polynucleotides, *Adv. Drug Deliv., Rev. 5*, 163-187.

12. Monsigny, M., Midoux, P., Roche, A.-C. (1993) Perspectives ex vivo et in vivo, pour la thérapie génique, de la transfection sélective à l'aide de complexes plasmide-polylysine ciblés, *Médecine/Sciences, 9*, 441-449.

13. Behr, J.P. (1993) Synthetic gene transfer vectors, *Acc. Chem. Res., 26*, 274-278.

14. Felgner, P.L., Gadek, T.R., Holm, M., Roman, R., Chan, H.W., Wenz, M., Northrop, J.P., Ringold, G.M., Danielsen, M. (1987) Lipofection: a highly efficient, lipid-mediated DNA-transfection procedure, *Proc. Natl. Acad. Sci. U.S.A., 84*, 7413-7417.

15. Malone, R.W., Felgner, P.L., Verma, I.M. (1989) Cationic liposome-mediated RNA transfection, *Proc. Natl. Acad. Sci. U.S.A., 86*, 6077-6081.

16. Brigham, K.L., Meyrick, B., Christman, B., Berry, L.C., Jr., King, G. (1989) Expression of a prokaryotic gene in cultured lung endothelial cells after lipofection with a plasmid vector, *Am. J. Respir. Cell Mol. Biol., 1*, 95-100.

17. Lu, L., Zeitlin, P.L., Guggino, W.B., Craig, R.W. (1989) Gene transfer by lipofection in rabbit and human secretory epithelial cells, *Pflügers Arch., 415*, 198-203.

18. Welsh, N., Öberg, C., Hellerström, C., Welsh, M. (1990) Liposome-mediated in vitro transfection of pancreatic islet cells, *Biomed. Biochim. Acta, 12*, 1157-1164.

19. Antonelli, N.M., Stadler, J. (1990) Genomic DNA can be used with cationic methods for highly efficient transformation of maize protoplasts, *Theor. Appl. Genet., 80*, 395-401.

20. Innes, C.L., Smith, P.B., Langenbach, R., Tindall, K.R., Boone, L.R. (1990) Cationic liposomes (Lipofectin) mediate retroviral infection in the absence of specific receptors, *J. Virol., 64*, 957-961.

21. Rippe, R.A., Brenner, D.A., Leffert, H.L. (1990) DNA-mediated gene transfer into adult rat hepatocytes in primary culture, *Mol. and Cell. Biol., 10*, 689-695.

22. Muller, S.R., Sullivan, P.D., Clegg, D.O., Feinstein, S.C. (1990) Efficient transfection and expression of heterologous genes in PC12 cells, *DNA Cell Biol., 9*, 221-229.

23. Brant, M., Nachmansson, N., Norrman, K., Regnell, Bredberg, A. (1991) Shuttle vector plasmid propagation in human peripheral blood lymphocytes facilitated by liposome-mediated transfection, *DNA Cell Biol., 10*, 75-79.

24. Spörlein, B., Koop, H.-U. (1991) Lipofectin: direct gene transfer to higher plants using cationic liposomes, *Theor. Appl. Genet., 83*, 1-5.

25. Parker-Ponder, K., Dunbar, R.P., Wilson, D.R., Darlington, G.J., Woo, S. (1991) Evaluation of relative promoter strength in primary hepatocytes using optimized lipofection, *Hum. Gene Ther., 2*, 41-52.

26. Jiang, C.-K., Connolly, D., Blumenberg, M. (1991) Comparison of methods for transfection of human epidermal keratinocytes, *J. Invest. Dermatol., 98*, 969-973.

27. Jarnagin, W.R., Debs, R.J., Wang, S.-S., Bissell, D.M. (1992) Cationic lipid-mediated transfection of liver cells in primary culture, *Nucl. Acids Res., 20*, 4205-4211.

28. Ray, J., Gage, F.H. (1992) Gene transfer into established and primary fibroplast cell lines: comparison of transfection methods and promoters, *Biotechniques, 13*, 598-603.

29. Yeoman, L.C., Danels, Y.J., Lynch, M.J. (1992) Lipofectin enhances cellular uptake of antisense DNA while inhibiting tumor cell growth, *Antisense Res. Develop., 2,* 51-59.

30. Li, A.P., Myers, C.A., Kaminski, D.L. (1992) Gene transfer in primary cultures of human hepatocytes, *In Vitro Cell. Dev. Biol., 28,* 373-375.

31. Bennett, C.F., Chiang, M.-Y., Chan, H., Shoemaker, J.E.E., Mirabelli, C.K. (1992) Cationic lipids enhance cellular uptake and activity of phosphorothioate antisense oligonucleotides, *Mol. Pharmacol., 41,* 1023-1033.

32. Behr, J.P. (1986) DNA strongly binds to micelles and vesicles containing lipopolyamines or lipointercalants, *Tet. Lett., 27,* 5861-5864.

33. Behr, J.P., Demeneix, B., Loeffler, J.P., Perez-Mutul, J. (1989) Efficient gene transfer into mammalian primary endocrine cells with lipopolyamine coated DNA, *Proc. Natl. Acad. Sci. U.S.A., 86,* 6982-6986.

34. Loeffler, J.P., Barthel, F., Feltz, P., Behr, J.P., Sassone-Corsi, P., Feltz, A. (1990) Lipopolyamine-mediated transfection allows gene expression studies in primary neuronal cells, *J. Neurochem., 54,* 1812-1815.

35. Demeneix, B.A., Fredriksson, G., Lezoual'ch, F., Daugeras-Bernard, N., Behr, J.P., Loeffler, J.P. (1991) Gene transfer into intact vertebrate embryos, *Int. J. Dev. Biol., 35,* 481-484.

36. Barthel, F., Boutillier, A.L., Giraud, P., Demeneix, B.A., Behr, J.P., Loeffler, J.P. (1992) Gene regulation analysis by lipopolyamine-mediated DNA transfer in primary neurons, in *Methods in Neurosciences,* Conn, P.M., Ed., Academic Press, New York, 291-312.

37. Loeffler, J.P., Behr, J.P. (1993) Gene transfer into primary and established mammalian cell lines with lipopolyamine-coated DNA, in *Methods in Enzymology: Recombinant DNA,* Wu, R., Ed., Academic Press, New York, 599-618.

38. Barthel, F., Remy, J.S., Loeffler, J.P., Behr, J.P. (1993) Gene transfer optimization with lipospermine-coated DNA, *DNA Cell Biol., 12,* 553-560.

39. Staedel, C., Hua, Z., Broker, T.R., Chow, L.T., Remy, J.S., Behr, J.P. (1994) High efficiency transfection of primary human keratinocytes with positively charged lipopolyamine:DNA complexes, *J. Invest. Dermatol., 5,* 768-772.

40. Demeneix, B.A., Abdel-Taweb, H., Benoit, C., Seugnet, I., Behr, J.P. (1994) Temporal and spatial expression of lipospermine-compacted genes transferred into chick embryos in vivo, *Biotechniques, 16,* 436.

41. Demeneix, B.A., Kley, N., Loeffler, J.P. (1990) Differentiation to a neuronal phenotype in bovine chromaffin cells is repressed by protein kinase C and is not dependent on c-Fos oncoproteins, *DNA Cell Biol., 9,* 335-345.

42. Boutillier, A.L., Sassone-Corsi, P., Loeffler, J.P. (1991) The protooncogene c-Fos is induced by corticotropin-releasing factor and stimulates proopiomelanocortin gene transcription in pituitary cells, *Mol. Endocrinol., 5,* 1301-1310.

43. Giraud, P., Kowalski, C., Barthel, F., Bequet, D., Renard, M., Grinau, M., Boudouresque, F., Loeffler, J.P. (1991) Striatal proenkephalin turnover and gene transcription are regulated by cAMP and protein kinase C related pathways, *Neuronscience, 43,* 67-79.

44. Schweighoffer, F., Barlat, I., Chevallier-Multon, M.C., Tocque, B. (1992) Implication of GAP in Ras-dependent transactivation of a polyoma enhancer sequence, *Science, 256,* 825-827.

45. Schenborn, E., Goiffon, V. (1992) Greatly increased transfection efficiency of NIH 3T3 and HeLa cells using Transfectam reagent, *J. NIH Res., 4,* 79.

46. Bejanin, S., Habert, E., Berrard, S., Dumas Milne Edwards, J.B., Loeffler, J.P., Mallet, J. (1992) Promoter elements of the rat choline acetyltransferase gene allowing nerve growth factor inducibility in transfected primary cultured cells, *J. Neurochem., 58,* 1580-1583.

47. Boutillier, A.L., Barthel, F., Loeffler, J.P., Hassan, A., Demeneix, B.A. (1992) Genetic analysis in neurons and neural crest-derived post mitotic cells, *Prog. Neuro-Psychopharmacol. Biol. Psychiat., 16,* 959-968.

48. Boutillier, A.L., Barthel, F., Roberts, J.L., Loeffler, J.P. (1992) β-adrenergic stimulation of c-Fos via protein kinase A is mediated by CREB dependent and tissue specific CREB independent mechanisms in corticotrope cells, *J. Biol. Chem., 267,* 23520-23526.

49. Kley, N., Chung, R.Y., Fay, S., Loeffler, J.P., Seizinger, B.R. (1992) Specific repression of the basal c-Fos promoter by wild-type p53, *Nucleic Acids Res., 20,* 4083-4087.

50. Barthel, F., Loeffler, J.P. (1993) Characterization and genetic analysis of functional corticotropin releasing hormone receptors in primary cerebellar cultures, *J. Neurochem., 60,* 696-703.

51. Lezoualc'h, F., Hassan, A.H.S., Giraud, P., Loeffler, J.P., Lee, S.L., Demeneix, B.A. (1992) Assignment of the β-thyroid hormone receptor to 3,5,3′-triiodothyronine-dependent inhibition of transcription from the thyrotropin-releasing hormone promoter in chick hypothalamic neurons, *Mol. Endocrinol., 6,* 1797-1804.

52. Bading, H., Ginty, D.D., Greenberg, M.E. (1993) Regulation of gene expression in hippocampal neurons by distinct calcium signalling pathways, *Science, 260,* 181-186.

53. Perron, H., Suh, M., Lalande, B., Gratacap, B., Laurent, A., Stoebner, P., Seigneurin, J.M. (1993) Herpes simplex virus ICP0 and ICP4 immediate early proteins strongly enhance expression of a retrovirus harboured by a leptomeningeal cell line from a patient with multiple sclerosis, *J. Gen. Virol., 74,* 65-72.

54. Ito, A., Miyazoe, R., Mitoma, J., Akao, T., Osaki, T., Kunitake, T. (1990) Synthetic cationic amphiphiles for liposome-mediated DNA transfection, *Biochem. Inter., 22,* 235-241.

55. Leventis, R., Silvius, J.R. (1990) Interactions of mammalian cells with lipid dispersions containing novel metabolizable cationic amphiphiles, *Biochim. Biophys. Acta, 1023,* 124-132.

56. Gao, X., Huang, L. (1991) A novel cationic liposome reagent for efficient transfection of mammalian cells, *Biochem. Biophys. Res. Commun., 179,* 280-285.

57. Zhou, X., Klibanov, A.L., Huang, L. (1991) Lipophilic polylysines mediate efficient DNA transfection in mammalian cells, *Biochim. Biophys. Acta, 1065,* 8-14.

58. Ballas, N., Zakai, N., Sela, I., Loyter, A. (1988) Liposomes bearing a quaternary ammonium detergent as an efficient vehicle for functional transfer of TMV-RNA into plant protoplasts, *Biochim. Biophys. Acta, 939,* 8-18.

59. Pinnaduwage, P., Schmitt, L., Huang, L. (1989) Use of a quaternary ammonium detergent in liposome mediated DNA transfection of mouse L-cells, *Biochim. Biophys. Acta, 985,* 33-37.

60. Koshizaka, T., Hayashi, Y., Yagi, K. (1989) Novel liposomes for efficient transfection of beta-galactosidase gene into COS-1 cells, *J. Clin. Biochem. Nutr., 7,* 185-192.

61. Rose, J.K., Buonocore, L., Whitt, M.A. (1991) A new cationic liposome reagent mediating nearly quantitative transfection of animal cells, *Biotechniques, 10,* 520-525.

62. Wilson, R.W., Bloomfield, V.A. (1979) Counterion-induced condensation of DNA. A light-scattering study, *Biochemistry, 18,* 2192-2196.

63. Perron, H., Bambilla, E., Unpublished results.

64. Haywood, A.M. (1975) 'Phagocytosis' of Sendai virus by model membranes, *J. Gen. Virol., 29,* 63-68.

65. Remy, J.S., Sirlin, C., Vierling, P., Behr, J.P., *Bioorg. Chem.,* in press.

Chapter 10

Cationic Liposome-Mediated Gene Transfer
In Vitro and *In Vivo*

Robert J. Debs

CONTENTS

I. INTRODUCTION

The transfer and expression of cloned genes in cultured cells and in animals has become increasingly important, both to analyze gene function in living systems and to develop gene-based therapies for a wide variety of human diseases. Cationic liposomes, liposomes bearing a net positive charge, have been shown to avidly complex to DNA, and to efficiently deliver cloned genes both *in vitro* and *in vivo*.[1,2] The efficiency and versatility of liposome-mediated DNA delivery, as well as the wide range of cationic DNA carrier molecules which have been synthesized and tested for gene delivery, have generated substantial interest in this approach to gene transfer.

II. CARRIER-BASED GENE TRANSFER INTO CULTURED CELLS

Previously, neutral liposomes or liposomes bearing a net negative charge,[3] including pH sensitive liposomes,[4] were studied as DNA delivery systems into cultured cells. Although some level of gene expression was seen in transfected cell types[3,4] the use of these liposomes is limited by their limited abilities to both capture DNA, and to deliver it intracellularly. More recently, preformed cationic liposomes have been shown to efficiently complex with DNA. These DNA-cationic liposome complexes have been shown to transfer and express cloned genes in a wide variety of cultured cells, including both transformed cell lines and primary cultures of human cells.[5-14] In addition to cationic liposomes, other cationic moieties, including polyamines[15] and polylysine-based DNA carrier systems,[16,17] have been shown to efficiently transfect a variety of cultured cell types. Uptake of ligand-polycation-DNA complexes appears to be via receptor-mediated endocytosis.[16,17] Agents which increase their egress from the endosomal-lysomal compartment into the cytoplasm can significantly increase transfection efficiency.[17]

III. CATIONIC LIPOSOME-MEDIATED GENE TRANSFER INTO CULTURED CELLS

The use of cationic liposomes as a DNA carrier system to transfect cultured cells was first reported by Felgner et al. using the synthetic cationic lipid, DOTMA, in a 1 to 1 molar ratio with DOPE.[5] Subsequently, a variety of novel synthetic cationic lipids have been reported to mediate efficient transfection of cultured cells. Some of these new cationic liposomes have been claimed to be superior to DOTMA/DOPE in their ability to transfect cells. However, in general, different cell types are transfected with different efficiencies by a given cationic liposome formulation. No single cationic lipid formulation appears to be uniformly superior to other available lipid compositions in their ability to transfect cultured cells. Rather, transfection conditions must be optimized individually for each cationic liposome formulation, in order to maximize the transfection of a given cell line or cultured primary cell type. Specifically, for each cationic liposome formulation tested, the following parameters should be optimized for the cell type to be transfected. These include: (1) the DNA to lipid ratio of the complex, (2) the total dose of DNA:lipid complex added, (3) the density of the cells at the time of transfection, (4) whether the cells are transfected in suspension or adherent, (5) the medium in which the cells are cultured (including whether or not serum is present initially, and at what concentration[18]), (6) the duration of exposure of the liposome:DNA complexes to the cells, (7) whether, when, and at what concentrations transfection enhancing agents, such as lysomotrophic agents[3] and/or agents thought to increase plasma membrane fusion or endocytosis[3] are added, and (8) the time points after addition of the DNA:liposome complexes when the cells are analyzed for expression of the transgene product. Optimization of some or all of these parameters may significantly improve transfection efficiency in cell types which are difficult to transfect using standard protocols.

IV. ADVANTAGES OF CATIONIC LIPOSOMES FOR TRANSFECTING CULTURED CELLS

The use of cationic liposomes as a DNA-carrier system for transfecting cultured cells offers several potential advantages over other cellular transfection systems. These include:

1. A variety of pre-made cationic liposomes are available commercially and different batches of the same formulation appear uniform.
2. Formation of liposome:DNA complexes is simple and straightforward.
3. At concentrations which transfect a wide variety of cell types, cationic liposomes appear minimally toxic.[5-14]
4. Since the DNA is complexed to preformed liposomes, there are theoretically very few limitations on the size of DNA that cationic liposomes can introduce intracellularly. Recently, approximately 150 kb YAC DNAs have been reported to be stably transfected into embryonic stem cells after cationic liposome-mediated transfer.[19,20]

V. CATIONIC LIPOSOME-MEDIATED GENE TRANSFER IN ANIMALS

Cationic liposomes have also been used to directly transfer and express a variety of cloned genes in individual tissues following administration to animals. Brigham et al. reported chloramphenicol acetyl/transferase (CAT) reporter gene expression in the lung for up to 7 days following intravenous (i.v.) or intratracheal administration of an SV40-CAT plasmid complexed to cationic liposomes composed of DOTMA:DOPE ("lipofectin").[21] Similar findings in the lung were subsequently reported by Hazinski et al.,[22] Yoshimura et al.,[23] and Hyde et al.[24] who instilled intratracheally lipofectin complexed to one of several different expression plasmids. Expression of biologically relevant transgenes, including the human cystic fibrosis transmembrane conductance regulator (CFTR) gene, was documented using this approach.[24] Localized expression of transgenes has also been reported following cationic liposome-mediated DNA injection either directly into the central nervous system[25-27] or into the arterial wall.[28]

VI. HUMAN GENE THERAPY TRIALS USING CATIONIC LIPOSOMES

Aerosol administration of DOTMA:DOPE liposomes complexed to a CMV-CAT expression plasmid into mice has been reported to produce CAT gene expression in the majority of cells present in the lungs. The CAT gene product was present in the lung for at least 3 weeks, as detected both by CAT enzymatic assay and by immunohistochemistry, which was used to detect CAT antigen.[29] No evidence of host toxicity was detected in treated animals. More recently, aerosol administration of cationic liposomes complexed to an SV40-human CFTR gene expression plasmid appeared to at least partially correct the chloride secretory defect (the biochemical hallmark of cystic fibrosis) in some of the CFTR gene knockout mice receiving this treatment.[30] Based on these results, a human clinical trial evaluating the effects in cystic fibrosis patients of intranasal instillation of DC-cholesterol:DOPE liposomes complexed to this SV40-CFTR expression plasmid is currently in progress in the U.K.[31]

In another on-going trial of cationic liposome-mediated human gene therapy, an expression plasmid containing the human HLA-B7 gene complexed to DC-cholesterol:DOPE liposomes has been injected directly into cutaneous tumors in human melanoma patients.[32] This approach is based on prior animal studies showing that local injection of these liposome:DNA complexes could produce anti-tumor effects in non-injected tumors remote from the injection site.[33] Thus, the expression of appropriate HLA genes in easily accessible tumor metastases may induce an immune-mediated anti-tumor response directed against tumors which are both distant from the injection site and do not express the transfected gene product. The mechanism of this remote anti-tumor effect is not well characterized. A preliminary report indicates that injection of cationic liposome-HLAB7 expression vector complexes into human melanoma patients appears non-toxic and has produced some evidence of anti-tumor activity in at least one patient.[32]

Recently, cationic liposomes have been shown to mediate transfection of most tissues in the body after i.v. injection of DOTMA:DOPE liposomes complexed to a CMV-CAT expression plasmid. Expression of the CAT gene was detected by immunohistochemistry several months after a single i.v. injection of cationic liposome:DNA complexes. The efficiency of transgene expression was related to the lipid formulation, the lipid to DNA ratio injected, the expression plasmid used, and the dose of liposome:DNA complexes injected.[34] Intravenous injection of cationic liposome:DNA complexes did not cause apparent host toxicity, either in this study or in another study examining the potential toxicity of cationic liposome:DNA complexes injected i.v.[34,35] The intraperitoneal injection of cationic liposome CMV-CAT complexes has also been reported to efficiently transfect a variety of visceral organs.[36] Different cationic liposome formulations complexed to the same CMV-CAT expression plasmid appeared to confer a degree of tissue-specific transgene expression *in vivo*. Here, efficient *in vivo* transfection of T lymphocytes in mice was demonstrated. Again no apparent host toxicity was produced by parenteral injection of cationic liposome:DNA complexes.

VII. SUMMARY

In summary, cationic liposomes have been shown to mediate efficient transfection of a variety of cultured cell types, including non-transformed primary cultures of rodent neurons, hepatocytes, pancreatic islet cells, alveolar type II cells, and alveolar macrophages, as well as a wide variety of transformed lines and tumor cell lines.[5-14] More recently, cationic liposomes have been shown to mediate transgene expression *in vivo*, expressed either in individual tissues or in a more generalized fashion. At least two human trials using cationic liposome-mediated gene transfer are currently in progress. Numerous others are under consideration by the regulatory agencies which are controlling the rapidly evolving area of gene transfer and gene therapy in human patients. As new and potentially more useful cationic DNA carrier molecules and DNA expression vectors are developed, and as the mechanisms of carrier-mediated gene transfer are better elucidated, cationic carrier-mediated gene transfer should become an even more powerful approach to human gene therapy.

REFERENCES

1. Felgner, P. L. and Rhodes, G., Gene therapeutics, *Nature*, 349, 351, 1991.
2. Hug, P. and Sleight, R. G., Review: liposomes for the transformation of eukaryotic cells, *Biochim. Biophys. Acta*, 1097, 1, 1991.
3. Fraley, R., Straubinger, R., Rule, G., Springer, L., and Papahadjopoulos, D., Liposome-mediated delivery of deoxyribonucleic acid to cells: enhanced efficiency of delivery related to lipid composition and incubation conditions, *Biochemistry*, 20, 6978, 1981.
4. Wang, C.-Y. and Huang, L., pH-sensitive immunoliposomes mediate target-cell-specific delivery and controlled expression of a foreign gene in mouse, *Proc. Natl. Acad. Sci. U.S.A.*, 84, 7851, 1987.
5. Felgner, P., Gadek, T., Holm, M., Roman, R., Chan, H. W., Wenz, M., Northrop, T. P., Ringold, G. M., and Danielson, M., Lipofection: a highly efficient, lipid-mediated DNA transfection procedure, *Proc. Natl. Acad. Sci. U.S.A.*, 84, 7413, 1987.
6. Leventis, R. and Silvius, J., Interactions of mammalian cells with lipid dispersions containing novel metabolizable cationic amphiphiles, *Biochim. Biophys. Acta*, 1023, 124, 1990.
7. Rose, J., Buonocore, L., and Whitt, M. A., A new cationic liposome reagent mediating nearly quantitative transfection of animal cells, *Biotechnology*, 10, 520, 1991.
8. Zhou, X., Klibanov, A., and Huang, L., Lipophilic polylysines mediate efficient DNA transfection in mammalian cells, *Biochim. Biophys. Acta*, 1065, 8, 1991.
9. Welsh, N., Oberg, C., Hellerstrom, C., and Welsh, M., Liposomes mediated in vitro transfection of pancreatic islet cells, *Biomedica Biochimica Acta*, 49, 1157, 1990.

10. Brigham, K. L., Meyrick, B., Christman, B., Berry, L. C., Jr., King, G., Expression of a prokaryotic gene in cultured lung endothelial cells after lipofection with a plasmid vector, *Am. J. Respir. Cell. Mol. Biol.*, 1, 95, 1989.

11. Jarnagin, W., Debs, R., and Bissell, M., Cationic lipid-mediated transfection of rat hepatocytes in primary culture, *Nucl. Acid Res.*, 20, 4205, 1992.

12. Malone, R., Felgner, P., and Verma, I., Cationic liposome-mediated RNA transfection, *Proc. Natl. Acad. Sci. U.S.A.*, 86, 6077, 1989.

13. Debs, R., Freedman, L., Edmunds, S., Gaensler, K. L., Duzgunes, N., and Yamamoto, K. R., Regulation of gene expression *in vivo* by liposome-based delivery of a purified transcription factor, *J. Biol. Chem.*, 265, 10189, 1990.

14. Debs, R., Pian, M., Gaensler, K., Clements, J., Friend, D., and Dobbs, L., Prolonged transgene expression in rodent lung cells, *Am. J. Resp. Cell. Mol. Biol.*, 7, 406, 1992.

15. Behr, J. P., Demeneix, B., Loeffler, J. P., and Perez-Mutul, J., Efficient gene transfer into mammalian primary endocrine cells with lipopolyamine-coated DNA, *Proc. Natl. Acad. Sci. U.S.A.*, 86, 6982, 1989.

16. Zenke, M., Steinlein, P., Wagner, E., Cotten, M., Beug, H., and Birnstiel, M. L., Receptor-mediated endocytosis of transferrin-polycation conjugates: an efficient way to introduce DNA into hematopoietic cells, *Proc. Natl. Acad. Sci. U.S.A.*, 87, 3655, 1990.

17. Curiel, D. T., Agarwal, S., Wagner, E., and Cotten, M., Adenovirus enhancement of transferrin-polylysine-mediated gene delivery, *Proc. Natl. Acad. Sci. U.S.A.*, 88, 8850, 1991.

18. Brunette, E., Stribling, R., and Debs, R., Lipofection does not require the removal of serum, *Nucl. Acids Res.*, 20, 1151, 1992.

19. Gnirke, A., Barnes, T. S., Patterson, D., Schild, D., Featherstone, T., and Olson, M. V., Cloning and *in vivo* expression of the human GART gene using yeast artificial chromosomes, *EMBO J.*, 10, 1629, 1991.

20. Strauss, W. M., Dausman, J., Beard, C., Johnson, C., Lawrence, J. B., and Jaenisch, R., Germ line transmission of a yeast artificial chromosome spanning the murine alpha 1 (I) collagen locus, *Science*, 259, 1904, 1993.

21. Brigham, K., Meyrick, B., Christman, B., Magnuson, M., King, G., and Berry, L. C., In vivo transfection of murine lungs with a functioning prokaryotic gene using a liposome vehicle, *Am. J. Med. Sci.*, 298, 278, 1989.

22. Hazinski, T., Ladd, P. A., and DeMatteo, C. A., Localization and induced expression of fusion genes in the rat lung, *Am. J. Resp. Cell. Mol. Biol.*, 4, 206, 1991.

23. Yoshimura, K., Rosenfeld, M. A., Nakamura, H., Scherer, E. M., Pavirani, A., Lecocq, J. P., and Crystal, R. G., Expression of the human cystic fibrosis transmembrane conductance regulator gene in the mouse lung after *in vivo* intratracheal plasmid-mediated gene transfer, *Nucl. Acids Res.*, 20, 3233, 1992.

24. Hyde, S. C., Gill, D. R., Higgins, C. F., Trezise, A. E., MacVinish, L. J., Cuthbert, A. W., Ratcliff, R., Evans, M. J., and Colledge, W. H., Correction of the ion transport defect in cystic fibrosis transgenic mice by gene therapy, *Nature*, 362, 250, 1993.

25. Holt, C. E., Garlick, N., and Cornel, E., Lipofection of cDNAs in the embryonic vertebrate central nervous system, *Neuron*, 4, 203, 1990.

26. Ono, T., Fujino, Y., Tsuchiya, T., and Tsuda, M., Plasmid DNAs directly injected into mouse brain with lipofectin can be incorporated and expressed by brain cells, *Neurosci. Lett.*, 117, 259, 1990.

27. Jiao, S., Acsadi, G., Jani, A., Felgner, P. L., and Wolff, J. A., Persistence of plasmid DNA and expression in rat brain cells *in vivo*, *Exp. Neurol.*, 115, 400, 1992.

28. Nabel, E., Plautz, G., and Nabel, G., Site-specific gene expression in vivo by direct gene transfer into the arterial wall, *Science*, 249, 1285, 1990.

29. Stribling, R., Brunette, E., Liggitt, D., Gaensler, K., and Debs, R., Aerosol gene delivery *in vivo*, *Proc. Natl. Acad. Sci. U.S.A.*, 89, 11277, 1992.

30. Alton, E. W. F. W., Middleton, P. G., Caplen, N. J., Smith, S. N., Steel, D. M., Munkonge, F. M., Jeffery, P. K., Geddes, D. M., Hart, S. L., Williamson, R., Fasold, K. I., Miller, A. D., Dickinson, P., Stevenson, B. J., McLachlan, G., Dorin, J. R., and Porteous, D. J., Non-invasive liposome-mediated gene delivery can correct the ion transport defect in cystic fibrosis mutant mice, *Nature Genetics*, 5, 135, 1993.

31. Anon.; United Kingdom scientists test liposome gene therapy technique in news, Nature, 365, 4, 1993.
32. Nabel, G. J., Nabel, E. G., Yang, Z.-Y., Fox, B. A., Plautz, G. E., Gao, X., Huang, L., Shu, S., Gordon, D., and Chang, A. E., Direct gene transfer with DNA-liposome complexes in melanoma: expression, biologic activity, and lack of toxicity in humans, *Proc. Natl. Acad. Sci. U.S.A.*, 90, 11307, 1993.
33. Plautz, G. E., Yang, Z.-Y., Wu, B.-Y., Gao, X., and Huang, L., Immunotherapy of malignancy by *in vivo* gene transfer into tumors, *Proc. Natl. Acad. Sci. U.S.A.*, 90, 4645, 1993.
34. Zhu, N., Liggitt, D., Liu, Y., and Debs, R., Systemic gene expression after intravenous DNA delivery into adult mice, *Science*, 261, 209, 1993.
35. Stewart, M. J., Plautz, G. E., Del Buono, L., Yang, Z.-Y., Xu, L., Gao, X., Huang, L., Nabel, E. G., and Nabel, G. J., Gene transfer *in vivo* with DNA-liposome complexes: safety and acute toxicity in mice, *Human Gene Therapy*, 3, 267, 1992.
36. Philip, R., Liggitt, D., Philip, M., Dazin, P., and Debs, R. J., *In vivo* gene delivery: efficient transfection of T lymphocytes in adult mice, *J. Biol. Chem.*, 268, 16087, 1993.

Chapter 11

Sterically Stabilized (Stealth®) Liposomes: Pharmacological Properties and Drug Carrying Potential in Cancer

Demetrios Papahadjopoulos and Alberto A. Gabizon

CONTENTS

I. CONVENTIONAL LIPOSOMES AS A DRUG CARRIER SYSTEM

The use of conventional liposomes as a carrier system for drug delivery has already shown promising results, but has been limited to specific applications because of their short circulation time in blood.[1-4] The rapid clearance of conventional liposomes from blood is mediated primarily through the scavenging function of phagocytic cells located in the liver, spleen and to a lesser extent in other tissues known as the reticuloendothelial system (RES) or as the mononuclear phagocytic system (MPS). The rapid uptake of liposomes by these cells has been the basis of most biomedical uses of liposomes so far,[5] including efficacy against intracellular parasites, activation of macrophages, and antigen presentation. These cells may also act as a slow release system for liposomally encapsulated Amphotericin B and Doxorubicin, which are currently being tested in clinical trials.

Because of the obvious advantages of liposomes that can circulate longer in blood and be able to reach a variety of cells in tissues other than the liver and spleen, there was a considerable research effort in that direction. Prior to the advent of the new formulations discussed below, attempts to reduce the uptake of liposomes by the RES had led to the use of small, rigid, neutral liposomes, which had shown the longest circulation time in blood.[4,6] Such liposomes have serious disadvantages as a delivery system, including small internal volume for drug encapsulation, size instability on prolonged storage, strict limitations for liposome composition, and a dose-dependent pharmacokinetic profile.[6]

The term "Stealth" is used to designate liposomes that circulate in blood for extended periods of time following i.v. injection and, in addition, show diminished uptake by the macrophage cells of liver and spleen, increased uptake by non-MPS tissues, and dose-independent blood clearance kinetics and tissue disposition. The term "sterically stabilized" is used to indicate liposomes that have long circulation times in blood irrespective of bilayer fluidity and surface charge density. "Stealth" has been registered as a trademark by Liposome Technology, Inc.

The term "conventional" is used to designate liposomes composed of various phospholipids, glycolipids, and possibly other lipids, without additional molecules such as monosialoganglioside GM1, phosphatidylinositol, polyethylene-glycol-derivatized phospholipids, etc., which confer the property of long circulation in blood and diminished recognition by mononuclear phagocyte cells.

0-8493-4569-3/95/$0.00+$.50
© 1995 by CRC Press, Inc.

II. INITIAL OBSERVATIONS ON STERIC STABILIZATION

A major advance was achieved with the observation that inclusion of certain specific lipids within the liposome bilayer can prolong their circulation time in blood, and decrease their uptake by the RES.[7-10] The term stealth was initially introduced[11] to refer to these liposomes and later the term steric stabilization was proposed[12] as a physicochemical description of the observed phenomena. In this communication we will use the abbreviation SL. The initial SL formulations were empirical, based on the lipid composition of the outer monolayer of red blood cells[7,9] and on a combination of earlier observations concerning the role of surface charge and bilayer fluidity.[8,10] Thus, it was established that the GM1 ganglioside, phosphatidylinositol and cerebroside sulfate, when added at a 5-10% mole ratio to the bulk neutral lipids used for making liposomes, would result in considerable prolongation of residence time ($T_{1/2}$) in blood. This effect was observable with fluid bilayers such as those composed of egg phosphatidylcholine (EPC) and cholesterol (Chol),[8] but the $T_{1/2}$ was even longer when the bulk phospholipid produced rigid bilayers such as those composed of sphingomyelin (SM)[7] or distearolyphosphatidlycholine (DSPC).[8]

One of the most important early observations with SL was that prolonged $T_{1/2}$ correlated with increased uptake by implanted tumors in mice.[8] Furthermore, conjugation on the liposome surface of antibodies recognizing antigens expressed on tumor cells produced a further increase in the uptake by the tumors.[10] At that time, we made the suggestion that the lipids which prolong circulation time carry a negative charge which is "shielded" by carbohydrate residues and is thus prevented from participating in direct interactions with proteins in blood involved in removal of foreign particles.[8] More recently, we have generalized this mechanism to include "steric stabilization" of the liposome surface with respect to both ionic interactions with the phospholipid head-groups, and hydrophobic interaction with the bilayer interior, by proteins either in blood or at the surface of cells.[12]

III. BLOOD CLEARANCE AND TISSUE LOCALIZATION

Recently, liposomes containing lipids derivatized with the hydrophilic polymer polyethyleneglycol (PEG-PE) have shown prolonged circulation independent of lipid composition.[12-22] The earlier SL formulations composed of glycolipids were more dependent upon a rigid bilayer than those with PEG derivatives.[8,19] The comparative behavior of liposomes of varying lipid composition following i.v. injection in mice is summarized in Table 1. These results show that the relation between blood circulation and MPS uptake of liposomes containing PEG-PE is similar to that found previously with liposomes containing glycolipids. Note that the results reported in the table for MPS uptake are the summation of liver and spleen only, and these tissues are different in their uptake mechanisms.[23-25]

Lipid dose and particle size are also important in determining the blood clearance rate of liposomes. As the lipid dose increases, the circulation time also increases for most conventional lipid formulations including those of DSPC or SM mixtures with Chol.[4,6] With SL containing GM1 or PEG-PE, such a dose response has not been observed over a wide range of lipid doses.[18,19,26] Perhaps as revealing as to the mechanism of uptake of liposomes containing PEG-PE is the finding that their clearance from blood can be represented by a single exponential rate over nearly two log units.[18,19,26] Increasing particle size typically decreases circulation time, but again, liposomes with PEG-PE show a smaller dependence. As the mean particle size distribution of PEG-PE/PC/Chol liposomes is increased from 100 nm to 250 nm the blood levels after 24 hr decrease only by about a factor of two.[19] For comparison, increasing the particle size of liposomes composed of DSPC and Chol from 30-70 nm to only 200 nm results in negligible blood levels after 24 hr, even with a saturating lipid dose.[4] The uptake by liver and spleen can be modulated widely in relation to the particle size of SL. For example large size SL are taken up preferentially by the spleen,[23] while liver shows increased uptake for very small SL.[24]

One of the earlier observations with SL was their increased accumulation in tumors.[8,10] This has been substantiated in several implanted murine tumors[18,27-31,34,35] and more recently in biopsies from human patients.[36,37] The enhanced accumulation of SL in implanted tumors has been attributed to a combination of prolonged circulation in blood in conjunction with extravasation of the small particles at sites of "leaky" vasculature.[18,27-30] Because of steric stabilization, the PEG-coated liposomes have a reduced tendency to adsorb onto biological surfaces or cells during circulation, improving their ability to penetrate through fenestrations or other endothelial gaps,[29,30] and thus increasing their rate of extravasation into tumors.[29-31] It is possible that such passage and increased accumulation can also occur in sites of other pathological conditions such as inflammation and infections.[32,33] Skin is the tissue with the highest

Table 1   **Tissue distribution of liposomes 24 hr after IV injection in mice:[a]
effect of headgroup and bilayer fluidity**

Lipid Composition	% Injected Dose (24 hr)			Ratio MPS/Blood
	Blood	MPS	Total	
PG:PC:Chol 1:9:5	0.3 ± 0.1	37.0 ± 3.6	49.8 ± 3.9	123.3
PC:Chol 10:5	3.0 ± 0.8	25.8 ± 1.3	44.4 ± 2.9	8.6
DSPC:Chol 10:5[b]	7.6 ± 6.1	23.7 ± 4.3	56.4 ± 3.4	3.1
PI:PC:Chol 1:9:5	4.1 ± 1.3	14.8 ± 3.4	37.4 ± 6.3	3.6
GM1:PC:Chol 1:9:5	6.0 ± 0.4	17.9 ± 4.4	42.3 ± 1.5	2.9
DPPG:DSPC:Chol 1:10:5	6.5 ± 0.8	31.2 ± 3.2	59.9 ± 5.5	4.8
HPI:DSPC :Chol 1:10:5[c]	12.2 ± 3.4	27.0 ± 3.5	61.6 ± 4.0	2.2
PEG-DSPE:PC:Chol 1:10:5	29.3 ± 2.8	20.6 ± 2.6	63.4 ± 3.3	0.7
GM1:DSPC:Chol 1:10:5	16.0 ± 3.8	16.5 ± 1.4	66.3 ± 4.2	1.0
PEG-DSPE:DSPC:Chol 1:10:5	23.6 ± 3.1	27.7 ± 2.2	68.1 ± 3.4	1.2

[a]Except where noted, data from Reference 12.
[b]Data from Reference 19.
[c]Data from Reference 8.

concentration of SL after liver, spleen, and tumors[8] and the vascular area most open to their passage is the post-capillary venules.[30]

IV. MICROSCOPIC LOCALIZATION IN TUMORS AND OTHER TISSUES

The mechanism by which SL accumulate in tumors at a higher concentration (per g tissue) than all other tissues except liver and spleen has been studied by microscopy.[28-31] A method for producing colloidal gold particles[38] within liposomes was used recently to investigate the precise localization of liposomes in tumors and other tissues at the cellular level. This was accomplished both by direct electron microscopy of colloidal gold particles within liposomes in tissues[29,30] and also by silver enhancement of the gold particles and formation of silver grains visible in eosin-hematoxylin-stained tissue slices by optical microscopy.[29,30] These studies have revealed that liposomes can extravasate beyond the endothelial layer in tumors and also in skin. These microscopic observations at the cellular level indicated that liposomes in the liver are intracellular within the Kupffer cells of the MPS, while in a subcutaneously growing sarcoma they are localized extracellularly around tumor cells.[29] It has been suggested that in tumors, SL extravasate by passing through gaps between the endothelial cells in areas of neovascularization.[29] Similar results were obtained by the use of rhodamine-labeled dextran, following its encapsulation in SL,[18] reenforcing the conclusion based on the localization of gold particles. In skin lesions of transgenic mice resembling Kaposi's Sarcoma, the extravasating liposomes marked by colloidal gold were observed to transverse endothelial cells within large caveollae and also were localized within the "spindle"-like cells in endocytic vacuoles.[30]

V. MECHANISTIC AND TARGETING ASPECTS

Studies on the interaction of liposomes with cultured mammalian cells *in vitro* have revealed that SL are not taken up as avidly when compared to conventional liposomes.[47-49] This is not strictly related to the presence or absence of negative charge on the liposome surface, because SL have been shown to exhibit a negative surface charge, albeit reduced due to the presence of PEG-PE.[52] Studies on the interaction of blood proteins with liposomes of various compositions have shown that SL have a diminished capacity for protein adsorption,[54] pointing to a possible correlation between diminished adsorption and prolonged $T_{1/2}$ in blood. Presumably for the same reason, steric stabilization diminishes the binding of SL to cell surface receptors.

The term "targeted" has been used to define liposomes with attached surface ligands that are capable of specific interactions with other macromolecules including cell surface receptors. This has been achieved by conjugating immunoglobulins and other ligands on the liposome surface.[45] Such ligand-directed liposomes can be used to probe specific interactions with cell surface receptors and to target drugs and other macromolecules to specific cells and tissues in living organisms for either diagnostic or therapeutic purposes. The early work of antibody-conjugation to conventional liposomes proved the feasibility of the concept with cells *in vitro*, but was not efficient enough for *in vivo* targeting. This is because the immunoglobulin-conjugated conventional liposomes were removed non-specifically by the phagocytic cells of the liver and spleen, before they could bind specifically to their intended target cells. The advent of long-circulating, sterically stabilized liposomes has revitalized this field. It has now been demonstrated in several labs that antibody-directed targeting to cancer cells[10,42-44] and endothelial cells[40,41] can be achieved *in vivo* (mice) in the presence of sterically stabilizing groups on the liposome surface. Delivery of encapsulated drugs into the interior of the target cells in the case of non-endocytic receptors remains a challenging problem because the majority of cells are not phagocytic and fusion of most liposomes with cells is a rare phenomenon.[5] We anticipate a lot more work in this area with high potential for important medicinal applications.

VI. LIPOSOMES AS CARRIERS OF DOXORUBICIN

Drug carriers in cancer chemotherapy aim at altering tissue distribution and various pharmacokinetic parameters in such a way that toxicity can be reduced and efficacy increased.[55] Control over the distribution and release of drugs is a crucial element of any rational therapy. The possibility of associating a drug to carrier systems which could regulate tissue distribution as well as rate of release and elimination will probably have important implications on the pharmacologic effect and therapeutic outcome. This is especially relevant to anticancer cytostatic agents which have a narrow safety margin of dosage because of their serious toxic effects. Doxorubicin (DOX) is a potent anticancer drug with a broad spectrum of antitumor activity against a variety of human solid tumors and leukemias.[56] Reducing the toxic effects of DOX, especially its dose limiting cardiotoxicity, with an appropriate delivery system could result in a significantly improved therapeutic index. This approach attracted several groups of investigators in the early 1980s who attempted to validate the concept by using liposomal carriers in preclinical models.[57]

Liposomes offer certain advantages as carriers: (1) a large quantity of the cytotoxic agent is encapsulated in the carrier and can be transported to an individual target cell; (2) liposome encapsulation does not require chemical bonds between the targeting ligand and the drug; and (3) the liposomal particle protects the drug from metabolic degradation until it reaches its destination. The association of DOX with the lipid bilayer is enhanced considerably by the presence of negatively charged phospholipids.[58] Several laboratories, including ours, were able to show that conventional negatively charged liposomes with bilayer-intercalated DOX cause important changes in the tissue distribution of the drug, namely decreased levels in the heart muscle and increased and sustained levels in the liver and spleen. The results reported in the scientific literature point to a decreased toxicity of liposome-associated DOX (L-DOX) in rodents with regard to cardiotoxicity and other parameters: gastrointestinal toxicity, nephrotoxicity, immunosuppression, and skin damage caused by extravasation. Decreased cardiotoxicity has also been confirmed in dogs by the Georgetown University group.[59] A possible drawback of L-DOX is that it does not appear to offer any advantages over free DOX in terms of myelosuppression.[60] In mice, the LD_{50} of DOX is increased approximately twofold using conventional liposome formulations containing negatively charged phospholipids and bilayer-entrapped drug.[60-62]

In principle, the reduced toxicity of L-DOX, enabling escalation of the single and cumulative dose of DOX, could well be translated in to a net gain in the drug therapeutic index if no loss of the antitumor activity occurs. However, this is a complex issue since the shift of drug distribution toward the RES may lead to a reduced drug concentration in tumor tissue. Results of our own studies indicate that the anatomic site of tumor growth plays an important role in the antitumor effect observed.[63] Briefly, it was found that L-DOX is more active than free DOX on lymphomas infiltrating the liver and spleen, and equally effective on bone marrow-residing leukemic cells. In contrast, free DOX is more effective than L-DOX when tested by the i.v. route against ascitic and SC-implanted tumors. The antitumor effect correlates well with differences in drug levels in the relevant anatomic areas, suggesting that the therapeutic advantage of L-DOX results from drug release in the RES and its vicinity, followed by death of bystander tumor cells. Our observations are in general agreement with those of Mayhew and Rustum.[64] The superior therapeutic effect of L-DOX over free DOX has also been demonstrated on liver metastases of mouse colon carcinomas.[65] The question of whether anatomical barriers will determine differential access of liposome-delivered drugs, thus limiting the antitumor activity in specific body locations, raised a major objection to the applicability of liposomal carriers in cancer chemotherapy. Pharmacokinetic and imaging studies in humans using L-DOX liposome formulations suggested indeed that while toxicity can be somewhat reduced, the predominant uptake of drug by the RES may result in suboptimal tumor exposure to the drug.[66]

The ability of long-circulating liposomes of small size (<100 nm diameter) to accumulate in tumors[8,27,67,68] makes them very attractive candidates for the delivery of cytotoxic drugs. A natural byproduct of the design of sterically stabilized liposomes (SL) was the application of this technology to the delivery of anthracyclines. This approach, coupled with the development of efficient methods to load anthracyclines into the liposome water phase[69-71] and the use of high phase transition temperature phospholipids (>37°C) as the main bilayer component, have enabled the production of a stable drug carrier system with reduced affinity of the RES and minimal leakage in circulation.[72]

VII. PHARMACOKINETICS OF SL-ENCAPSULATED DOX AND OTHER ANTHRACYCLINES

In 1989, we reported striking changes in the pharmacokinetics of DOX in mice, using an early formulation of sterically stabilized liposomes composed of hydrogenated phosphatidylinositol/phosphatidylcholine/cholesterol (HPI/HPC/Chol) liposomes as carriers.[72] The drug was shown to circulate in liposome-associated form with a half-life of 15.5 hr. In contrast, using a "conventional" liposome formulation based on egg-derived phosphatidylglycerol/phosphatidylcholine/cholesterol (EPG/EPC/Chol), the half-life of liposome-associated drug was only 1 hr. When the pharmacokinetic parameters of DOX in HPI/HPC/Chol liposomes were compared to those of free DOX, the increase in AUC and reduction in clearance are greater than 200-fold, thus stressing the ability of liposomes to manipulate the pharmacokinetics of DOX in a substantial way. Recently, we have summarized our pharmacokinetic studies in rodents and dogs with DOX encapsulated in SL (SL-DOX) containing a PEG-derivatized phospholipid.[73] The $T_{1/2}$ for DOX in these liposomes in dogs' blood is approximately 30 hr. It was also found that: (1) PEG-containing liposomes confer to DOX a circulation time significantly longer than HPI-containing liposomes and also longer than liposomes of the same composition (HPC/Chol) without PEG or HPI; (2) when HPC is replaced with a phospholipid of lower phase transition temperature (DPPC,EPC), the leakage rate of DOX increases and the circulation time decreases even in the presence of PEG.

The pharmacokinetics of epirubicin (EPI, an anthracyclin similar to DOX) encapsulated in liposomes incorporating a PEG-derivatized phospholipid (SL-EPI) injected into rats were studied by Mayhew et al.[74] In agreement with the results obtained with SL-DOX, it was found that SL-EPI is slowly cleared from plasma with a half-life of 18 hr, while free EPI is rapidly distributed to peripheral tissues with an initial half-life of only 14 min. The differences in AUC and clearance between free EPI and SL-EPI were 200-fold,[74] a remarkable similarity with the mouse pharmacokinetic study with DOX mentioned above.[72] In both studies, the clearance of the liposome-encapsulated drug was best described by a bi-exponential curve. The first phase of clearance is faster but usually accounts for a minor fraction of the dose. The second phase is slower and dominates the pattern of clearance of the liposomal drug. The meaning of the first exponential is unclear, but it could reflect either a small saturable compartment or a pool of drug leaking from liposomes.

Bally et al.[75] have observed long circulation times of liposome-encapsulated DOX in mice, using a high dose of drug (20 mg/kg) and small (100 nm) neutral EPC/Chol or DSPC/Chol conventional liposomes devoid of PEG or any other sterically stabilizing lipid. As much as 30 to 80% of the injected dose remained in circulation after 24 hr. This observation may be due in part to a toxic effect, such as RES blockade, since the authors also found that the circulation time of drug-carrying liposomes was two- to threefold longer than that of plain liposomes after i.v. injection in mice. The authors did not investigate whether this effect holds for subtoxic doses of DOX (5–10 mg/kg). In the case of drug-free SL, no RES blockade or dose-dependent saturation have been detected.[76] However, a side-by-side comparison of the pharmacokinetics of plain and DOX-loaded PEG liposomes has not been done.

Reports on two other cytotoxic drugs encapsulated in sterically stabilized liposomes (Ara-C, Vincristine) did not include pharmacokinetic data.[77,78]

VIII. BIODISTRIBUTION IN TUMOR-BEARING ANIMALS

Since sterically stabilized liposomes localize in tumors in substantial amounts,[8,18,27,29,79] one would expect a significant enhancement in the concentration of a liposome-encapsulated drug such as DOX, as compared to conventional free drug administration. Using a solid implant of the J6456 lymphoma in the mouse thigh, the peak drug levels obtained after injection of DOX encapsulated in HPI/HPC/Chol liposomes were about 4-fold higher than those after free DOX injection. In addition, the peak tumor drug level occurs late after drug injection, stressing the fact that liposome accumulation in tumors is a slow process requiring a long circulation time. In contrast to tumor, there was no difference between DOX levels in a normal tissue such as muscle when comparing DOX in HPI/HPC/Chol and free DOX.[34] Similar observations have been made with implants of the C-26 colon carcinoma and the M-109 lung carcinoma using PEG liposomes as carriers of DOX in mice.[79,80] Using an ascitic tumor model, we have found that there was a gradual process of extravasation of SL-DOX from plasma into ascitic fluid.[18,34] Between 10 to 20% of the injected dose was recovered in the ascitic fluid 48 to 72 hr after injection, where it remained mostly in liposome-associated form. No evidence of liposome uptake by ascitic tumor cells was found. This suggests that the drug is first released from liposomes in the extracellular fluid and then gains access to tumor cells. Of interest, no significant accumulation of SL-DOX was observed in peritoneal washes from tumor-free mice.

These studies provide evidence that SL-DOX extravasates with relative selectivity into tumor areas. Furthermore, it appears that SL can significantly boost drug levels in tumors, regardless of the type of tumor and site of growth. In fact, a recent communication indicates that the DOX concentration in a brain-implanted tumor can be increased by nearly 20-fold using PEG-stabilized liposomes as a delivery system.[81] A substantial increase in the tumor drug levels has also been obtained with daunorubicin encapsulated in small-sized vesicles composed of DSPC/Chol.[82] Although these liposomes do not contain sterically stabilizing components, they display a relatively long circulation time.

Regarding drug levels in normal tissues such as liver and heart, two important observations have been made:[8,72,73] (1) the peak level in liver tissues is roughly similar for free DOX and SL-DOX; however, in the latter case the peak is achieved at 24 hr after injection, while in the former, the peak is already attained at 30 min following injection; (2) the peak concentration in the heart muscle seen shortly after injection of free DOX and believed to be correlated with cardiotoxicity is not observed after SL-DOX injection.

Altogether, the current knowledge on the biodistribution of SL-DOX is consistent with the morphological data pointing at extravasation of SL through the abnormally permeable vessels characteristic of many tumors, and their accumulation in the interstitial fluid.[79] Drug delivery to tumor cells depends on the rate of release of drug from these extravasated liposomes. In vitro studies of cell drug uptake and cytotoxicity[83] also support the contention that extracellular release of drug from liposomes is the key step to allow drug access to the intracellular compartment. On the one hand, this may be a favorable aspect for liposomal systems since the released drug will diffuse across the tumor cell layers and extracellular matrix without the physical constraints imposed on macromolecules and particulate systems. On the other hand, if drug release becomes very slow, there might be a problem of drug bioavailability which could limit the therapeutic effect of SL-DOX.

Consequently, factors which accelerate the drug efflux rate of liposomes or the cellular uptake of liposomes should enhance the pharmacologic and antitumor effects. Two respective examples are the use of hyperthermia[84-86] and the use of antibody-targeted liposomes.[10,43,44,87,88] The latter have been shown to decrease the in vitro IC_{50} of DOX, suggesting an increase of liposome-mediated drug uptake by cells.

IX. ANTITUMOR ACTIVITY IN ANIMAL MODELS

A number of studies indicate that the therapeutic activity of DOX or EPI encapsulated in sterically stabilized liposomes is superior to that of the respective free drugs in various tumor models including: mouse colon carcinoma (C-26) implanted SC,[74,79] mouse lymphoma (J-6456) inoculated i.p.,[18,34] mouse mammary carcinomas (MC2A, MC2B, MC19, MC65) implanted SC or intramammary,[89] lung metastases of mouse M-109 lung carcinoma,[80] and human squamous cell carcinoma of lung implanted in SCID mice.[90] In these studies, the therapeutic gain was achieved at subtoxic doses and using the i.v. route against tumors residing outside the RES compartment of liver, spleen, and bone marrow. Thus, unlike "conventional" liposomes, the improvement in therapeutic activity with long-circulating liposomes does not depend on toxicity buffering and on the proximity of tumor cells to the RES. The mechanism of action, as discussed above, appears to be related to selective extravasation in tumor sites with gradual release of the drug. The fact that small liposomes (<100 nm) are required to obtain the optimal antitumor effect[34] and the delay in optimal antitumor effects[79] are consistent with the importance of extravasation.

Forssen et al.[82] have reported a superior antitumor effect for daunorubicin encapsulated in DSPC/Chol liposomes over free daunorubicin using the mouse MA16C mammary adenocarcinoma and P-1798 lymphosarcoma implanted SC. Their results are consistent with the tumor localizing properties of these long-circulating vesicles. There are no published studies comparing side-by-side the therapeutic activity of DOX in HPC/Chol or DSPC/Chol, with or without inclusion of a PEG-derivatized lipid.

It should be stressed that, for kinetic reasons, experiments with long-circulating liposomes in fast-growing mouse tumor models (e.g., L1210 and P388 leukemias) may be rather misleading. These tumors grow so rapidly that their doubling times are shorter than the circulation half-life of the liposomal drug. Thus, the rates of liposome distribution and drug release may be too slow to be effective in these systems.

X. TOXICITY STUDIES

Recent pharmacological studies of SL-DOX in rodents demonstrate reduced toxicity concomitant with increased tumor localization. Toxicity studies point to improved tolerance of SL-DOX, with significantly less mortality. The LD_{50} of DOX in mice is between 20 to 30 mg/kg for HPI/HPC/Chol and PEG-PE/HPC/Chol liposomes, and between 10 to 15 mg/kg for free DOX.[34,80] This is a modest but significant increase. For EPI encapsulated in PEG-PE/HPC/Chol liposomes, the LD_{50} is 9 mg/kg in 3 weekly injections, while that of free EPI is 6 mg/kg using the same schedule.[74] In rat studies, treatment with a PEG-PE stabilized liposome preparation of DOX (referred to as Doxil®) resulted in less cardiotoxicity, nephrotoxicity, and myelotoxicity than treatment with free DOX.[91] These results are valuable since encapsulation in SL could conceivably allow us the use of higher cumulative doses of DOX in a clinical setting. One additional observation is that the vesicant effect caused by DOX after intradermal injection is prevented by encapsulation in PEG-PE/HPC/Chol liposomes.[80]

The fact that acute toxicity is substantially reduced by liposomal delivery despite the slow clearance of SL-DOX is worth noting. This seemingly paradoxical point can be explained by the fact that most of the circulating drug is in liposome-associated form and is therefore not bioavailable. Thus, SL appear to offer a double advantage as an anticancer drug delivery system: toxicity buffering as with conventional liposome formulations, and selective tumor accumulation.

XI. CLINICAL STUDIES: PRELIMINARY RESULTS

We have recently completed a clinical study in which the pharmacokinetics of DOX encapsulated in PEG-PE/HPC/Chol liposomes (Doxil®) were examined and compared to that of free DOX in cancer patients.[92] This pilot study clearly established that the drug pharmacokinetic properties are significantly changed by encapsulation in SL. DOX was cleared very slowly with a distribution phase median half-life of 42–46 hr, as opposed to the rapid distribution of free DOX ($T_{1/2}$ = 5 min). The AUC and clearance of DOX were increased and decreased, respectively, by a 200-fold factor. With Doxil®, approximately 100% of the injected dose was recovered in the patients' plasma volume 5 min after injection, and nearly all the drug measured in plasma was accounted for by liposome-associated drug through 7 days after injection. There was a linear increase in plasma drug concentration when the dose of Doxil® was raised from 25 to 50 mg/m², with no major effect on the pharmacokinetic parameters. Thus, the pharmacokinetic observations in rodents and dogs were confirmed in humans.[18]

In a total number of 15 patients receiving 53 courses, Doxil® was generally well tolerated, stomatitis being the most significant side effect. Dose escalation studies of Doxil® are currently ongoing within the frame of a phase I study. It should be pointed out that no substantial increase of the human maximal tolerated dose of DOX has been obtained so far in any of the phase I studies reported with various formulations of liposomal anthracyclines.[93-96] Despite this, liposomes may still increase the dose intensity of the encapsulated drug since the pharmacokinetic changes caused by liposome delivery may lead to increased body drug retention, which will ultimately result in greater dose intensity than an equal dose of free drug.

Doxil® is also being extensively studied in AIDS-related Kaposi Sarcoma. A recent communication suggests that Doxil® is highly active against Kaposi Sarcoma lesions.[97] Of note, stealth liposomes have been shown to extravasate very efficiently in Kaposi Sarcoma-like lesions of mice.[30]

The ability of SL to extravasate into tumor compartments is supported by our recent clinical data.[92] There was a slow drug accumulation in malignant effusions, reaching its peak concentration around one week after injection of Doxil®. In 3 patients with malignant pleural effusions in whom a direct comparison was made between free DOX and Doxil®, the pleural fluid drug levels were four- to tenfold greater than after free DOX treatment.

These results are consistent with preclinical observations and indicate that Doxil® pharmacokinetics are controlled by the liposome carrier and differ substantially from those of free DOX. The observed drastic changes in the drug pharmacokinetic pattern are likely to result in a change in drug biodistribution and pharmacodynamics, which should ultimately be translated into an improved therapeutic index in humans.

REFERENCES

1. Gregoriadis, G., Ed., *Liposomes as Drug Carrier Recent Trends and Progress,* John Wiley and Sons, New York, 1988.
2. Lopez-Berestein, G., and Fidler, I. J., Eds., *Liposomes in the Therapy of Infectious Diseases and Cancer,* Alan R. Liss, New York, 1989.
3. Nassander, U. K., Storm, G., Peters, P. A. M., and Crommelin, D. J. A., Liposomes, in *Biodegradable Polymers as Drug Delivery Systems,* Chasin, M. and Langer, R., Eds., Marcel Dekker, New York, 1990.
4. Senior, J. H., Fate and behavior of liposomes *in vivo*: a review of controlling factors, *Crit. Rev. Ther. Drug Carrier Syst.,* 3, 123, 1987.
5. Papahadjopoulos, D., A personal perspective of liposomes as a drug carrier system, *J. Liposome Res.,* 2(3), 3, 1992.
6. Hwang, K. J., Liposome pharmacokinetics, in *Liposomes from Biophysics to Therapeutics,* Ostro, M. J., Ed., Marcel Dekker, New York, 1987, chap. 4.
7. Allen, T. M., and Chonn, A., Large unilamellar liposomes with low uptake by the reticuloendothelial system, *FEBS Lett.,* 223, 1987.
8. Gabizon, A., and Papahadjopoulos, D., Liposome formulations with prolonged circulation time in blood and enhanced uptake by tumors, *Proc. Natl. Acad. Sci. U.S.A.,* 85, 6949, 1988.
9. Allen, T. M., Hansen, C., and Rutledge, J., Liposomes with prolonged circulation times: factors affecting uptake by reticuloendothelial and other tissues, *Biochim. Biophys. Acta,* 981, 27, 1989.
10. Papahadjopoulos, D., and Gabizon, A., Targeting of liposomes to tumor cells *in vivo, Ann. N.Y. Acad. Sci.,* 507, 64, 1987.
11. Allen, T. M., "Stealth" liposomes: avoiding reticuloendothelial uptake, in *Liposomes in the Therapy of Infectious Diseases and Cancer,* Lopez-Berestein, G. and Fidler, I., Eds., Alan R. Liss, New York, 1989, 405.
12. Lasic, D. D., Martin, F. J., Gabizon, A., Huang, S.-K., and Papahadjopoulos, D., Sterically stabilized liposomes: a hypothesis on the molecular origin of the extended circulation times, *Biochim. Biophys. Acta,* 1070, 187, 1991.
13. Woodle, M. C., Newman, M., Collins, L., Redemann, C., and Martin, F. J., Improved long circulating liposomes ("Stealth") using synthetic lipids, *Proc. Int. Symp. Control. Rel. Bioact. Mater.,* 1990, 1777.
14. Klibanov, A. L., Maruyama, K., Torchilin, V.P., and Huang, L., Amphipathic polyethyleneglycols effectively prolong the circulation time of liposomes, *FEBS Lett.,* 268, 235, 1990.
15. Blume, G., and Cevc, G., Liposomes for the sustained drug release *in vivo, Biochim. Biophys. Acta,* 1029, 91, 1990.
16. Senior, J., Delgado, C., Fisher, D., Tilcock, C., and Gregoriadis, G., Influence of surface hydrophilicity of liposomes on their interaction with plasma protein and clearance from the circulation: studies with poly(ethylene glycol)-coated vesicles, *Biochim. Biophys. Acta,* 1062, 77, 1991.
17. Allen, T. M., Hansen, C., Martin, F. J., Redemann, C., and Yau-Young, A., Liposomes containing synthetic lipid derivatives of polyethylene glycol show prolonged circulation half-lives *in vivo, Biochim. Biophys. Acta,* 1062, 29, 1991.
18. Papahadjopoulos, D., Allen, T. M., Gabizon, A., Mayhew, E., Matthay, K. K., Huang, S.-K., Lee, K.-D., Woodle, M. C., Lasic, D. D., Redemann, C., and Martin, F. J., Sterically stabilized liposomes: improvements in pharmacokinetics, and anti-tumor therapeutic efficacy, *Proc. Natl. Acad. Sci. U.S.A.,* 88, 11460, 1991.

19. Woodle, M. C., Matthay, K. K., Newman, M. S., Hadiyat, J. E., Collins, L. R., Redemann, C., Martin, F. J., and Papahadjopoulos, D., Sterically stabilized liposomes: versatility of lipid compositions with prolonged circulation, *Biochim. Biophys. Acta*, 1105, 193, 1992.

20. Klibanov, A. L., Maruyama, K., Beckerleg, A.-M., Torchilin, V. P., and Huang, L., Activity of amphipathic poly(ethylene glycol) 5000 to prolong the circulation time of liposome depends on the liposome size and is unfavorable for immunoliposome binding to target, *Biochim. Biophys. Acta*, 1062, 142, 1991.

21. Mori, A., Klibanov, A. L., Torchilin, V. P., and Huang, L., Influence of the steric barrier activity of amphipathic poly(ethyleneglycol) and ganglioside GM_1 on the circulation time of liposomes and on the target binding of immunoliposomes *in vivo*, *FEBS Lett.*, 284, 2, 263, 1991.

22. Maruyama, K., Yuda, T., Okamoto, A., Ishikura, C., Kojima, S., and Iwatsuru, M., Effect of molecular weight in amphipathic polyethylenegylcol on prolonging the circulation time of large unilamellar liposomes, *Chem. Pharm. Bull.*, 39(6), 1620, 1991.

23. Liu, D., Mori, A., and Huang, L., Large liposomes containing ganglioside GM_1 accumulate effectively in spleen, *Biochim. Biophys. Acta*, 1066, 159, 1991.

24. Liu, D., Mori, A., and Huang, L., Role of liposome size and RES blockade in controlling biodistribution and tumor uptake of GM1-containing liposomes, *Biochim. Biophys. Acta*, 1104, 95, 1992.

25. Patel, H. M., Serum opsonins and liposomes: their interaction and opsonophagocytosis, *Crit. Rev. Ther. Drug Carrier Syst.*, 9(1), 39, 1992.

26. Allen, T. M., and Hansen, C., Pharmacokinetics of stealth versus conventional liposomes: effect of dose and composition, *Biochim. Biophys. Acta*, 1068, 133, 1991.

27. Gabizon, A., Price, D. C., Huberty, J., Bresalier, R. S., and Papahadjopoulos, D., Effect of liposome composition and other factors on the targeting of liposomes to experimental tumors: biodistribution and imaging studies, *Cancer Res.*, 50, 6371, 1990.

28. Huang, S. K., Hong, K., Lee, K.-D., Papahadjopoulos, D., and Friend, D. S., Light microscopic localization of silver-enhanced liposome entrapped colloidal gold in mouse tissues, *Biochim. Biophys. Acta*, 1069, 117, 1991.

29. Huang, S.-K., Lee., K.-D., Hong, K., Friend, D. S., and Papahadjopoulos, D., Microscopic localization of sterically stabilized liposomes in colon carcinoma-bearing mice, *Cancer Res.*, 52, 5135, 1992.

30. Huang, S.-K., Martin, F. J., Jay, G., Vogel, J., Papahadjopoulos, D., and Friend, D. S., Extravasation and transcytosis of liposomes in Kaposi's sarcoma-like dermal lesions of transgenic mice bearing the HIV tat gene, *Am. J. Pathol.*, 143(1), 10, 1993.

31. Wu, N. Z., Da, D., Rudoll, T., Needham, D., Whortin, R. A., and Dewhirst, M. W., Increased microvascular permeability contributes to preferential accumulation of stealth liposomes in tumor tissue, *Cancer Res.*, 53, 3765, 1993.

32. Bakker-Woudenberg, I. A. J. M., Lokersee, A. F., and Storm, G., Enhanced localization of liposomes with prolonged blood circulation time in infected lung tissue, *Biochim. Biophys. Acta*, 1138, 318, 1992.

33. Bakker-Woudenberg, I. A. J. M., Lokersee, A. F., ten Kate, M. T., Moulton, J. W., Woodle, M. C., and Storm, G., Liposomes with prolonged blood circulation times and selective localization in Klebstiella pneumoniae infected lung tissue, *J. Infect. Dis.*, 168, 164, 1993.

34. Gabizon, A. A., Selective tumor localization and improved therapeutic index of anthracyclines encapsulated in long circulating liposomes, *Cancer Res.*, 52, 891, 1992.

35. Liu, D., Mori, A., and Huang, L., Role of liposome size and RES blockade in controlling biodistribution and tumor uptake of GM_1-containing liposomes, *Biochim. Biophys. Acta*, 1104, 95, 1992.

36. Northfelt, D., Kaplan, L., Russell, J., Volberding, P., Martin, F. J., Anderson, M., and Lang, J., Single dose pharmacokinetics, safety, and tumor localization of Doxil in AIDS patients with Kaposi's Sarcoma, *J. Liposome Res.*, 2, 411, 1992.

37. Gabizon, A., Catane, R., Uzieli, B., Kaufman, B., Safra, T., Barenholz, Y., and Huang, A., A pilot study of DOX encapsulated in long-circulating (Stealth) liposomes in cancer patients, *Proc. Am. Soc. Clin. Oncol.*, 11, 129, 1992.

38. Hong, K., Friend, D. S., Glabe, C. G., and Papahadjopoulos, D., Liposomes containing colloidal gold are a useful probe of liposome-cell interactions, *Biochim. Biophys. Acta*, 732, 320, 1983.

39. Kronberg, B., Dahlman, A., Carlfors, J., and Karsson, J., Preparation and characterization of sterically stabilized liposomes, *J. Pharm. Sci.*, 79, 667, 1990.

40. Maruyama, K., Holmberg, E., Kennel, S. J., Klibanov, A., Torchilin, V. P., and Huang, L., Characterization of *in vivo* immunoliposome targeting to pulmonary endothelium, *J. Pharm. Sci.*, 79(11), 978, 1990.

41. Maruyama, K., Kennel, S. J., and Huang, L., Lipid composition is important for highly efficient target binding and retention of immunoliposomes, *Proc. Natl. Acad. Sci. U.S.A.*, 87, 5744, 1990.

42. Ahmad, I., Longenecker, M., Samuel, J., and Allen, T. M., Antibody-targeted delivery of Doxorubicin entrapped in sterically stabilized liposomes can eradicate lung cancer in mice, *Cancer Res.*, 53, 1484, 1993.

43. Ahmad, I., and Allen, T. M., Antibody-mediated specific binding and cytotoxicity of liposome-entrapped Doxorubicin in lung cancer cells *in vitro*, *Cancer Res.*, 52, 4817, 1992.

44. Park, J. W., Hong, K., Carter, P., Kotts, C., Shalaby, R., Giltinian, D., Wirth, C., Asgari, H., Wood, W. I., Papahadjopoulos, D., and Benz, C., Development of anti-her-2 immunoliposomes for breast cancer therapy, Abstract, *Proc. Am. Soc. Clin. Oncol.*, 1993.

45. Gregoriadis, G., *Liposome Technology*, Vol. III, Gregoriadis, G., Ed., CRC Press, Boca Raton, FL, 1992, chap. 1, 9-18.

46. Allen, T. M., Williamson, P., and Schlegel, R. A., Phosphatidylserine as a determinant for reticuloen-dothelial recognition of model erythrocyte membranes, *Proc. Natl. Acad. Sci. U.S.A.*, 85, 8067, 1988.

47. Allen, T. M., Austin, G. A., Chonn, A., Lin, L., and Lee, K. C., Uptake of liposomes by cultured mouse bone marrow macrophages: influence of liposome composition and size, *Biochim. Biophys. Acta*, 1061, 56, 1990.

48. Lee, K.-D., Hong, K., and Papahadjopoulos, D., Recognition of liposomes by cells: *in vitro* binding and endocytosis mediated by specific lipid headgroups and surface charge density, *Biochim. Biophys. Acta*, 1103, 185, 1992.

49. Lee, K.-D., Nir, S., and Papahadjopoulos, D., Quantitative analysis of liposome-cell interactions *in vitro*: rate constants of binding and endocytosis with suspension and adherent J774 cells and human monocytes, *Biochemistry*, 32, 889, 1993.

50. Connor, J., Bucana, C., Fidler, I. J., and Schoit, A. J., Differentiation-dependent expression of phosphatidylserine in mammalian plasma membranes: quantitative assessment of outer-leaflet lipid by prothrombinase complex formation, *Proc. Natl. Acad. Sci. U.S.A.*, 86, 3184, 1989.

51. Burkhanov, S. A., Kosykh, V. A., Repin, V. S., Saatov, T. S., and Torchilin, V. P., Interaction of liposomes of different phospholipid and ganglioside composition with rat hepatocytes, *Int. J. Pharm.*, 46, 31, 1988.

52. Woodle, M. C., Collins, L. R., Sponsler, E., Kossovsky, N., Papahadjopoulos, D., and Martin, F. J., Sterically stabilized liposomes: reduction in electrophoretic mobility but not electrostatic surface potential, *Biophys. J.*, 61, 902, 1992.

53. Huang, L., Stealth liposomes, ninja liposomes or cryptosomes: are they really sterically stabilized liposomes?, *J. Liposome Res.*, 2(3), 451, 1992.

54. Chonn, A., Semple, S. C., and Cullis, P. R., Association of blood proteins with large unilamellar liposomes *in vivo*, *J. Biol. Chem.*, 267(26), 18759, 1992.

55. Mayhew, E. M., and Papahadjopoulos, D., Therapeutic applications of liposomes, in *Liposomes*, Ostro M. J., Ed., Marcel Dekker, New York, 1983, 289.

56. Young, R. C., Ozols, R. F., and Myers, C. E., The anthracycline antineoplastic drugs, *New Engl. J. Med.*, 305, 139, 1981.

57. Szoka, F. C., Liposomal drug delivery: current status and future prospects, in *Membrane Fusion*, Wilschut J. and Hoekstra D., Eds., Marcel Dekker, New York, 1991, 845.

58. Gabizon, A., Liposomes as a drug delivery system in cancer chemotherapy, in *Drug Carrier Systems (Horizons in Biochemistry and Biophysics Series, Vol. 9)*, Roerdink, F. H. and Kroon, A. M., Eds., John Wiley & Sons, Chichester, 1989, 185.

59. Herman, E. H., Rahman, A., Ferrans, V. J., Vick, J. A., and Schein, P. S., Prevention of chronic Doxorubicin cardiotoxicity in Beagles by liposomal encapsulation, *Cancer Res.*, 43, 5427, 1983.

60. Gabizon, A., Meshorer, A., and Barenholz, Y., Comparative long-term study of the toxicities of free and liposome-associated Doxorubicin in mice after intravenous administration, *J. Natl. Cancer Inst.*, 7, 459, 1986.

61. Mayer, L. D., Tai, L. C., Ko, D. S. C., Masin, D., Ginsberg, R. S., Cullis, P. R., and Bally, M. B., Influence of vesicle size, lipid composition, and drug-to-lipid ratio on the biological activity of liposomal doxorubicin in mice, *Cancer Res.*, 49, 5922, 1989.

62. Olson, F., Mayhew, E. M., Maslow, D., Rustum, Y., and Szoka, F., Characterization, toxicity and therapeutic efficacy of Adriamycin encapsulated in liposomes, *Eur. J. Cancer Clin. Oncol.*, 18, 167, 1982.

63. Gabizon, A., Goren, D., and Barenholz, Y., Investigations on the antitumor efficacy of liposome-associated doxorubicin in murine tumor models, *Isr. J. Med. Sci.*, 24, 512, 1988.

64. Mayhew E. M., and Rustum Y. E., The use of liposomes as carriers of therapeutic agents, *Prog. Clin. Biol. Res.*, 172B, 301, 1985.

65. Mayhew, E. M., Goldrosen, M. H., Vaage, J., and Rustum, Y. M., Effects of liposome-entrapped Doxorubicin on liver metastases of mouse colon carcinomas 26 and 38, *J. Natl. Cancer Inst.*, 78, 707, 1987.

66. Gabizon, A., Chisin R., Amselem, S., Druckmann, S., Cohen, R., Goren, D., Fromer, I., Peretz, T., Sulkes, A., and Barenholz, Y., Pharmacokinetic and imaging studies in patients receiving a formulation of liposome-associated Adriamycin, *Br. J. Cancer*, 64, 1125, 1991.

67. Proffitt, R. T., Williams, L. E., Presant, C. A., Tin, G. W., Uliana, J. A., Gamble, R. C., and Baldeschwieler, J. D., Tumor imaging potential of liposomes loaded with In-1-NTA: biodistribution in mice, *J. Nucl. Med.*, 24, 45, 1983.

68. Ogihara-Umeda, I., and Kojima, S., Increased delivery of Gallium-67 to tumors using serum-stable liposomes, *J. Nucl. Med.*, 29, 516, 1988.

69. Mayer, L. D., Tai, L. C., Bally, M. B., Mitilenes, G. N., Ginsberg, R. S., and Cullis P. R., Characterization of liposomal systems containing Doxorubicin entrapped in response to pH gradients, *Biochim. Biophys. Acta*, 1025, 143, 1990.

70. Haran, G., Cohen, R., Bar, L. K., and Barenholz, Y., Transmembrane ammonium sulfate gradients in liposomes produce efficient and stable entrapment of amphipathic weak bases, *Biochim. Biophys. Acta*, 1151, 201, 1993.

71. Lasic, D. D., Frederik, P. M., Stuart, M. C. A., Barenholz., Y. and McIntosh, T. J., Gelation of liposome interior: a novel method for drug encapsulation, *FEBS Lett.*, 312(2,3), 255, 1992.

72. Gabizon, A., Shiota, R., and Papahadjopoulos, D., Pharmacokinetics and tissue distribution of doxorubicin encapsulated in stable liposomes with long circulation times, *J. Natl. Cancer Inst.*, 81, 1485, 1989.

73. Gabizon, A., Barenholz Y., and Bialer M., Prolongation of the circulation time of Doxorubicin encapsulated in liposomes containing a polyethylene glycol derivatized phospholipid: pharmacokinetic studies in rodents and dogs, *Pharm.. Res.*, 10, 703, 1993.

74. Mayhew, E. M., Lasic, D. D., Babbar, S., and Martin, F. J., Pharmacokinetics and antitumor activity of epirubicin encapsulated in long-circulating liposomes, *Int. J. Cancer*, 51, 302, 1992.

75. Bally, M. B., Nayar, R., Masin D., Cullis, P. R., and Mayer, L. D., Liposomes with entrapped Doxorubicin exhibit extended blood residence times, *Biochim. Biophys. Acta*, 1023, 133, 1990.

76. Allen, T. M., and Hansen, C., Pharmacokinetics of stealth versus conventional liposomes: effect of dose and composition, *Biochim. Biophys. Acta*, 1068, 133, 1991.

77. Allen, T. M., Mehra, T., Hansen, C., and Chin, Y. C., Stealth liposomes: an improved release system for 1-β-D-Arabinofuranosylcytosine, *Cancer Res.*, 52, 2431, 1992.

78. Vaage, J., Donovan, D., Mayhew, E. M., Uster, P., and Woodle, M., Therapy of mouse mammary carcinomas with vincristine and Doxorubicin encapsulated in sterically stabilized liposomes, *Int. J. Cancer*, 54, 959, 1993.

79. Huang, S. K., Mayhew, E. M., Gilani, S., Lasic, D. D., Martin, F. J., and Papahadjopoulos, D., Pharmacokinetics and therapeutics of sterically stabilized liposomes in mice bearing C-26 colon carcinoma, *Cancer Res.*, 52, 6774, 1992.

80. Gabizon, A., Pappo, O., Goren, D., Chemla, M., Tzemach, D., and Horowitz, D., Preclinical studies with doxorubicin encapsulated polyethylene-glycol-coated liposomes, *J. Liposome Res.*, 3, 517, 1993.

81. Siegal, T., Lossos, A., and Gabizon, A. A., Selective tumor localization of doxorubicin encapsulated in long-circulating liposomes in an experimental brain tumor model, *Neurology*, 43, A360, 1993.

82. Forssen, E. A., Coulter, D. M., and Proffitt, R. T., Selective *in vivo* localization of Daunorubicin small unilamellar vesicles in solid tumors, *Cancer Res.*, 52, 3255, 1992.

83. Horowitz, A. T., Barenholz, Y., and Gabizon, A., *In vitro* cytotoxicity of liposome-encapsulated doxorubicin: dependence on liposome composition and drug release, *Biochim. Biophys. Acta*, 1109, 203, 1992.

84. Maruyama, K., and Iwatsuru, M., Doxorubicin encapsulated in long-circulating thermo-sensitive liposomes, *J. Liposome Res.*, 4, 513, 1994.

85. Huang, S. K., Stauffer, P. R., Hong, K., Guo, J. W. H., Phillips, T. L., Huang, A., and Papahadjopoulos, D., Liposomes and hyperthermia in mice: increased tumor uptake and therapeutic efficacy of Doxorubicin in sterically stabilized liposomes, *Cancer Res.*, 54, 2186, 1994.

86. Maruyama, K., Unezaki, S., Takahashi, N., and Iwatsuru, M., Enhanced delivery of Doxorubicin to tumor by long-circulating thermosensitive liposomes and local hyperthermia, *Biochim. Biophys. Acta*, 1149, 209, 1993.

87. Matthay, K. K., Abai, A. A., Cobb, S., Hong, K., Papahadjopoulos, D., and Straubinger, R. M., Role of ligand in antibody-directed endocytosis of liposomes by human T-leukemia cells, *Cancer Res.*, 49, 4879, 1989.

88. Heath, T. D., Montgomery, J. A., Piper, J. R., and Papahadjopoulos, D., Antibody-targeted liposomes: increase in specific toxicity of methotrexate-γ-aspartate, *Proc. Natl. Acad. Sci. U.S.A.*, 80, 1377, 1983.

89. Vaage, J., Mayhew, E. M., Lasic, D. D., and Martin, F., Therapy of primary and metastatic mouse mammary carcinomas with Doxorubicin encapsulated in long-circulating liposomes, *Int. J. Cancer*, 51, 942, 1992.

90. Williams, S. S., Alosco, T. R., Mayhew, E. M., Lasic, D. D., Martin, F. J., and Bankert, R. B., Arrest of human lung tumor xenograft in severe combined immunodeficient mice using Doxorubicin encapsulated in sterically stabilized liposomes, *Cancer Res.*, 53, 3964, 1993.

91. Working, P. K., Newman, M. S., Huang, S. K., Mayhew, E. M., Vaage, J., and Lasic, D. D., Pharmacokinetics, biodistribution and therapeutic efficacy of Doxorubicin encapsulated in Stealth® liposomes (Doxil®), unpublished data, 1994.

92. Gabizon, A., Catane, R., Uziely, B., et al., unpublished data, 1994.

93. Gabizon, A., Peretz, T., Sulkes, A., Amselem, S., Ben-Yosef, R., Ben-Baruch, N., Catane, R., Biran, S., and Barenholz, Y., Systemic administration of doxorubicin-containing liposomes in cancer patients: a phase I study, *Eur. J. Cancer Clin. Oncol.*, 25, 1795, 1989.

94. Rahman, A., Treat, J., Roh, J. K., Potkul, L. A., Alvord, W. G., Forst, D., and Wolley, P. V., A phase I clinical trial and pharmacokinetic evaluation of liposome-encapsulated doxorubicin, *J. Clin. Oncol.*, 8, 1093, 1990.

95. Pestalozzi, R., Schwendener, R., and Sauter, C., Phase I/II study of liposome-complexes mitoxantrone in patients with advanced breast cancer, *Ann. Oncol.*, 3, 445, 1992.

96. Cowens, J. W., Creaven, P. J., Greco, W. R., Brenner, D. E., Tung, Y., Ostro, M., Pilkiewicz, F., Ginsberg, R., and Petrelli, N., Initial clinical (Phase I) trial of TLC D-99 (Doxorubicin encapsulated in liposomes), *Cancer Res.*, 53, 2796, 1993.

97. Szelenyi, H., Jablonowski, H., Armbrecht, C., Mauss, S., Niederau, C., and Strohmeyre, G., Liposomal Doxorubicin — a new formulation for the treatment of Kaposi Sarcoma. A study on safety and efficacy in AIDS patients, *Int. Conf. AIDS*, 9, 397, 1993.

Chapter 12

Liposomes as Carriers of Antigens

Martin Friede

Liposomes exert an immunoadjuvant activity on associated protein, enhancing both humoral and cellular immune responses against the protein. This activity arises out of an enhanced delivery of the protein to antigen presenting cells and, under certain circumstances, to a liposome-mediated cytoplasmic delivery of the antigen. Based on these properties, liposomes are ideal candidates as adjuvants for new-generation vaccines. In contrast to proteins, short peptides lacking T-epitopes are not rendered immunogenic by liposomes. This can, however, be overcome by the inclusion of an immunostimulant, monophosphoryl lipid A, in the liposomal membrane. Immunogenicity of the resulting conjugates is dependent on the mode of association of the peptide with the liposomes, and appears to arise out of a targeting by the peptide of immunostimulant directly to lymphocytes recognizing the peptide. In addition, peptides representing T-epitopes can be added, thus inducing highly specific B- and T-cell responses. Such systems permit the design of peptide-based vaccines.

CONTENTS

I. INTRODUCTION

When liposomes were first hailed several decades ago as being the panacea to a host of medical problems, it was assumed by some that being composed only of lipid, liposomes would protect entrapped protein from any immunological response by the body. The discovery, therefore, by Allison and Gregoriadis[1] in 1974 that associating a protein with liposomes actually augmented its immunogenicity, spelled the demise of several liposome projects (a message not yet received by all researchers) yet opened up the flourishing field of using liposomes to generate an immune response.

This field has been given tremendous impetus by the urgent need to find acceptable adjuvants for new-generation vaccines aimed at combating diseases for which conventional immunization strategies are either inapplicable or have untoward side effects.[2] These new-generation vaccines will, in the most part, be composed of recombinant proteins or peptides, which alone tend to be very poor immunogens. It has therefore been the object of a large body of research to render such proteins or peptides sufficiently immunogenic, and also to induce the most relevant immune response in terms of cellular or humoral immunity by combining the proteins with novel carriers or vehicles. Currently, aluminum salts are the only adjuvants widely licensed for use in humans, and these are not always suitable nor without side effects. The fact that liposomes augment the immunogenicity of associated proteins and are biodegradable and non-toxic has made liposomes appear as ideal candidates as adjuvants for such vaccines. The number of parameters involved in preparing immunogenic liposome-protein conjugates, however, is vast and the

0-8493-4569-3/95/$0.00+$.50
© 1995 by CRC Press, Inc.

parameters may play a significant role on the nature of response induced. As we begin to understand how liposomes interact with the immune system and with the body in general, the choice of liposomal parameters can be made on a scientific basis. Furthermore, as we dissect the interaction of liposomes with the immune system we are able to learn more about the immune system itself.

Several detailed reviews have appeared on these topics in recent years.[3-5] Rather than repeating what has been said in those papers, this chapter will attempt to explain how liposomes can be used to induce an immune response against proteins and peptides and the promise such an approach offers towards the development of synthetic vaccines. Since the immune response toward proteins and peptides is different, these will be treated separately.

II. LIPOSOMES AS PROTEIN CARRIERS

The research surrounding the use of liposomes to augment the immunogenicity of proteins has followed two non-exclusive approaches: on one hand there has been the development of an understanding of how liposomes present associated protein to the immune system, and on the other hand, to achieve an optimal formulation with the aim of achieving a vaccine formulation.

A. THE PHYSIOLOGICAL BASIS OF LIPOSOME ADJUVANTICITY FOR PROTEINS

The induction of a classical, T-dependent humoral response requires that the protein antigen be taken up via a phagocytotic or endocytotic mechanism by an antigen presenting cell (APC). The protein is then degraded in the lysozome to small fragments, some of which are subsequently presented on the surface of the cell in association with the MHC II glycoprotein complex. Recognition of this peptide-MHC II complex by T-helper lymphocytes (CD4) results in the secretion of a number of lymphokines by the T-lymphocyte. These lymphokines lead to proliferation of the T-lymphocyte recognizing the peptide and also to activation and proliferation of B-lymphocytes, leading ultimately to secretion of antibody recognizing the antigen that was originally taken up by the APC.

A number of cells are able to act as APCs, the most important of which are macrophages, B-lymphocytes, and the dendritic and Langerhans cells. Of these, the macrophages have appeared as the most likely key components in the induction of an immune response towards liposomal antigen, since it is well known that liposomes are avidly phagocytosed by macrophages. A number of experiments in vitro[6,7] and in vivo[8,9] have now proven that macrophages are indeed the fulcrum of liposome adjuvanticity. van Rooijen has demonstrated the key role played by macrophages in vivo by eliminating the macrophages with toxic liposomes[10] and subsequently studying the immune response toward free and liposomal antigen. He showed that in the absence of macrophages, the response toward liposomal antigen is significantly reduced, yet the response against certain soluble antigens may be enhanced. Two mechanisms are therefore proposed:[11]

1. Soluble antigen, when rendered particulate by encapsulation in, or binding to, liposomes is much more efficiently taken up and processed by macrophages than in the free form. This results in efficient recruitment of T-lymphocytes which are then able to stimulate B-lymphocytes recognizing the antigen (Figure 1). It is also possible that the macrophages are able to transfer fragments of the antigen directly to B-lymphocytes or dendritic cells, which may in turn present the antigen to T-lymphocytes.[12]

2. Liposomes (not necessarily with protein associated) block macrophages by occupying their phagocytotic activity. These macrophages may otherwise suppress certain immune responses. Once the macrophages are blocked, soluble antigen is free to interact with other cells of the immune system and induce an immune response.

It is also possible that for certain liposome types (primarily small liposomes, see later) the B-lymphocytes are able to act as APC, and there is also evidence that dendritic cells play an important role in presenting liposomal antigen.[13] The adjuvanticity of liposomes may also arise in part from a slow-release effect. Liposome associated antigen has been shown to remain at the site of injection for a considerable time,[14] during which time gradual leakage will result in continual stimulation of the immune system (Figure 1).

B. INDUCTION OF A HUMORAL RESPONSE

On the above evidence it appears that rendering the antigen particulate by simply associating it with liposomes is sufficient to enhance macrophage uptake and immune stimulation. A number of parameters, however, affect the adjuvanticity of the liposomes (reviewed in Reference 3). Parameters that have been

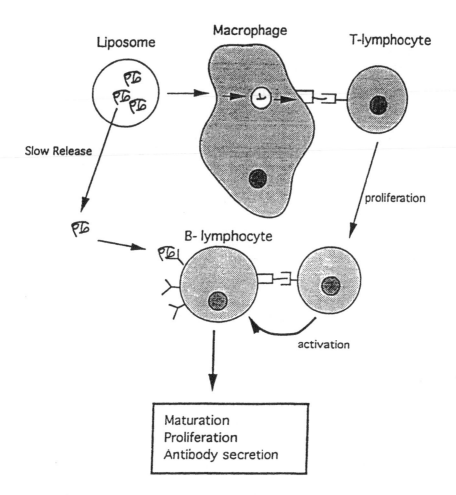

Figure 1 A simplified illustration of the proposed mechanism of liposomal adjuvanticity. Liposome encapsulated protein is phagocytosed by macrophages, leading to protein degradation in the lysosomes. Fragments of the protein are then presented on the surface of the macrophage associated with the MHCII complex, leading to specific T-helper cell stimulation. Lymphokine secretion, and direct interaction of the T-cells with B-cells which have bound and endocytosed free antigen leads to specific B-cell stimulation and antibody secretion.

investigated include, inter alia, the size of the liposomes,[15] the lipid:protein ratio,[16] the lipid transition temperature,[17] and the lipid charge.[17] While all of these factors play a crucial role on the intensity of the immune response, ranging from complete abolition of the response to 500-fold increases in intensity, it is difficult to draw any general conclusions regarding an optimal formulation. It is probably necessary to optimize these parameters for each different antigen, animal model, and site of injection, and also for the type of immune response desired, since what is deleterious for one antigen may be beneficial for another. One parameter that appears to play a role not on the intensity but rather on the type of immune response generated is the localization of antigen in or on the liposomes. While the intensity of the response is roughly the same for the same quantity of antigen encapsulated in or bound to the surface of liposomes, the ratio of IgG to IgM is significantly affected, with surface-bound antigen generating a high ratio of IgM to IgG.[18,19] This suggests that surface-bound antigen may, apart from inducing immunity via macrophages, interact directly with B-lymphocytes and activate B-lymphocytes via an alternative mechanism.

To enhance the intensity of the humoral response beyond that brought about by simply associating the antigen with liposomes, a number of authors have investigated the co-inclusion of immunostimulants with the liposomes. Significant improvements in the immune response have been obtained with lipid A and its non-toxic derivative monophosphoryl lipid A (MPLA),[20,21] as well as Il2,[22,23] and muramyl-dipeptide.[24] In these cases the liposomes act to concomitantly present the antigen and immunostimulant to the same cell or site, leading either to a more intense response or a switch of the immune response to a desired isotype.[24] This ability to easily co-present two or more different molecules in the same structure may in fact be one of the most appealing features of liposomes.

C. INDUCTION OF CELLULAR IMMUNITY

When an antigen presenting cell takes up antigen via an endocytotic or phagocytotic mechanism, the antigen (or a fragment of the antigen) is presented on the surface associated with MHC II. This leads ultimately to the activation of B-lymphocytes and production of antibodies. If, however, the antigen, instead of entering an endosomal route, is delivered into the cytoplasm, the antigen undergoes a different processing and is presented on the surface associated with the MHC I complex. This leads to the proliferation of a different group of T-lymphocytes, cytotoxic T-lymphocytes (CTL), specific for the antigen that was in the cytoplasm. These CTL form another branch of the immune response called cell mediated immunity (CMI) which is often essential for immunity against viral pathogens, and hence it is of prime importance that certain vaccines be capable of inducing CTL. Conventional adjuvants such as alum or Freund's adjuvant, generally speaking, are not capable of inducing significant CTL, yet liposomes, and in particular pH sensitive liposomes, appear able to induce significant CTL for a wide variety of encapsulated antigens.

When liposomal antigen is taken up by cells (primarily macrophages) liposomes pass from the endosome to lysosome where degradation occurs, and antigen is then recycled to the endosomal compartment and associated with MHC II.[25] It has, however, been shown in vitro that if the antigen is incorporated in pH sensitive liposomes, the liposome disrupts in the acidic endosomal compartment and fuses with the endosome, thus releasing part of the entrapped protein directly into the cytoplasm. Once in the cytoplasm the antigen is processed and presented associated with the MHC I complex.[26] While this effect is not observed with normal liposomes in vitro, in vivo it has been demonstrated that both types of liposomes are able to induce a CTL response, with pH sensitive liposomes possibly giving a slightly better response.[27] The difference between the in vitro and in vivo results could be explained by a destabilization of pH dependent liposomes in vivo, and possibly also to the trafficking of antigen fragments from macrophages to dendritic cells in vivo. A number of other liposomal parameters affecting CMI induction have been investigated including using liposomes in which the surface charge varied,[28] which has a negligible effect, or varying lipid transition temperature,[29] which appears to have an important role, at least for a trans-membrane protein.

In the above experiments a large amount of liposomal antigen was generally required to induce CMI. More recently, however, it has been shown that the inclusion of an immunostimulator, monophosphoryl lipid A (MPLA), in the liposomal membranes provides a co-stimulatory signal, permitting the induction of enhanced CMI with very little liposomal antigen.[30] In this case, it appears that the MPLA provides a co-stimulatory signal to the APC (probably macrophage), thus permitting the induction of a CTL response with only a few MHC I complexes occupied by the antigen. In the absence of such co-stimulation, a high density of MHC I occupancy is probably required to stimulate T-cells, thus necessitating a large antigen dose.

Liposomes appear therefore to meet the promise of being widely applicable immunoadjuvants for proteins. Protein antigens can be associated either by entrapment or by surface conjugation, and both a humoral and cellular response are induced. The intensity of the response can be further augmented by the inclusion of an immunostimulant either into the liposomal membrane for molecules such as lipid A, or entrapped, for soluble molecules such as Il2. Finally, being composed only of lipid the liposomal constructs are non-toxic, and may even reduce the toxicity of associated compounds. The tremendous potential of such constructs as vaccines is demonstrated by the recent human trials of a liposomal vaccine against malaria.

III. PRESENTATION OF HAPTENS AND PEPTIDES TO THE IMMUNE SYSTEM BY LIPOSOMES

In contrast to proteins, for antigens lacking T-epitopes an enhanced delivery of the antigen to APC is of very little utility since T-helper lymphocyte recruitment cannot occur. Examples of such antigens are short peptides and non-peptidic molecules termed haptens, e.g., di- and trinitrophenol. Generally, to induce an immune response against such molecules they have to be coupled to a carrier. If the carrier is non-proteinaceous, such as polysaccharide, the carrier serves to present the antigen multimerically and the immune response is characterized by being primarily IgM with no induction of immunological memory. Such a response is termed T-independent (TI) and can occur in the absence of T-lymphocytes. In order to induce a T-dependent response with an isotype switch to IgG and induction of memory, the carrier has to be a protein. In this case the carrier serves not only to present the antigen, but also as a source of T-epitopes, permitting T-lymphocyte recruitment.

A. LIPOSOMES AS CARRIERS OF HAPTENS

It was recognized very early in the development of liposomes that they could act as carriers of haptens in much the same way that polysaccharides were used. Initial experiments demonstrated that, as expected, haptens such as dinitrophenol when bound to the surface of liposomes induced a T-independent response.[31] What was interesting was that the intensity of the response could be modulated both by varying the lipid transition temperature,[32] a low transition temperature giving a poor response, and by varying the length of the linker between the hapten and the lipid to which it was linked.[33] Subsequent experimentation demonstrated that these effects arose out of a physical interaction between the hapten and the liposomal membrane, yielding a low accessibility of the hapten on fluid liposomes or when linked via a short linker.[34] The induction of a T-independent response could be significantly enhanced and switched to a T-dependent type by the inclusion of lipid A in the liposomes.[35] Under these circumstances the lipid A probably triggers the B-lymphocytes to behave as if they had received lymphokines from T-lymphocytes.

An alternative method of enhancing the response against haptens, and also switching the response to a T-dependent type, was demonstrated in an elegant series of experiments performed by Garçon and Six.[36] These authors prepared mixed liposomes containing a fragment of influenza hemagglutinin (HA) in the liposomal membrane and the dinitrophenyl hapten at the surface of the liposomes. Therefore the construct contained both T-epitopes (HA) and separate B-epitopes (nitrophenol), and in mice induced a classical T-dependent IgG response. This indicates that T-epitopes in the same construct, although not directly coupled to the B-epitope, are able to induce the recruitment of T-lymphocytes and a mature immune response. In principle, therefore, by using liposomes one should be able to induce specific T- and B-cell responses, and also to make use of existing T-cell memory to augment a B-cell response.

B. LIPOSOMES AS CARRIERS OF PEPTIDE ANTIGENS

While the experiments with chemical haptens demonstrated the principle of using liposomes to render haptens immunogenic, these molecules are by themselves of very little interest medically. In contrast, peptide antigens, which should behave immunologically as haptens, have significant medical potential.

As a basis for vaccines, peptides may offer several advantages over proteins in that they are produced synthetically, thus bypassing problems of protein purification or hazard associated with proteins from recombinant sources, and very specific epitopes could in principle be mimicked by peptides, thus inducing very specific immune responses. For example, it might be possible to immunize with a peptide representing an epitope common to all strains of a particular virus, whereas immunization with entire proteins may very well induce an immune response directed primarily against a variable region of the protein.

Because of the great potential of peptide-based vaccines, a lot of research has been done to identify what kind of peptides can be used to induce antibodies that recognize the native protein and confer protection.[37] One stumbling block to the implementation of peptide based vaccines is that, in general, short peptides (less than 15 amino acids) tend not to be immunogenic, and in order to induce antibodies the peptides need to be coupled to a carrier[38] which has normally been a protein such as tetanus toxoid, or keyhole limpet hemacyanin. The protein carrier serves to provide T-epitopes, often absent in short peptides, as well as to present the peptide multimerically. It is also possible that conjugation to the carrier serves to stabilize a conformation of the peptide[39] which, being short, generally will not have a defined structure in solution, and thus provides a structure against which antibodies can be more readily induced.

Protein carriers, however, do present a number of drawbacks; they themselves are immunogenic, and this may have undesired consequences if the same carrier is used for several different peptides.[40] The process of conjugation of the peptide to a protein often generates adducts on the protein which are highly immunogenic and against which an immune response may be directed.[41] Additionally, conjugation to a protein may induce changes in the conformation of the peptide, arising from interaction of the peptide with the heterogeneous protein surface, thus yielding a number of different conformations of the peptide, or perhaps the wrong conformation.[42] Furthermore, the use of a non-synthetic protein carrier negates to some extent the advantage to be gained from using synthetic peptides. Because of these, and a number of other drawbacks associated with protein or polysaccharide carriers, liposomes have been investigated as potential vehicles to render peptides immunogenic in the absence of proteins. The advantages hoped to be gained are to have a non-immunogenic carrier which is completely synthetic, and also one into which additional immunostimulants can be incorporated.

In our laboratories we have investigated the influence of liposomes on the immunogenicity of a number of short peptides, which by themselves are non-immunogenic. The first peptide used had the sequence Ile-Arg-Gly-Glu-Arg-Ala, and corresponds to the C-terminal hexapeptide of histone H3. By itself this peptide is of no interest for vaccines, but it has been shown that antibodies against this peptide recognize the parent protein, a prerequisite for peptide-based vaccines.

The addition of an N-terminal Cys-Gly sequence permitted coupling of the peptide to liposomes containing a maleimido-derivatized lipid in their membrane, by virtue of the reaction between the Cys-thiol and the maleimide function. While a number of techniques exist for coupling peptides to the surface of liposomes (see Chapter 2 of this volume), the technique described above has the advantage of being irreversible and defined, with only one possible orientation for the peptide. It also has wide applicability since cysteine can be readily added to a peptide during synthesis, and is a residue not often found in the middle of linear epitopes.

Initial results using this peptide associated with liposomes and injected into mice demonstrated that the peptide, whether surface-bound or entrapped in the liposomes, did not induce any detectable immune response. This, however, changed when the immunostimulant monophosphoryl lipid A was included in the liposomes, in which case surface-bound, but not entrapped peptide induced an IgG response that cross reacted with the parent protein.[43] It was later shown that if the MPLA was in liposomes other than those to which the peptide was bound, no response was elicited.[44] Since the peptide appeared unable to activate T-lymphocytes, the process therefore involves a T-independent B-cell activation. MPLA is known to be able to activate resting B-lymphocytes in mice,[45] and it therefore seems that the mechanism whereby surface-bound peptide induced an immune response is one in which liposomes bind preferentially to lymphocytes whose surface immunoglobulin (sIg) recognizes the peptide. Since MPLA is associated with the same liposome, this will result in preferential activation of these B-lymphocytes, leading to secretion of antibody against the peptide (Figure 2).

The actual mechanism whereby the activation of murine B-lymphocytes by MPLA takes place has not been fully elucidated. It is known that B-lymphocytes have surface receptors for lipopolysaccharide (LPS), of which MPLA is a component, and interaction of LPS with this receptor leads to a complex series of events resulting in B-cell activation. However, it is possible that endocytosis of the liposomes is required to elicit full activation.[46] Indeed, we observed that small liposomes (SUV, 100 nm diameter) bearing MPLA and surface-bound peptide induced a response with a much higher ratio of IgG to IgM, and of a longer duration than that observed with large liposomes (REV, 400 nm diameter).[43] This difference could be explained by the fact that large liposomes, unlike liposomes of 100 nm diameter, cannot be endocytosed by B-lymphocytes, and if endocytosis of the MPLA is required for full activation small liposomes should then give a superior response.

The immune response induced by small liposomes is also dependent on the net charge of the liposomes, neutral liposomes inducing a more intense and longer lasting response than negatively charged liposomes.[44] Neutral liposomes tend to have a longer half-life in the body than charged liposomes[47] and also have greater tendency to accumulate in the spleen. Both of these factors probably play a role in the enhanced immune response toward neutral liposomes. It is also possible that the difference described above in the immune response toward small and large liposomes is likewise due to the differing half-lives in the circulation of small and large liposomes.

As described previously, dinitrophenylated liposomes are able to induce an IgM response against the hapten even in the absence of MPLA, although the latter augments the response and induces a switch to IgG. The fact that the same behavior is not observed with peptides warns against extrapolating mechanisms elucidated for small haptens to a peptide-based system where one is looking for antibodies capable of recognizing the parent protein. The difference between the two systems probably arises out of differing interactions of the antigens with lipid membranes: small phenolic haptens are thought to interact with lipidic membranes in the fluid state, resulting in them not being sufficiently exposed to elicit an immune response (even in the presence of MPLA), and being exposed on solid membranes in a manner able to induce IgM secretion by B-lymphocytes. The results presented in Figure 3 suggest that, in contrast to phenolic haptens, the peptide of sequence CGIRGERA is exposed on fluid liposomes, and in the presence

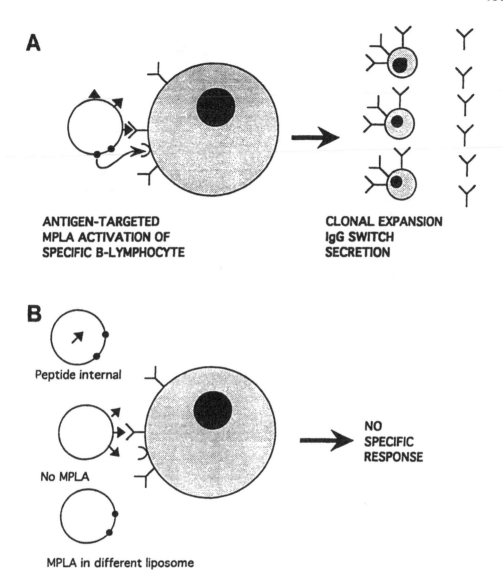

Figure 2 An illustration demonstrating the requirement for peptidic antigen to be surface bound, and for the adjuvant monophosphoryl lipid A (MPLA) to be present in the same liposome, in order to achieve antibody induction. **(A)** Surface-bound peptide targets the liposome to B-lymphocytes recognizing the peptide. MPLA in the membrane then activates the B-lymphocyte. **(B)** If the peptide is internally entrapped, no targeting of the liposome to B-lymphocytes occurs, and if the MPLA is in a different liposome there is no peptide to target it to B-lymphocytes.

of MPLA antibodies are induced so that they recognize in ELISA both the peptide and the cognate protein, histone H3. On solid liposomes, however, while the antibodies that are induced recognize the peptide in ELISA, they do not recognize histone H3. Hence, for this particular peptide the use of solid liposomes is detrimental for the specificity of the antibodies. This result appears to be due to the peptide lying flat on the surface of the solid liposomes, inducing antibodies directed primarily against the Ile-Arg-Glu- region, which is less accessible in the parent protein than when the peptide is coupled to a carrier.

C. A LIPOSOME-BASED INFLUENZA VACCINE MODEL
To test the applicability of the above system to different peptides, and in the framework of a vaccine model, two cyclic peptides representing a nine-residue sequence from influenza hemagglutinin cyclized

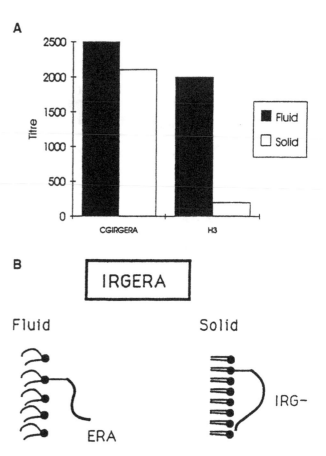

Figure 3 The specificity antibodies raised against peptide-liposome conjugates can vary depending on the nature of the liposomes. **(A)** The peptide of sequence CGIRGERA (corresponding to the C-terminal hexapeptide of histone H3) coupled to solid or fluid liposomes containing MPLA induces antibodies recognizing the peptide in ELISA (coupled to a protein carrier). The cognate protein (histone H3) is only recognized by antibodies raised using the fluid liposomes. **(B)** This appears to be due to the C-terminus of the peptide being less accessible on solid liposomes and hence the response is directed towards the N-terminus, a region on the cognate protein less accessible to antibodies.

two different ways (Figure 4) were coupled to SUV containing MPLA.[42] It had previously been demonstrated that both peptides, when conjugated to protein carriers, were able to confer significant (70%) protection in mice against a nasal challenge with live influenza virus. However, the protection achieved with protein carriers required Freund's adjuvant, thus reducing the applicability of such a system for human use.[48] Conjugated to liposomes, however, one of the peptides was able to confer 80% protection against a viral challenge, and this in the absence of protein carrier or Freund's adjuvant. Such a system therefore represents a synthetic, non-toxic, immunogen with potential for use as a human vaccine.

The fact that when coupled to proteins both peptides conferred immunity, whereas only one of the peptides was able to do so when coupled to the liposomes, appeared to be due to both peptides adopting a similar range of conformations when coupled to proteins, but unique conformations when coupled to liposomes and only one of these conformations resembling the epitope on the virus.[42] This is not

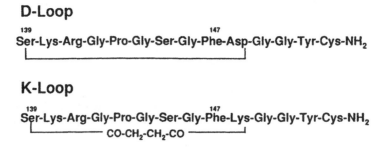

Figure 4 Composition of cyclic peptides corresponding to the region 139–147 of the hemagglutinin of influenza virus (strain X31) and with a terminal cys residue to permit coupling to liposomes. The D-loop when coupled to liposomes induces protective immunity against influenza, the K-loop does not.

unexpected when one considers that the surface of a protein carrier is heterogeneous with regard to charge localization, whereas a liposomal surface is homogeneous. Therefore, coupled to a protein a peptide will be forced by the environment to adopt a number of conformations. Depending on the peptide, this may or may not be advantageous: if the peptide does not have a structure identical to the parent epitope, the heterogeneity of proteins may induce a certain proportion of the peptides to adopt the correct conformation. On the other hand if the peptide does have the correct conformation, the homogeneity of the liposomal surface will not alter it.

An additional feature which differentiates the use of protein carriers with Freund's adjuvant and the liposomes containing MPLA is the subtype of antibodies produced. The protein carrier induces a response which is primarily IgG1, whereas the liposomes induce a significantly stronger IgG2a and IgG2b response.[49] For certain diseases, especially those in which complement fixation is important, such a switch in isotype may be critical to the efficacy of a vaccine.

IV. PERSPECTIVES

The preceding paragraphs demonstrate that liposomes can be used very effectively as carriers of peptidic antigens. Not only is a humoral response generated against the peptides, but this response can be directed specifically against a defined structure without inducing antibodies against the carrier or structures induced by the carrier. Additionally, these liposomal conjugates can be lyophilized and stored,[49] a significant advantage over aluminum-based adjuvants for vaccines.

It should also be possible to use such liposomes to generate very specific T-lymphocyte activation. The source of the T-epitopes experiment described previously by Garçon and Six[36] was a fragment from influenza hemagglutinin. In principle, any widely recognized T-epitope, such as the universal T-epitope peptide from tetanus toxoid,[50] could be used. Initial experiments from our laboratories suggest that the synthetic peptide representing residues 50–67 of influenza hemagglutinin, a highly conserved region of the protein which is recognized as a T-epitope in a wide population of mice,[51] when coupled to liposomes induces the proliferation of T-lymphocytes directed against the virus. We are currently investigating whether these lymphocytes confer protection against a challenge with the virus.

Liposomes as carriers of synthetic peptides thus lead us to the ideal vaccine: a defined B-epitope, a defined T-epitope, a chemically defined adjuvant, and all these held together in an easily formed, non-toxic, biodegradable carrier.

REFERENCES

1. Allison, A.C. and Gregoriadis, G., Liposomes as immunological adjuvants, *Nature,* 252, 252, 1974.
2. Warren, H.S. and Chedid, L.A., Future prospects for vaccine candidates, *CRC Crit. Rev. Immunol.,* 8, 83-101, 1988.
3. Gregoriadis, G., Immunological adjuvants, a role for liposomes, *Immunol. Today,* 11, 89-98, 1990.
4. Alving, C., Immunological aspects of liposomes: presentation and processing of liposomal protein and phospholipid antigens, *Biochim. Biophys. Acta,* 1113, 307-372, 1992.
5. Buiting, A., Van Rooijen, N., and Claasen, E., Liposomes as antigen carriers and adjuvants in vivo, *Res. Immunol.,* 143, 471-586, 1992.
6. Dal Monte, P. and Szoka, F., Effect of liposome encapsulation on antigen presentation in vitro: comparison of presentation by peritoneal macrophages and B cell tumours, *J. Immunol.,* 142, 1437-1440, 1989.
7. Szoka, F., The macrophage as the principal antigen presenting cell for liposome encapsulated antigen, *Res. Immunol.,* 143, 186, 1992.
8. Su, D. and Van Rooijen, N., The role of macrophages in the immunoadjuvant action of liposomes: effects of elimination of spleenic macrophages on the immune response against intra-venously injected liposome associated albumin antigen, *Immunology,* 66, 466, 1989.
9. Van Rooijen, N., Macrophages as accessory cells in the in vivo humoral response: from processing of particulate antigen to regulation by suppression, *Semin. Immunol.,* 4, 237, 1992.
10. Van Rooijen, N., The liposome mediated macrophage suicide technique, *J. Immunol. Methods,* 124, 1, 1989.
11. Van Rooijen, N., Immunoadjuvant activities of liposomes: two different macrophage mediated mechanisms, *Vaccine,* 11, 1170, 1993.

12. Van Rooijen, N., Antigen processing and presentation in vivo. The microenvironement as a crucial factor, *Immunol. Today*, 11, 436, 1990.
13. Huang, L., Reddy, R., Nair, S., Zhou, F., and Rouse, B., *Res. Immunol.*, 143, 192-196, 1992.
14. Cohen, S., Bernstein, H., Hewes, C., Chow, M., and Langer, R., The pharmocokinetics of, and humoral response to, antigen delivered by microencapsulated liposomes, *Proc. Natl. Acad. Sci. U.S.A.*, 88, 10440-10444, 1991.
15. Francis, M.J., Fry, C.M., and Rowlands, D., *J. Gen. Virol.*, 66, 2347-2354, 1985.
16. Davis, D. and Gregoriadis, G., *Immunology*, 68, 277-82, 1989.
17. Gregoriadis, G., Davis, D., and Davies, A., Liposomes as immunological adjuvants: antigen incorporation studies, *Vaccine*, 5, 145-151, 1987.
18. Therien, H., Lair, D., and Shahum, E., Liposomal vaccine: influence of antigen association on the kinetics of the humoral response, *Vaccine*, 8, 558-562, 1990.
19. Therien, H., Shahum, E., and Fortuin, A., Liposome adjuvanticity: influence of dose and protein:lipid ratio on the humoral response to encapsulated and surface linked antigen, *Cell. Immun.*, 136, 402-413, 1991.
20. Richards, R., Hayne, M., Hockmeyer, W., and Alving, C., Liposomes, lipid A and Alum enhance the immune response to a synthetic malaria sporozite antigen, *Infect. Immun.*, 56, 682-686, 1988.
21. Fries, L., Gordnay, D., Richards, R., et al., Liposomal malaria vaccine in humans: a safe and potent adjuvant strategy, *Proc. Natl. Acad. Sci. U.S.A.*, 89, 358-362, 1992.
22. Tan, L. and Gregoriadis, G., Effect of IL2 on the immunoadjuvant activity of liposomes, *Biochem. Soc. Trans.*, 17, 693-694, 1988.
23. Ho, R., Burke, R., and Merigen, T., Liposome formulated IL2 as an adjuvant of recombinant HSV glycoprotein for the treatment of recurrent genital HSV in guinea pigs, *Vaccine*, 10, 209-213, 1992.
24. Phillips, N. and Emili, A., Enhanced antibody response to liposome-associated protein antigens: preferential stimulation of IgG2a/production, *Vaccine*, 10, 151-158, 1992.
25. Harding, C., Collins, D., SLot, J., Genze, H., and Unanue, E., Liposome encapsulated antigens are processed in lysosomes, recycled and presented to T-cells, *Cell*, 64, 393-401, 1991.
26. Reddy, R., Zhou, F., Huang, L., Carbone, F., Bevan, M., and Rouse, B., pH sensitive liposomes provide an efficient means of sensitizing target cells to class I restricted CTL recognition of a soluble protein, *J. Immunol. Methods*, 141, 157-163, 1991.
27. Reddy, R., Zhou, F., Nair, S., Huang, L., and Rouse, B., In vivo cytotoxic T lymphocyte induction with soluble proteins administered in liposomes, *J. Immunol.*, 148, 1585-1589, 1992.
28. Lopes, L. and Chain, B., Liposome mediated delivery stimulates a class I-restricted cytotoxic T cell response to a soluble antigen, *Eur. J. Immunol.*, 22, 287-290, 1992.
29. Garnier, F., Forquet, F., Bertolino, P., and Gerlier, D., Enhancement of an in vivo and in vitro T cell response against measles virus haemagglutinin after its incorporation into liposomes: effect of the phospholipid composition, *Vaccine*, 9, 340-345, 1991.
30. Zhou, F. and Huang, L., Monophosphoryl lipid A enhances specific CTL induction by a soluble protein antigen entrapped in liposomes, *Vaccine*, 11, 1139-1141, 19—.
31. Kochibe, N., Nicolotti, R., Davie, J., and Kinsky, S., Stimulation and inhibition of anti-hapten response in guinea pigs immunised with hybrid liposomes, *Proc. Natl. Acad. Sci. U.S.A.*, 72, 4582-4586, 1975.
32. Dancey, G.F., Yasuda, T., and Kinsky, S., Effect of liposomal model membrane composition on immunogenicity, *J. Immunol.*, 120, 1109-1114, 1978.
33. Dancey, G.F., Isakson, P.C., and Kinsky, S., Immunogenicity of liposomal model membranes sensitised with dinitrophenylated phosphatidylethanolamine derivatives containing different length spacers, *J. Immunol.*, 122, 638-642, 1979.
34. Okayama, N., Hamano, T., Hamakawa, N., Inagaki, K., and Nakanishi, M., Membrane fluidity and lipid hapten structure of liposomes affect calcium signals in antigen specific B-cells, *Biochemistry*, 30, 11154-11156, 1991.
35. Dancey, G.F., Yasuda, T., and Kinsky, S., Enhancement of liposomal model mebrane immunogenicity by incorporation of lipid A, *J. Immunol.*, 119, 1868-1871, 1977.
36. Garçon, N. and Six, H., Universal vaccine carrier: liposomes that provide T-dependent help to weak antigens, *J. Immunol.*, 146, 3697-3702, 1991.
37. Arnon, R., Synthetic peptides as the basis for vaccine design, *Molec. Immunol.*, 28, 209-215, 1991.
38. Katz, D.H., Adaptive differentiation of lymphocytes. Theoretical implications for cell-cell recognition and regulation of the immune response, *Adv. Immunol.*, 29, 137, 1980.

39. Macquaire, F., Balex, F., Giacobbio, E., Huynh-Ginh, T., Neumann, J., and Sanson, A., *Biochemistry*, 31, 2576-2582, 1992.

40. Herzenberg, L.A. and Tokuhisha, T., Carrier priming leads to hapten specific suppression, *Nature,* 285, 664-666, 1980.

41. Briand, J.P., Muller, S., and Van Regenmortel, M.H.V., Synthetic peptides as antigens: pitfalls of conjugation methods, *J. Immunol. Methods,* 78, 59-69, 1985.

42. Friede, M.H., Muller, S., Briand, J.P., Plaué, S., Fernandes, I., Frisch, B., Schuber, F., and van Regenmortel, M.H.V., Induction of protective immunity to influenza in mice with a synthetic cyclic peptide coupled to liposomes, *Vaccine,* 1993 in press.

43. Frisch, B., Muller, S., Briand, J.P., van Regenmortel, M.H.V., and Schuber, F., Parameters affecting the immmunogenicity of liposome associated synthetic hexapeptide antigen, *Eur. J. Immunol.,* 21, 185-193, 1991.

44. Friede, M., Muller, S., Briand, J.P., van Regenmortel, M.H.V., and Schuber, F., Induction of immune response against a short synthetic peptide antigen coupled to small neutral liposomes containing monophosphoryl lipid A, *Molec. Immunol.,* 30, 539-547, 1993.

45. Hiernaux, J.R., Stasha, P., Cantrel, J., Rudbach, J., and Baker, P., Immunomodulatory activity of monophosphoryl lipid A in C3H/HeJ and C3H/HeSnJ mice, *Infect. Immun.,* 57, 1483-1490, 1989.

46. Verma, J.N., Mangala, R., Amselem, S., Krzych, U., Alving, C., Green, S., and Wassef, N., Adjuvant effects of liposomes containing lipid A: enhancement of liposome antigen presentation and recruitment of macrophages, *Infect. Immun.,* 60, 2438-2444, 1992.

47. Senior, J., Crawley, J., and Gregoriadis, G., Tissue distribution of liposomes exhibiting long half-lives in the circulation after i.v. injection, *Biochim. Biophys. Acta,* 839, 1-7, 1985.

48. Muller, S., Plaué, S., Samama, J.P., Valette, M., Briand, J.P., and van Regenmortel, M.H.V., Antigenic properties and protective capacity of a cyclic peptide corresponding to site A of influenza haemagglutinin, *Vaccine,* 8, 308-314, 1990.

49. Friede, M., van Regenmortel, M.H.V., and Schuber, F., Lyophilised liposomes as shelf items for the preparation of immunogenic liposome-peptide conjugates, *Anal. Biochem.,* 211, 117-122, 1993.

Chapter 13

pH-Sensitive Liposomes as Tools for Cytoplasmic Delivery

David Collins

CONTENTS

I. INTRODUCTION

pH-Sensitive liposomes are stable at neutral pH, but unstable and fusion-active at acidic pH values. The earliest proposed use of pH-sensitive liposomes was to take advantage of the finding that the pH in the proximity of metastatic tumors was lower than that of normal tissue.[1] Liposomes which were pH-sensitive were envisioned to release encapsulated drug as they passed through these regions of low pH.

As it became clear that liposomes were taken up by cells through the endocytic pathway,[2] several laboratories began constructing liposomes which could release drugs into cells following endocytic uptake. Viruses of a number of families including vesicular stomatitis virus, influenza virus, and Semliki virus have membrane glycoproteins which undergo conformational changes at low pH and allow endocytosed virions to fuse with cellular membranes.[3,4] Reconstitution of such viral proteins into liposomes promotes acid-induced liposome-cell fusion and delivery of the encapsulated contents into the cell cytoplasm.[5] However, reconstitution is a time-consuming and cumbersome task, so efforts in several laboratories became directed toward producing pH-sensitive liposome systems without using viral fusion proteins.

Several pH-sensitive liposome systems have been described. Excluding those constructed using viral proteins, they fall into two main categories: intrinsically pH-sensitive liposomes and those which utilize an external non-lipid trigger. Intrinsically pH-sensitive liposomes are constructed by combining phosphatidylethanolamine (PE) with one of a number of acidic amphiphiles (Figure 1). This type of liposome is the focus of this chapter. Externally triggered pH-sensitive liposomes combine an otherwise stable liposome with an external soluble component such as a titratable polymer[6,7] or a titratable synthetic peptide[8] which undergoes a conformational change upon acidification. These systems are interesting and potentially useful, but successful delivery using this type of acid-sensitivity has yet to be demonstrated.

Following a brief review of the physical/chemical characteristics of pH-sensitive liposomes, several examples of cellular delivery using these liposomes will be presented.

0-8493-4569-3/95/$0.00+$.50

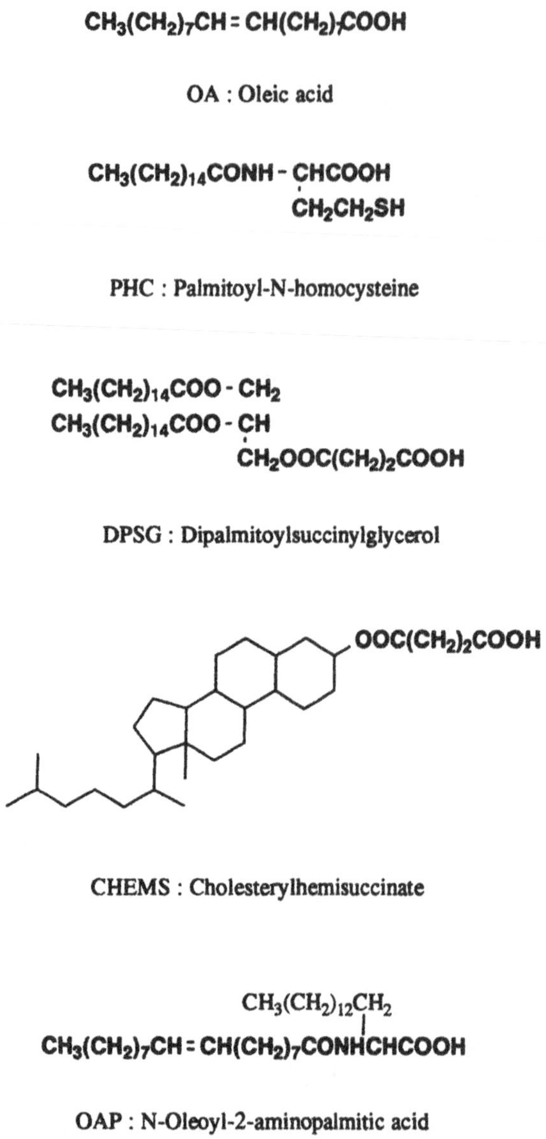

$$CH_3(CH_2)_7CH = CH(CH_2)_7COOH$$

OA : Oleic acid

$$CH_3(CH_2)_{14}CONH - \underset{|}{C}HCOOH$$
$$\phantom{CH_3(CH_2)_{14}CONH - }CH_2CH_2SH$$

PHC : Palmitoyl-N-homocysteine

$$CH_3(CH_2)_{14}COO - CH_2$$
$$CH_3(CH_2)_{14}COO - \underset{|}{C}H$$
$$\phantom{CH_3(CH_2)_{14}COO - }CH_2OOC(CH_2)_2COOH$$

DPSG : Dipalmitoylsuccinylglycerol

$$OOC(CH_2)_2COOH$$

CHEMS : Cholesterylhemisuccinate

$$CH_3(CH_2)_{12}CH_2$$
$$CH_3(CH_2)_7CH = CH(CH_2)_7CON\overset{|}{H}CHCOOH$$

OAP : N-Oleoyl-2-aminopalmitic acid

Figure 1 Examples of titratable amphiphiles used to stabilize PE and prepare pH-sensitive liposomes. The protonated form of the amphiphiles is shown.

II. STABILIZATION AND DESTABILIZATION OF PH-SENSITIVE LIPOSOMES

Construction of pH-sensitive liposomes takes advantage of the polymorphic phase behavior of unsaturated PE which forms the inverted hexagonal (H_{II}) phase[9-11] (see also Chapter 7) rather than bilayers under physiological conditions of pH and temperature. PE vesicles can be formed at high pH (≥ 9.0) due to the deprotonation of the amine group. This leads to an overall negative charge and the generation of electrostatic repulsions between adjacent PE molecules and between PE bilayers,[12-14] inhibiting H_{II} phase formation. PE can also form bilayers at temperatures well below the H_{II} phase transition temperature (T_H).[13] Increasing the temperature of these samples above the T_H leads to collapse into H_{II} phase. The ability to make pure PE liposomes is limited by the extreme conditions required for the feat and such liposomes are of limited usefulness due to their inherent instability.

Stabilization of PE into bilayers at neutral pH and construction of pH sensitive liposomes can be achieved by using one of several titratable, acidic lipids which are negatively charged at neutral pH values. Figure 1 shows the structure of some of the lipids which can be used to stabilize PE and form pH-sensitive liposomes. In their deprotonated forms, the stabilizers interfere with H_{II} phase formation by inhibiting hydrogen bonding between PE molecules and by increasing interbilayer repulsions. When the pH of the media is reduced, the headgroups are protonated. This neutralizes the negative charge and

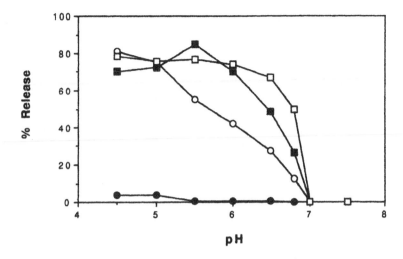

Figure 2 Destabilization of pH-sensitive liposomes of different compositions. Liposomes containing [^{125}I]-HEL were incubated 1 min. at the indicated pH values. After ultracentrifugation (100,000 × g) to pellet liposomes, the% release was determined by counting the supernatants and pellets from liposomes composed of (□) DOPE:OA, (○) DOPE:DPSG, and (■) DOPE: PHC . All mol ratios are 4:1 (DOPE:amphiphile). The leakage from DOPC:DOPS (●) pH-insensitive liposomes is shown for comparison.

permits close apposition of the bilayers, allowing the PE to adopt the H_{II} phase. The intermediates in the transition from bilayers to H_{II} phase are thought to be involved in the process of membrane fusion.[12-14]

pH-sensitive liposomes have been prepared using a variety of techniques including reverse phase evaporation (REV),[12] dehydration-rehydration (DRV),[15] sonication,[16] detergent dialysis,[17] freeze-thaw,[18] and extrusion.[19] In general, the interested investigator should choose the method most compatible with the drug to be encapsulated. The most important precaution to be taken is to monitor the pH of the liposome samples during preparation. Failure to form liposomes or low drug capture will result if a pH ≥7.0 is not maintained. Another parameter which is important is the divalent cation concentration. It has been shown that pH-sensitive liposomes are destabilized by divalent cations such as Ca^{2+} and Mg^{2+}, for this reason inclusion of EDTA or EGTA in the preparation buffer is recommended.

In general, forming pH-sensitive liposomes requires from 15–30 mol% of an amphiphilic lipid, like those shown in Figure 1, combined with unsaturated PE. As mentioned above, various stabilizers have been used which impart different properties on the liposomes. For example, double chain stabilizers can be used to prepare plasma stable pH-sensitive liposomes (see below). Amphiphiles with different pKs can be exploited to prepare liposomes which destabilize at different pH values (Figure 2).

Liposomes prepared using dioleoyl PE (DOPE) and palmitoyl homocysteine (PHC) were first described by Connor et al.[16] The liposomes were prepared by sonication and contained 20 mol% PHC. Fusion of these liposomes at low pH was demonstrated using gel filtration, electron microscopy, and membrane mixing. By assaying calcein leakage, the authors showed that the acid-induced fusion was a leaky process. Inclusion of cholesterol made the fusion less leaky.[16]

The PE:cholesterylhemisuccinate (CHEMS) system is another well-characterized pH-sensitive liposome system. Acid-induced lipid mixing of TPE:CHEMS membranes was shown to occur at pH ≤5.5 using the NBD-PE/Rh-PE resonance energy transfer assay.[20] Ellens et al.[21] analyzed the destabilization process of TPE:CHEMS at low pH and found that aggregation and interbilayer contact occurred prior to leakage and destabilization. By differential scanning calorimetry (DSC) and ^{31}P nmr it was shown that at concentrations of ≥16 mol% CHEMS abolishes the T_H of PE at neutral pH.[20] By contrast, at acidic pH values CHEMS actually enhances H_{II} phase formation. No acid-induced mixing of aqueous contents could be demonstrated for TPE:CHEMS, however divalent cations could induce liposome fusion at neutral pH.[12] Ca^{2+} and Mg^{2+} induced fusion was abolished at pH 4.5, where the CHEMS is presumably uncharged.[12]

Fatty acids, such as oleic acid (OA),[22] have also been combined with PE to prepare pH-sensitive liposomes. PE:OA (7:3) liposomes were shown to undergo changes in turbidity, leakage, content mixing (fusion), and membrane mixing at pH ≤6.5.[22] The leakage rate and the rate of lipid mixing of PE:OA in acidified medium was faster than the fusion (content mixing) rate. This indicates that both leakage and lipid mixing are the first destabilization processes which occur in this system upon aggregation, and that they compete with and occur prior to significant fusion. PE:OA membranes did not fuse with PC:OA membranes, even when PC:OA liposomes were in excess.

Nayer and Schroit[23] synthesized a carboxylated derivative of PE, N-succinyldioleoyl-phosphatidylethanolamine (COPE) using succinic anhydride. COPE liposomes were stable at pH 4.0, but leaky at pH 7.4. This behavior appears to arise from packing defects induced by the deprotonated succinyl groups at pH 7.4, yielding a net charge of −2. Unlike pure COPE vesicles, DOPE:COPE (7:3) liposomes were leaky in acidic conditions and stable at neutral pH values. In contrast to other types of pH-sensitive liposomes, leakage from DOPE:COPE liposomes was independent of aggregation, fusion, and lipid mixing, indicating that acid-induced rearrangements of the liposomal lipids were responsible for the leakage observed.

Double chain amphiphiles, such as N-oleoyl aminopalmitic acid (OAP)[24] and 1,2-dipalmitoyl-succinylglycerol (1,2-DPSG),[25] have been synthesized and used in combination with PE to form pH-sensitive liposomes. Depending on the amphiphile, stable liposomes can be formed by including 10–25 mol% of the synthetic lipids in the membranes. The liposomes are destabilized at low pH and exhibit leakage, lipid mixing, and content mixing under acidic conditions. DSC analysis of the PE:amphiphile mixtures failed to demonstrate a correlation between H_{II} phase formation and destabilization at low pH.[24,25] One exception to this conclusion was found when 1,2-DPSG was shown to slightly increase the T_H of DEPE at pH 7.4 but decrease the T_H at low pH.[25]

III. CYTOPLASMIC DELIVERY WITH pH-SENSITIVE LIPOSOMES

A. ENDOCYTOSIS

Liposomes enter cells via endocytosis. This vesicular pathway is the main route for cellular uptake of nutrients and exogenous macromolecules and is composed of compartments which are kinetically and biochemically distinct. After binding to the cell surface, liposomes are internalized into endosomes (Figure 3),where they encounter a pH more acidic than the external milieu. Early endosomes, reached within 5 min, generally have an internal pH of ~6.5.[26,27] Through either maturation or vesicular fusion, the early endosome contents are transferred to a more acidic environment as endocytosis proceeds. The late endosome environment, with an internal pH of 5.5–6.0, is reached after 10–15 min of uptake. The last endocytic compartment, the lysosome, is further acidified to pH values of 5.0 or lower, and is reached at uptake times longer than 20 min. The lysosome is the main degradative compartment in the endocytic pathway and contains a wide range of proteolytic enzymes as well as disulfide reducing activity in some cell types.[28]

Conventional, pH-insensitive liposomes are delivered to lysosomes via endocytosis and degraded.[2] Some drugs, in particular macromolecules, are extremely sensitive to lysosomal degradation and may be inactivated if delivered into this hydrolytic compartment. By using acid sensitive liposomes, encapsulated molecules can be released prior to arriving in lysosomes.

B. UPTAKE OF pH-SENSITIVE LIPOSOMES

In general, pH-sensitive liposomes are taken up more efficiently than pH-insensitive liposomes.[29,30] The reason for this phenomenon is unknown, but may be related to the propensity of PE-based liposomes to adhere to each other due to the poor hydration of the PE headgroup. PE-containing liposomes may therefore have a higher affinity for cell membranes than PC-based pH-insensitive liposomes. Regardless of the exact mechanism, one must consider the relative uptake of pH-sensitive vs. pH-insensitive liposomes when designing and interpreting in vitro delivery experiments.

C. DELIVERY OF FLUORESCENT DYES

Using DOPE:PHC liposomes, Connor and Huang demonstrated cytosolic delivery of encapsulated calcein to murine L-929 fibroblasts.[31] They used liposomes targeted with monoclonal antibodies to the mouse major histocompatability complex (MHC) class I antigen (H-2K^k) haplotype of L-929 cells. No liposome uptake was demonstrated for murine A-31 cells (H-2K^d).[31] While incubation of cells with acid-insensitive, PC immunoliposomes yielded only punctate fluorescence, cells incubated with DOPE:PHC immunoliposomes showed diffuse, cytoplasmic fluorescence. Endosome acidification was required for

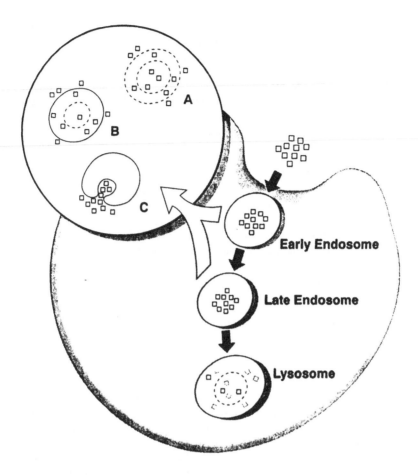

Figure 3 Uptake and delivery by pH-sensitive liposomes. Liposomes are endocytosed through either coated or uncoated pits in the cell surface. The liposomes pass through early endosomes, late endosomes, and finally lysosomes as endocytosis proceeds. Liposomes which are unstable in the pH range of the endosomes can have three fates, each leading to cytoplasmic delivery of encapsulated drug: (**A**) pH-sensitive liposomes destabilize or rupture endosome membranes, (**B**) leak into the endosome lumen, (**C**) or fuse with the endosome membrane. Liposomes which are pH-insensitive can pass through endosomes intact (see Reference 30) but are degraded in lysosomes.

liposome release as demonstrated by the fact that ammonium chloride and chloroquine blocked delivery of calcein. Despite efficient cell surface binding, no intracellular release was observed at 4°C where endocytosis is inhibited.[31]

Liposomes composed of TPE:OA (7:3) were shown to be capable of mediating cytoplasmic delivery of calcein in CV-1 cells, while uptake of control liposomes composed of either PS:PE or PC:OA yielded only punctate fluorescence in the cells.[19] Cytosolic delivery of calcein by PE:OA liposomes was not inhibited significantly by the inclusion of cholesterol (up to 30 mol%) in the liposome membranes, however weak bases such as ammonium chloride were shown to reduce cytoplasmic delivery of calcein by PE:OA liposomes, indicating the importance of endosome acidification for delivery.[19] When calcein was replaced with fluoresceinated dextrans (18 and 40 kDa), only punctate fluorescence was noted, even for pH-sensitive PE:OA liposomes. Cytosolic delivery of the dextrans could be achieved if cells which had endocytosed liposomes were osmotically shocked using glycerol. This finding suggests that cytoplasmic delivery of large molecules by liposomes is less efficient than delivery of small molecules such as calcein and that liposome-endosome fusion in this system is an infrequent event.

In a very thorough study,[29] the delivery of fluorescent compounds using DOPE:CHEMS (3:2) was compared with delivery using DOPC:CHEMS. DOPE:CHEMS was able to deliver calcein, FITC dextran, and HPTS into the cytoplasm of macrophage-like cell lines (P-388 and RAW 264.7). Compared to liposomes composed of DOPE:OA, DOPE:CHEMS liposomes were more efficient for cytosolic delivery of calcein. By using 8-hydroxypyrene-1,3,6-trisulfonate (HPTS), the cytosolic location of the delivered dye was unambiguously shown. The fluorescence spectrum of HPTS is pH sensitive.[32] Using an excitation wavelength of 450 nm, the fluorescence of the dye increases dramatically as the pH increases from 6 to 8. The isobestic point for the pH-dependent phenomenon is 413 nm. Therefore, if the ratio of fluorescence emission using 450 nm excitation is compared to that obtained after excitation at 413 nm (R = F450/F413), an indication of the pH experienced by the probe is obtained. Cells which had been incubated with HPTS-containing DOPE:CHEMS liposomes were examined using an excitation filter in the range of 450–490 nm and exhibited diffuse cytoplasmic fluorescence, indicating that the dye was not at an acidic pH. If an excitation filter in the range of 350–410 nm was used, punctate as well as cytoplasmic fluorescence was observed.

In contrast to the findings above, Brown and Silvius[33] observed no cytoplasmic fluorescence in CV-1 cells incubated with pH-sensitive liposomes containing calcein or FITC-dextrans. They compared delivery using PE:OA and PE:OAP to delivery with pH-insensitive liposomes composed of PC:PG:cholesterol. Cells incubated with either pH-sensitive or pH-insensitive liposomes showed only punctate fluorescence, indicating that the liposomes were unable to mediate cytoplasmic delivery under the conditions used. The liposomes in this study were conjugated to transferrin to increase their uptake by the cells. However, the uptake pathway taken by liposome-conjugated transferrin may be significantly different than that of free protein. In some cases, liposomes which are endocytosed by cells can be seen in non-acidified 'sacs' near the periphery of cells.[30] If transferrin conjugation directs liposomes to these sacs, pH-sensitive liposomes cannot be expected to release their contents or mediate cytoplasmic delivery of dye. This remains a possibility in this study, since acidification of the liposomes after uptake was not demonstrated.

D. DELIVERY OF ARA C

The anti-tumor drug ara-C was encapsulated in pH-sensitive liposomes composed of DOPE:OA which had been prepared to contain a fatty acid-derivatized antibody directed toward the murine H-2K^k MHC antigen.[33] These pH-sensitive immunoliposomes increased the cytotoxic effect of the drug toward target L-929 cells over that of free ara C, while they were non-toxic towards non-target A-31 cells (H-2K^d). Antibody-free liposomes were less effective than free drug. DOPC liposomes and immunoliposomes containing ara C were toxic only at very high drug concentrations. The delivery by the pH-sensitive immunoliposomes was blocked by pretreating the cells with ammonium chloride and chloroquine, indicating that acidification of the liposomes was required for delivery.

Ara C delivery by pH-sensitive and pH-insensitive liposomes bearing surface conjugated transferrin has also been investigated.[32] Ara C encapsulated in liposomes composed of TPE:OAP-serine:EPC:chol:SATA-PE (pH-sensitive) and EPC:TPG:chol:SATA-PE (pH-insensitive) liposomes were more cytotoxic towards CV-1 cells than free drug. The liposomes used in this study were less efficient for ara C delivery than those used by Connor and Huang.[34] In the latter case, 50% inhibition of thymidine incorporation occurred at liposomal ara C concentrations of ~2 μM using liposomes targeted with monoclonal antibodies. By contrast, Brown and Silvius report 50% inhibition at 30 μM ara C encapsulated in pH-sensitive liposomes targeted using transferrin.[33] The less efficient delivery of ara C may be due to less efficient cell binding by the conjugated transferrin or could arise from different ara C sensitivities for the cells used. The mechanism of delivery of ara C by the different liposome compositions may also differ. In the case of DOPE:OA it is clear that endosome acidification is required as well as release from a site earlier than the cellular lysosome.[19,34] For the transferrin-conjugated liposomes, the route of delivery is much less clear. Delivery by some of the liposome preparations is inhibited by the nucleoside transport inhibitor, nitrobenzothioinosine, suggesting that active transport of drug may be involved in liposomal delivery of ara C.[32] The putative transporter may be located in either endosomes or lysosomes and was suggested to be the main route by which liposomal ara C gains access to the cytoplasm. The intracellular events involved in release of ara C from the liposomes are also unknown. Specifically, it is unclear whether acidification of the liposomes after uptake by CV-1 cells is required for delivery.

E. DELIVERY OF A BACTERIAL TOXIN

Cytoplasmic delivery of protein toxins using pH-sensitive liposomes has been well documented.[15,25,29,36] DTA is the cytotoxic fragment of diphtheria toxin which inhibits eukaryotic protein synthesis by catalyzing the ADP-ribosylation of elongation factor 2. DTA alone is not toxic to cells, since it is unable to cross lipid membranes in the absence of the binding and membrane translocation activity located in the B fragment of the toxin.[35] Furthermore, murine cells are resistant to diphtheria toxin even though they bind and endocytose the toxin and possess a DTA-sensitive elongation factor.[35] The resistance of murine cells to diphtheria toxin arises from a block in the translocation of DTA across the endosome membrane. DTA was encapsulated in pH-sensitive immunoliposomes composed of DOPE:OA (4:1) and bearing surface-conjugated anti-H-2K^k antibody. These immunoliposomes were able to bypass the normal block in DTA translocation and kill L929 cells (H-2K^k), with an IC$_{50}$ for the liposomes of ~1.4 × 10^{-9} M DTA.[15] The liposomes were shown to be target-specific since toxicity of the liposomes was only seen in L929 cells and not A31 cells (H-2K^d) or Vero cells. The liposome-mediated toxicity was dependent on endosome acidification since ammonium chloride and chloroquine inhibited cytoplasmic delivery of the toxin. pH-insensitive liposomes composed of DOPC were non-toxic.[15]

DOPE:CHEMS (2:1) liposomes were shown to be more efficient than DOPE:OA for cytoplasmic delivery of DTA.[29] P388 D1 cells were killed by DOPE:CHEMS liposomes containing DTA in a process which involved a low pH pathway. The efficiency of delivery of DTA by DOPE:CHEMS was calculated to be at least 0.01% of cell-associated liposomal DTA and an IC$_{50}$ of 7 × 10^{-10} M was demonstrated for this composition. This difference in efficiency between DOPE:OA and DOPE:CHEMS may be due to the macrophage-like nature of the P388 cells as compared to L929 cells. However, the enhanced stability of DOPE:CHEMS liposomes in the presence of cells as compared to the liposomes composed of DOPE:OA may also play a role in the apparently more efficient delivery obtained with the former composition.

Liposomes composed of DOPE:DPSG and DOPE:DOSG were also shown to be able to deliver DTA to L929 cells.[25,36] DOPE:DPSG liposomes were more efficient for DTA delivery than DOPE:DOSG (IC$_{50}$ ~1 × 10^{-9} M vs. 1 × 10^{-8} M). The reason for this difference is not known, but may be related to the greater stability of DOPE:DPSG liposomes compared to DOPE:DOSG.

F. PROBING THE ENDOCYTIC PATHWAY WITH pH-SENSITIVE LIPOSOMES

The different vesicular organelles of the endocytic pathway differ in their internal pH.[26,27] pH-sensitive liposomes can be prepared to have different sensitivities to acid in the range of pH 5-6.9 and are therefore a useful tool for studying endocytosis. This is illustrated in Figure 2 for liposomes composed of DOPE:OA, DOPE:PHC, and DOPE:DPSG. For DOPE:OA the pH at which 50% release occurs (pH$_{50}$) is approximately 6.8, while for DOPE:PHC and DOPE:DPSG the pH$_{50}$ values are ~6.5 and 5.7, respectively. Immunoliposomes of these compositions were prepared to contain DTA and the t$_{1/2}$ (time required for the DTA immunoliposomes to inhibit protein synthesis by 50%) was assayed.[36] Immunoliposomes bearing anti-H-2K^k antibodies were incubated with L929 cells at 4°C, under conditions where binding but not uptake takes place, then endocytosis was initiated by rapidly warming the cells to 37°C. NH$_4$Cl was added at various times after warm-up to inhibit endosome acidification and thus release from pH-sensitive liposomes.[15,36] Liposomes composed of DOPE:PHC exhibited a t$_{1/2}$ for delivery of ~5 min, as NH$_4$Cl was unable to block the toxicity of DTA at longer time points.[36] By contrast, DTA delivery by DOPE:DPSG liposomes took longer (t$_{1/2}$ ~25 min). These data suggest that liposomes reach pH ~6.5 within 5 min after uptake while ~25 min of uptake is required to reach pH ~5.5–5.7. Using an independent method for determining the average pH encountered by endocytosed liposomes, we found that delivery of DTA occurred faster than the acidification of the liposomes.[36] This indicates that DTA delivery may occur from the subpopulation of liposomes which first reaches a compartment with sufficient acidity. These results also suggest that it is very difficult to "synchronize" liposome uptake.

DOPE:OA liposomes exhibit a pH$_{50}$ of ~6.8-6.9 but a t$_{1/2}$ for DTA delivery of 15 min, which is intermediate between that of DOPE:DPSG and DOPE:PHC.[36] This result was obtained due to cell-induced alterations of the DOPE:OA liposomes. Specifically, 29% of liposomal OA became cell associated after 30 min incubation at 37°C. Besides liposome-to-cell transfer of OA other alterations occurred upon incubation of DOPE:OA with cells. After cell incubation, the liposomes became leaky and exhibited an alteration in pH$_{50}$ (from 6.9 to 6.0) which was probably responsible for the unexpected t$_{1/2}$ for DTA delivery. The data in this study are consistent with a model in which DOPE:PHC liposomes release at early endosomes and DOPE:DPSG liposomes release their contents at late endosomes (Figure 3). Liposomes composed of DOPE:OA, after being altered by interactions with the cells, release in the late endosome.

Early endosomal release has been confirmed for DOPE:PHC using an independent technique.[30] Anti-FITC antibody and FITC-dextran-biotin were encapsulated in separate sets of DOPE:PHC liposomes and incubated with peritoneal macrophages. Release of internalized liposomal contents into endosome or lysosomes allowed formation of immune complexes. The cells were then lysed in the presence of an excess of FITC-dextran and any immune complexes which formed were precipitated using protein A Sepharose. The precipitated material was then assayed using an enzymatic assay (avidin-B-galactosidase) which detected immune complexes containing biotinylated FITC-dextran. Intracellular release by DOPE:PHC was detectable within 5–10 min of uptake, consistent with release in early endosomes. By contrast, pH-insensitive liposomes composed of DOPC:dioleoylphosphatidylserine (DOPS) showed significant release only after prolonged uptake (30–120 min), a result consistent with lysosomal release. The sites of release were further defined by homogenizing macrophages which had been allowed to endocytose liposomes and fractionating the cell homogenates on Percoll density gradients to separate endosomes (light density) from lysosomes (heavy density). The subcellular fractions were then assayed for liposome release. DOPE:PHC liposomes released within 5–10 min in light density fractions which contained endosomes, while pH-insensitive liposomes exhibited significant release only in heavy density, lysosomal fractions which were reached after long (>30 min) incubations.

Using the same assay, DOPE:CHEMS liposomes were shown to release their contents within minutes of cellular uptake,[37] suggesting that this composition, like DOPE:PHC, released in early endosomes. Furthermore, DOPE:CHEMS liposomes endocytosed by macrophage cells in a 5 min pulse exhibited virtually no increase in release over 25 min of incubation at 37°C.[37] This result showed that the DOPE:CHEMS liposomes released within 5 min, consistent with early endosomal release.

Using immunoelectron microscopy, the intracellular site of release by DOPE:PHC and DOPC:DOPS liposomes was confirmed.[30] Hen egg lysozyme (HEL) was encapsulated in liposomes and incubated with peritoneal macrophages for 10–30 min at 37°C. Following fixation and embedding of the cells, HEL was located using anti-HEL antibodies. DOPE:PHC liposomes which were located in endosomes were swollen and disrupted and contained only low amounts of lysozyme, while DOPC:DOPS liposomes remained intact in endosomes. DOPC:DOPS liposomes in lysosomes appeared as large onion-like structures, consistent with their destabilization in the lysosomal environment. Interestingly, both types of liposomes were found in large intracellular sac-like structures near the surface of the macrophage cells. These sacs are not extensively acidified, since DOPE:PHC liposomes remain intact in these structures.[30]

G. pH-SENSITIVE LIPOSOMES IN ANTIGEN PROCESSING

Protein antigens are not presented intact by the immune system. Instead, they are partially catabolized into peptide fragments and these fragments are presented on MHC molecules for recognition by T cells.[38] In general, presentation of peptide on class II MHC occurs after endocytosis of antigen. Peptide-MHC II complexes stimulate CD4+ T cells and a humoral immune response. Presentation in the class I MHC pathway occurs after cytosolic processing of endogenously synthesized antigen. Peptide-MHC I complexes stimulate CD8+ T cells and a cellular immune response.

Endocytosis is a key step in antigen processing for class II presentation. Liposomes, and in particular pH-sensitive liposomes, have been useful tools for dissection of this important pathway. As previously demonstrated, liposomes of differing acid sensitivity have been shown to release at different endosome/lysosome sites after uptake.[36] Harding et al.[30] took advantage of this finding to determine the intracellular site of antigen processing for class II presentation by peritoneal macrophages (see above). Despite higher uptake by cells, the pH-sensitive composition, DOPE:PHC, was less effective than pH-insensitive DOPC:DOPS liposomes at delivering antigen into the antigen processing compartment for presentation on class II MHC. These data suggest that the macrophage lysosome is involved in antigen processing in the class II pathway, and that a retrograde pathway from lysosomes to the cell surface is responsible for delivery of antigenic peptides to the site of MHC II-T cell recognition.[30]

The majority of antigen encapsulated within liposomes, either pH-sensitive or pH-insensitive, is released and processed within cellular endosomes and presented on class II MHC. However, some fraction of encapsulated antigen undoubtedly reaches the cytosol when pH-sensitive liposomes are used. Accordingly, two groups working independently have shown that pH-sensitive liposomes can deliver exogenous antigen into the cytosolic class I antigen processing pathway in vitro.[39,40] The in vitro class I processing and presentation of liposome-encapsulated ovalbumin has been demonstrated for DOPE:PHC,[39] DOPE:CHEMS,[37] DOPE:DPSG,[37] and DOPE:DOSG.[37,40] By contrast, pH-insensitive liposomes composed of DOPC:DOPS[37,39] and DOPC:DOPS:CHOL[40] did not deliver antigen into the class I pathway in these in vitro studies. In the case of DOPE:PHC-encapsulated ovalbumin, class I processing and presentation was much less efficient than class II processing and presentation of the same antigen.[39] This suggests that cytosolic delivery of antigen by DOPE:PHC is an inefficient process. However, this conclusion must be made with caution since the relative efficiencies of the two pathways and the sensitivities of the different T hybridomas have not been quantitated and may differ significantly.

The successful delivery of antigen into the class I MHC pathway *in vitro* led the two groups to investigate the *in vivo* activity of pH-sensitive liposome-encapsulated antigen. *In vitro*, MHC class I processing and presentation were seen for ovalbumin encapsulated in pH-sensitive but not pH-insensitive liposomes. However, the *in vivo* story was different, since both pH-sensitive and pH-insensitive liposomes containing ovalbumin were able to elicit a cytotoxic T cell response.[37,41] The response was ovalbumin specific, directed toward EL4 cells transfected with ovalbumin (E.G7-OVA), but not untransfected EL4.[37,41] The response was MHC class I restricted since the lysis of the cells was blocked by antibody to class I MHC and EL4 is a MHC class II negative cell line.[37] The response was not due to anti-ovalbumin antibody production.[41] Zhou et al.[41] found that ovalbumin covalently conjugated to the liposome membrane was as effective as encapsulated antigen for stimulating a cytotoxic T cell response *in vivo*. Macrophage cells were responsible for the processing of liposomal antigen *in vivo*, since the macrophage-specific killer, liposomal dichloromethylene-diphosphonate, abolished the response to liposomal ovalbumin.[41] Zhou et al. also demonstrated that immunization of mice with ovalbumin-containing pH-sensitive liposomes protected mice against tumors induced by E.G7-OVA cells.[42] Vaccination with liposome-encapsulated antigen prolonged the survival of mice injected with E.G7-OVA cells but not untransfected EL4 cells.

H. DELIVERY OF GENETIC MATERIAL

Gene therapy, the correction of genetic disorders by introduction of exogenous genes, has become an important field of research with great potential. Tremendous advances in molecular biology in recent years have made it possible to isolate and clone a large number of genes that may have therapeutic benefit. However, obtaining a particular gene is only part of the problem. Also important are efficient, non-toxic delivery methods for introducing genes into the desired cell or tissue type.

Retroviruses have been suggested as possible gene carriers. While the efficiency of retroviral gene transfer can be quite high, the safety of the technique is still an open question. By contrast, pH-sensitive liposomes exhibit low toxicity, and may be suitable alternatives to retroviral gene carriers.

Liposomes composed of DOPE:OA:chol were shown to be capable of transfecting mouse Ltk- (cells lacking thymidine kinase) with an exogenous thymidine kinase (TK) gene.[43] The plasmid delivered by the liposomes, pPCTK-6A, contained the TK gene under the control of a cAMP-dependent promoter, allowing controlled expression of the enzyme. In this study, pH-sensitive liposomes were 8-fold more efficient in delivery of the gene than pH-insensitive liposomes. Under optimal conditions, the efficiency of transformation by pH-sensitive liposomes was 47% in short term assays (24 hr) and 2% for long term transformation (up to 30 days). These levels were much higher than those attainable by the calcium phosphate method. This system has also been exploited *in vivo* for targeted delivery of a bacterial gene to ascites tumors.[17] The liposomes were shown to be target-specific and an absolute requirement for the pH-sensitive liposome composition was also demonstrated.

Liposomes composed of DOPE:CHEMS have also been shown to mediate gene transfer *in vitro*.[44] Legendre and Szoka[44] compared the transfection efficiency mediated by pH-sensitive, pH-insensitive (PS:chol and DOPC:CHEMS) and cationic (DOPE: N-[1-(2-3 dioleoyloxy)]-propyl-N,N,N-trimethylammonium chloride (DOTMA)) liposomes using two different genes and five different cell types. For all cell types investigated, cationic liposomes mediated the highest transfection level, while pH-sensitive liposomes mediated gene transfer efficiencies of 1-30% of that obtained with DOPE:DOTMA liposomes. pH-insensitive liposomes did not induce transfection in any case. While less efficient than cationic liposomes, pH-sensitive liposomes still may offer some advantages. Cationic liposomes are more toxic than pH-sensitive liposomes and the ability to target cationic liposomes has not been demonstrated.

Pancreatic islet cells have been transfected with a viral oncogene (v-src) *in vitro* using pH-sensitive liposomes.[46] In this study, DNA was not encapsulated, but was associated with the surface of liposomes composed of either PC:chol:stearylamine (pH-insensitive), PE:chol:OA (pH-sensitive), or PE:DOTMA (Lipofectin®, cationic). After exposure to the liposome-DNA complexes, the expression of the src gene was assessed. Cationic Lipofectin® liposomes were more efficient in transfecting dispersed islet cells than were pH-sensitive liposomes. pH-insensitive liposomes also mediated transfection of islet cells, but not intact islets.[46] By contrast, Lipofectin® and pH-sensitive liposomes were equally effective for transfecting intact islets.

Antisense oligonucleotides, short lengths of DNA or RNA complementary to a specific gene or mRNA, offer the exciting possibility of inhibiting the expression of a targeted gene without affecting the function of other genes. The delivery of antisense oligonucleotides is therefore an important goal. Since cytoplasmic delivery of antisense oligonucleotides is necessary for activity, pH-sensitive liposomes may play an important role in the full realization of the technology's potential. The delivery of antisense DNA by pH-sensitive liposomes has recently been demonstrated.[45] In this study, an antisense oligonucleotide directed against the env gene of Friend retrovirus was delivered using DOPE:OA:chol liposomes. The ED50 for the antisense DNA was lowered significantly when encapsulated in pH-sensitive liposomes in NT3 cells which were chronically infected with the retrovirus.[45]

The double-stranded RNA, polyriboinosinic acid-polyribocytidylic acid (Poly IC) induces interferon alpha and beta genes.[47] As such, poly IC-like molecules may have powerful anti-tumor[48] and anti-viral[49] effects. Since the site of poly IC action may be the cytoplasm, pH-sensitive liposomes were used to deliver poly IC into L929 cells and Hela cells.[50] Poly IC encapsulated in pH-insensitive liposomes was unable to stimulate interferon production in either cell type, however pH-sensitive liposomes stimulated interferon production with minimal toxicity to the cells. The stimulation of interferon occurred at much lower doses with poly IC in pH-sensitive liposomes as compared to free (unencapsulated) poly IC.[50]

I. MECHANISM OF DELIVERY

The possible mechanisms for delivery used by pH-sensitive liposomes are illustrated in Figure 3. Association of the liposomes (or immunoliposomes) leads to their uptake via endocytosis. Upon encountering the acidified environment of the endosomal lumen the liposomes are destabilized. The liposomes then (A) destabilize or rupture the endosome membrane, (B) release their contents into the endosomal lumen, or (C) fuse with the endosome membrane. While no mechanism has been conclusively ruled out, liposome-endosome fusion is unlikely. In support of this view, pH-sensitive liposomes were shown to be unable to fuse with pH-insensitive membranes.[22] The fact that DNA can be delivered when associated with the surface of pH-sensitive liposomes[43,46] supports the idea that endosome destabilization may be involved. However, the alternate mechanisms illustrated in Figure 3 are not mutually exclusive and all may contribute to the observed cytosolic delivery.

IV. pH-SENSITIVE LIPOSOMES *IN VIVO*

Although pH-sensitive liposomes have great potential as drug carriers, the poor *in vivo* stability of the liposomes has hampered their widespread development and use. Early studies using large, unilamellar DOPE:OA liposomes (prepared by REV) showed that these liposomes leak rapidly (within 5 min) in plasma at 37°C[51] and are rapidly cleared by spleen and liver after intravenous injection. DOPC:OA pH-insensitive liposomes were also quite leaky in plasma, but to a lesser extent.[51]

By contrast, small (<0.2 μm) DOPE:OA liposomes are stable in plasma at 37°C but unstable in PBS.[52] Furthermore, while albumin could lyse small DOPE:OA vesicles in PBS, pre-incubation of the vesicles in plasma abolished the albumin-induced destabilization.[52] Stabilization of liposomes by plasma occurred very rapidly (~1 min) and was absolutely dependent on the liposome size: liposomes >0.2 μm were not stabilized by plasma. The mechanism of liposome stabilization was determined to consist of insertion of plasma proteins into the highly curved, small liposome membranes and extraction of the OA (~70% of the original OA is extracted from the liposomes after 1 hr in plasma).[53] The liposome lipid composition was further altered by the incorporation of plasma lipids into the liposome membranes.[53] Associated with the changed lipid composition is a loss of acid sensitivity upon pre-incubation of DOPE:OA liposomes in plasma.[53] After pre-incubation in plasma, no release of encapsulated contents was observed down to pH 3.0.

If cholesterol is included in DOPE:OA membranes the liposomes are more stable in plasma, however some loss of pH-dependent fusion activity occurs.[54] The mechanism by which cholesterol enhances DOPE:OA stability in serum is independent of OA transfer to plasma: liposomes prepared with 40% cholesterol showed greater OA extraction by plasma than liposomes without cholesterol.[54] Cholesterol seems to inhibit the insertion of plasma proteins into the liposome bilayer by altering lipid packing.

Based on these findings, it would seem that a titratable stabilizer which resisted extraction in the presence of plasma proteins would be ideal for producing pH-sensitive liposomes with high *in vivo* stability. Leventis et al. examined the effects of lipid membranes, albumin, and serum on liposomes composed of TPE:OAP.[24] POPC:DOPG sonicated vesicles caused very little leakage from TPE:OAP liposomes. In the presence of serum albumin, faster initial leakage rates were obtained in the first minute of incubation. However, the leakage rate fell to zero after release of about 10% of the contents. In the presence of serum OAP remains associated with the liposomes and is not extracted by serum proteins as is the case with OA. Despite the lack of stabilizer extraction, TPE:OAP liposomes are unstable in serum, exhibiting a leakage rate of ~5–7%/min in 25% serum and a cumulative release of ~40% at incubation times longer than 2 min.[24]

By contrast, liposomes prepared with PE and DPSG are stable in plasma.[25] Like OAP, DPSG is relatively resistant to extraction by serum proteins when compared to fatty acids. Liu and Huang[55] show that ~40% of DPSG is removed from DOPE:DPSG (4:1) liposomes upon incubation in plasma, however, since the total lipid recovery was not assayed this may not be a conclusive finding. Even if extraction of this amount of amphiphile does occur, sufficient DPSG remains to stabilize DOPE into bilayers, since stabilization of DOPE into bilayers requires only 10 mol% DPSG.[25] DOPE:DPSG and DOPE:DOSG liposomes remain acid sensitive even after prolonged incubation in plasma.[25] This finding suggests that though some plasma-induced alteration of the liposomes may occur, the changes are not as drastic as for DOPE:OA liposomes. In contrast to liposomes containing cholesterol, DOPE:DPSG liposomes allow serum stability without sacrificing pH sensitivity.

Stability in plasma is only part of the problem facing the scientist interested in using pH-sensitive liposomes for *in vivo* drug delivery. The rapid clearance and accumulation of most types of liposomes in spleen, liver and lung after i.v. injection presents a major barrier to their widespread use *in vivo*.[56-59] An important recent advance is the development of liposomes which exhibit reticuloendothelial system (RES) avoidance and have been named Stealth® liposomes.[56,57] Stealth® liposomes are constructed using either GM_1 or one of several polyethylene glycol (PEG) conjugated lipids. While the exact mechanism of liposome 'stealth' activity is unknown, inhibition of liposome opsonization has been demonstrated for Stealth® liposomes constructed using PEG-PE. The large volume occupied by the flexible PEG chain tethered to the liposome membrane may be responsible for inhibiting opsonin binding. GM_1 seems to work by a different, unknown mechanism.[57]

As mentioned above, pH-sensitive liposomes are rapidly cleared by RES. Liu and Huang successfully combined the "stealth" concept with standard pH-sensitive liposome technology to make long-circulating pH-sensitive liposomes.[55] The liposomes were constructed using DOPE and DPSG, a composition known to be plasma stable,[25] and various amounts of GM_1. At up to 5 mol% GM_1, the acid sensitivity of DOPE:DPSG was not affected.[55] Liposomes which contained GM_1 were no less stable in plasma than their counterparts lacking GM_1 and showed less loss of acid sensitivity after pre-incubation in plasma. *In vivo*, GM_1 was shown to enhance the blood concentration of pH-sensitive liposomes for the first few hours after i.v. injection . This effect was correlated with a decreased affinity of the liposomes for the cells of the RES.[55] While other pH-insensitive "stealth" liposomes showed longer periods of blood residence, the delay in RES uptake exhibited by DOPE:DPSG:GM_1 may be long enough to enhance target-specific binding *in vivo*.

Liposomes prepared using PEG-PE seem to have more "stealth" activity than those prepared with GM_1.[56,58] It is not known whether pH-sensitivity can be retained if PEG-PE is incorporated in DOPE:DPSG liposomes. Klibanov et al. demonstrated that streptavidin binding to liposomes containing biotinylated PE was inhibited by >2% PEG(5000)DSPE.[59] In addition, avidin-induced aggregation of biotinylated liposomes was abolished by inclusion of 7.4% PEG(5000)DSPE.[59] Since destabilization of most types of acid-sensitive liposomes depends on aggregation and inter-liposomal collisions, PEG(5000)DSPE may decrease or even abolish acid sensitivity. Using smaller PEG groups which are still able to give the "stealth" effect[56,58] may be one approach to making long-circulating pH-sensitive liposomes. To date, the optimal formulation which exhibits pH-sensitivity, serum stability, and long circulation remains a tantalizing goal yet to be achieved.

V. CONCLUSIONS

Although their full potential as drug delivery vehicles has not yet been realized, pH-sensitive liposomes have become useful tools for scientists with quite diverse interests. For workers interested in liposome fusion, pH-sensitive liposomes permit precise regulation and control of the fusion process, and allow the importance of membrane contact and the possible involvement of non-bilayer phases in fusion to be studied.[12-14,16] pH-sensitive liposomes are useful for studies of endosome acidification[36] and have played a major role in the development of our current understanding of antigen processing in the class I[37,39,40-42] and class II[30] pathways used by antigen presenting cells. The demonstrated ability of pH-sensitive liposomes to mediate cytoplasmic delivery of genetic material[43-46,50] suggests that the liposomes may be particularly useful for delivering antisense therapeutics.[60]

The problems associated with all types of liposomes are *in vivo* stability, targeting, and RES avoidance. If these hurdles can be overcome, pH-sensitive liposomes may someday be as useful *in vivo* as they are *in vitro*.

ACKNOWLEDGMENTS

I wish to thank Jennifer Liu and Jennifer Keysor for preparing the figures used in this chapter. Thanks also to Alan Habberfield for helpful criticism and advice during the preparation of this manuscript.

REFERENCES

1. Yatvin, M. B., Kreutz, W., Horwitz, B. A., and Shinitzky, M., pH-Sensitive liposomes: possible clinical implications, *Science*, 210, 1253, 1980.
2. Straubinger, R. M., Hong, K., Friend, D. S., and Papahadjopoulos, D., Endocytosis of liposomes and intracellular fate of encapsulated molecules: encounter with a low pH compartment after internalization in coated vesicles, *Cell*, 32, 1069, 1983.
3. Marsh, M., The entry of enveloped viruses into cells by endocytosis, *Biochem. J.*, 218, 1, 1984.
4. White, J., and Helenius, A., pH-Dependent fusion between the Semliki Forest Virus membrane and liposomes, *Proc. Natl. Acad. Sci. U.S.A.*, 77, 3273, 1980.
5. Uchida,T., Kim, J., Yamaizumi, M., Miyake, Y., and Okada, Y., Reconstitution of lipid vesicles associated with HVJ (Sendai virus) spikes, *J. Cell Biol.*, 80, 10, 1979.
6. Tirrell, D. A., Takigawa, D., Y., and Seki, K., pH-Sensitization of phospholipid vesicles via complexation with synthetic poly(carboxylic acid)s, *Ann. N.Y. Acad. Sci.*, 446, 237, 1985.
7. Uster, P. S., and Deamer, D., pH-Dependent fusion of liposomes using titratable polycations, *Biochemistry*, 24, 1, 1985.
8. Parente, R. A., Nir, S., and Szoka, F. C., pH-Dependent fusion of phosphatidylcholine small vesicles: induction by a synthetic amphiphilic peptide, *J. Biol. Chem.*, 263, 4724, 1988.
9. Cullis, P. R., and DeKruijff, B., Lipid polymorphism and the functional roles of lipids in biological membranes, *Biochim. Biophys. Acta*, 559, 399, 1979.
10. Gruner, S. M., Cullis, P. R., Hope, M. J., and Tilcock, C. P. S., Lipid polymorphism: the molecular basis of nonbilayer phases, *Ann. Rev. Biophys. Biophys. Chem.*, 14, 211, 1985.
11. Cullis, P. R., Tilcock, C. P., and Hope, M. J., Lipid polymorphism, in *Membrane Fusion*, Wilshut, J. and Hoekstra, D., Eds., Marcel Dekker, New York, 1990, 35.
12. Ellens, H., Bentz, J., and Szoka, F. C., H^+- and Ca^{2+}- induced fusion and destabilization of liposomes, *Biochemistry*, 24, 3099, 1985.
13. Ellens, H., Bentz, J., and Szoka, F. C., Destabilization of phosphatidylethanolamine liposomes at the hexagonal phase transition temperature, *Biochemistry*, 25, 285, 1986.
14. Ellens, H., Bentz, J., and Szoka, F. C., Fusion of phosphatidylethanolamine-containing liposomes and mechanism of the L - H_{II} phase transition, *Biochemistry*, 25, 4141, 1986.
15. Collins, D., and Huang, L., Cytotoxicity of diphtheria toxin A fragment to toxin-resistant murine cells delivered by pH-sensitive immunoliposomes, *Cancer Res.*, 47, 735, 1987.
16. Connor, J., Yatvin, M. B., and Huang, L., pH-Sensitive liposomes: acid-induced liposome fusion, *Proc. Natl. Acad. Sci. U.S.A.*, 81, 1715, 1984.
17. Wang, C.-Y., and Huang, L., pH-Sensitive liposomes mediate target cell-specific delivery and controlled expression of a foreign gene in mouse, *Proc. Natl. Acad. Sci. U.S.A.*, 84, 7851, 1987.
18. Zhou, F., Rouse, B. T., and Huang, L., An improved method of loading pH-sensitive liposomes with soluble proteins for class I-restricted antigen presentation, *J. Immunol. Methods*, 145, 143, 1991.
19. Straubinger, R. M., Duzgunes, N., and Papahadjopoulos, D., pH-Sensitive liposomes mediate cytoplasmic delivery of encapsulated macromolecules, *FEBS Lett.*, 179, 148, 1985.
20. Lai, M.-Z., Vail, W. J., and Szoka, F. C., Acid- and calcium-induced structural changes in phosphatidylethanolamine membranes stabilized by cholesterol hemisuccinate, *Biochemistry*, 24, 1654, 1985.
21. Ellens, H., Bentz, J., and Szoka, F. C., pH-Induced destabilization of phosphatidylethanolamine-containing liposomes: role of bilayer contact, *Biochemistry*, 23, 1532, 1984.

22. Düzgünes, N., Straubinger, R. M., Baldwin, P. A., Friend, D. S., and Papahadjopoulos, D., Proton-induced fusion of oleic acid-phosphatidylethanolamine liposomes, *Biochemistry*, 24, 3091, 1985.

23. Nayer, R., and Schroit, A. J., Generation of pH-sensitive liposomes: use of large unilamellar vesicles containing N-succinyldioleoylphosphatidylethanolamine, *Biochemistry*, 24, 5467, 1985.

24. Leventis, R., Diacovo, T., and Silvius, J. R., pH-Dependent stability and fusion of liposomes combining protonatable double-chain amphiphiles with phosphatidylethanolamine, *Biochemistry*, 26, 3267, 1987.

25. Collins, D., Litzinger, D., and Huang, L., Structural and functional comparisons of pH-sensitive liposomes composed of phosphatidylethanolamine and three different diacylsuccinylglycerols, *Biochim. Biophys. Acta*, 1025, 234, 1990.

26. Mellman, I., Fuchs, R., and Helenius, A., Acidification of the endocytic and exocytic pathways, *Ann. Rev. Biochem.*, 55, 663, 1986.

27. Schmid, S. L., Fuchs, R., Male, P., and Mellman, I., Two distinct subpopulations of endosomes involved in membrane recycling and transport to lysosomes, *Cell*, 52, 73, 1988.

28. Collins, D., Unanue, E. R., and Harding, C. V., Reduction of disulfide bonds in lysosomes is a key step in antigen processing, *J. Immunol.*, 147, 4054, 1991.

29. Chu, C.-J., Dijkstra, J., Lai, M.-Z., Hong, K., and Szoka, F. C., Efficiency of cytoplasmic delivery by pH-sensitive liposomes to cells in culture, *Pharm. Res.*, 7, 824, 1990.

30. Harding, C. V., Collins, D., Slot, J. W., Geuze, H. J., and Unanue, E. R., Liposome-encapsulated antigens are processed in lysosomes, recycled and presented to T cells, *Cell*, 64, 393, 1991.

31. Connor, J., and Huang, L., Efficient cytoplasmic delivery of a fluorescent dye by pH-sensitive immunoliposomes, *J. Cell Biol.*, 101, 582, 1985.

32. Wolfbeis, O. F., Furlinger, E., Kroneis, H., and Marsoner, H., Fluorometric analysis. I. A study on fluorescent indicators for measuring near neutral ("physiological") pH-values, *Fresenius Z. Anal. Chem.*, 314, 119, 1983.

33. Brown, P. M., and Silvius, J. R., Mechanism of delivery of liposome-encapsulated cytosine arabinoside to CV-1 cells in vitro. Fluorescence-microscopic and cytotoxicity studies, *Biochim. Biophys. Acta*, 1023, 341, 1990.

34. Connor, J., and Huang, L., pH-sensitive liposomes as efficient and target specific carriers for antitumor drugs, *Cancer Res.*, 46, 3431, 1986.

35. Gill, D. M., and Pappenheimer, A. M., Structure-activity relationships in diphtheria toxin, *J. Biol. Chem.*, 246, 1492, 1971.

36. Collins, D., Maxfield, F. R., and Huang, L., Immunoliposomes with different acid sensitivities as probes for the cellular endocytic pathway, *Biochim. Biophys. Acta*, 987, 47, 1989.

37. Collins, D., Findlay, K., and Harding, C. V., Processing of exogenous liposome-encapsulated antigens in vivo generates class I MHC-restricted T cell responses, *J. Immunol.*, 148, 3336, 1992.

38. Harding, C. V., and Unanue, E. R., Cellular mechanisms of antigen processing and the function of class I and class II major histocompatability complex molecules, *Cell Regulation*, 1, 449, 1990.

39. Harding, C. V., Collins, D. S., Kanagawa, O., and Unanue, E. R., Liposome-encapsulated antigens engender lysosomal processing for class II MHC presentation and cytosolic processing for class I presentation, *J. Immunol.*, 147, 2860, 1991.

40. Reddy, R., Zhou, F., Huang, L., Carbone, F., Bevan, M., and Rouse, B. T., pH-Sensitive liposomes provide an efficient means of sensitizing target cells to class I restricted CTL recognition of a soluble protein, *J. Immunol. Methods*, 141, 157, 1991.

41. Zhou, F., Rouse, B. T., and Huang, L., Induction of cytotoxic T lymphocytes in vivo with protein antigen entrapped in membranous vesicles, *J. Immunol.*, 149, 1599, 1992.

42. Zhou, F., Rouse, B. T., and Huang, L., Prolonged survival of thymoma-bearing mice after vaccination with a soluble protein antigen entrapped in liposomes: a model study, *Cancer Res.*, 52, 1, 1992.

43. Wang, C.-Y., and Huang, L., Highly efficient transfection mediated by pH-sensitive immunoliposomes, *Biochemistry*, 28, 9508, 1989.

44. Legendre, J.-Y., and Szoka, F. C., Delivery of plasmid DNA into mammalian cells using pH-Sensitive liposomes: comparison with cationic liposomes, *Pharm. Res.*, 9, 1235, 1992.

45. Ropert, C., Lavignon, M., Dubernet, C., Couvreur, P., and Malvy, C., Oligonucleotides encapsulated in pH-sensitive liposomes are efficient toward Friend retrovirus, *Biochem. Biophys. Res. Commun.*, 183, 879, 1992.

46. Welsh, N., Oberg, C., Hellerstrom, C., and Welsh, M., Liposome-mediated in vitro transfection of pancreatic islet cells, *Biomed. Biochim. Acta*, 49, 1157, 1990.

47. Whatlet, M. G., Clauss, I. M., Nols, C. B., Content, J., and Huez, G., New inducers revealed by the promoter sequence analysis of two interferon activated human genes, *Eur. J. Biochem.*, 169, 313, 1987.

48. Chapekar, M., and Glazer, R. I., Synergistic effect of human immune interferon and double-stranded RNA against human colon carcinoma cells in vitro, *Cancer Res.*, 45, 2539, 1985.

49. Montefiori, D. C., and Mitchell, W. M., Antiviral activity of mismatched double-stranded RNA against human immunodeficiency virus in vitro, *Proc. Natl. Acad. Sci. U.S.A.*, 84, 2985, 1987.

50. Milhaud, P. G., Compagnon, B., Bienvenue, A., and Philippot, J. R., Interferon production of L929 and Hela cells enhanced by polyriboinosinic acid-polyribocytidylic acid pH-Sensitive liposomes, *Bioconjugate Chem.*, 3, 402, 1992.

51. Connor, J., Norley, N., and Huang, L., Biodistribution of pH-sensitive liposomes, *Biochim. Biophys. Acta*, 884, 474, 1986.

52. Liu, D., and Huang, L., Small, but not large, unilamellar liposomes composed of dioleoylphos-phatidylethanolamine and oleic acid can be stabilized by human plasma, *Biochemistry*, 28, 7700, 1989.

53. Liu, D., Huang, L., Moore, M. A., Anantharamaiah, G. M., and Segrest, J. P., Interactions of serum proteins with small unilamellar liposomes composed of dioleoylphosphatidylethanolamine and oleic acid: high density lipoprotein, apolipoprotein A1, and amphipathic peptides stabilize liposomes, *Biochemistry*, 29, 3637, 1990.

54. Liu, D., and Huang, L., Role of cholesterol in the stability of pH-Sensitive, large unilamellar liposomes prepared by the detergent dialysis method, *Biochim. Biophys. Acta*, 981, 254, 1989.

55. Liu, D., and Huang, L., pH-sensitive, plasma-stable liposomes with relatively prolonged residence time in circulation, *Biochim. Biophys. Acta*, 1022, 348, 1990.

56. Allen, T. M., Stealth liposomes: five years on, *J. Liposome Res.*, 2, 289, 1992.

57. Klibanov, A. L., Maruyama, K., Torchillin, V. P., and Huang, L., Amphipathic polyethylene glycols effectively prolong the circulation time of liposomes, *FEBS Lett.*, 268, 235, 1990.

58. Allen, T. M., Hansen, C., Martin, F., Redemann, C., and Yan-Young, A., Liposomes containing synthetic lipid derivatives of poly(ethylene glycol) show prolonged circulation half-lives in vivo, *Biochim. Biophys. Acta*, 1066, 29, 1991.

59. Klibanov, A. L., Maruyama, K., Beckerleg, A. M., Torchilin, V. P., and Huang, L., Activity of amphipathic poly (ethylene glycol) 5000 to prolong circulation time of liposomes depends on the liposome size and is unfavorable for liposome binding to target, *Biochim. Biophys. Acta*, 1062, 142, 1991.

60. Akhtar, S., and Juliano, R. L., Cellular uptake and intracellular fate of antisense oligonucleotides, *Trends Cell Biol.*, 2, 139, 1992.

Chapter 14

Liposomes as Transporters of Oligonucleotides

Lee Leserman

CONTENTS

I. INTRODUCTION

The potential of liposomes as drug-carriers was appreciated and began to be exploited soon after it was realized that liposomes were closed structures. Drugs which have been encapsulated in liposomes include nucleic acid analogues, such as 5-fluorouracil, which inhibits thymidylate synthetase, and cytosine arabinoside, which inhibits DNA polymerase; others are conventional drugs which have as mechanisms of action intercalation within DNA or the prevention of its repair, such as actinomycin D and bleomycin (reviewed in Reference 1). Thus, there is nothing conceptually new about liposomes as carriers of nucleotides or of molecules destined to interact with nucleic acids.

What is new, however, is an increased appreciation of the ability of oligodeoxyribo- or oligoribonucleotides (about 10 to 30 nucleotides in length) of defined nucleotide sequence to form hybrid duplexes via hydrogen bonded base pairing with complementary DNA or RNA in a sequence-specific manner: these are called antisense oligonucleotides. Hybridization may result in an inhibition of expression of the target gene or mRNA by blocking access to DNA- or RNA-binding proteins necessary for transcription or translation; or if the antisense agent is composed of DNA the mRNA may be degraded by cellular enzymes (RNases H), which are specific for the RNA component of a DNA-RNA hybrid. Statistically, specific sequences of 15 or more nucleotides are likely to be expressed only once in the human genome. Consequently viral genes, or host genes whose sequences have been altered by mutation or translocation, may be specifically recognized and potentially blocked, leaving other genes and messages untouched (reviewed in Reference 2). It is now known that well-defined translocations or mutations of certain oncogenes, including point mutations, are etiologically related to malignant transformation, though the details of the process have not been completely elucidated.[3] The capacity of antisense oligonucleotides to distinguish normal and mutated *Ras* oncogenes, the sequences of which differ by a single nucleotide, has been demonstrated.[4] This antisense strategy necessarily requires that these agents have physical access to target DNA or RNA in the nucleus or cytoplasm.

II. POTENTIAL OF AND PROBLEMS WITH ANTISENSE OLIGONUCLEOTIDES

A. ADVANTAGES FOR DRUG SCREENING AND SYNTHESIS

The difference between DNA intercalating drugs, which have little or no sequence specificity, and the selective inhibition of particular genes or the mRNAs they encode corresponds to a fundamental change and reveals new perspectives in pharmacology. Since techniques for the determination of gene sequences are extremely rapid, and since the antisense sequence is complementary to this sequence, the structure of a "drug" with the potential to inhibit the expression of the gene is at hand when the gene is sequenced. Further, the synthesis of this drug is relatively simple and is little affected by differences in its nucleotide sequence. This is in marked contrast to conventional drugs, for which there is no obvious relation between the structure of, for example, an enzyme substrate, and an antagonist of the enzyme. Currently, drug development requires that appropriate inhibitors be found by screening or by modeling, which are inefficient and expensive, and which can yield molecules which, even if good enzyme inhibitors, may have poor solubility or bioavailability. The structures of these drugs are highly variable and their synthesis or purification require a large range of methods. Finally, the drug may have unacceptable levels of toxicity for reasons which may not be predictable by structural analogy with related compounds.

B. UNKNOWN ELEMENTS CONCERNING ACTIVITY *IN VIVO*

On the other hand, most of the appeal of the antisense strategy is based on theoretical considerations, and our enthusiasm for these agents has not yet been tested in confrontation with the difficulty of developing useful drugs. A detailed discussion of the mechanisms of action of antisense oligonucleotides and of the closely related ribozymes, which may be considered as antisense molecules with intrinsic nuclease activity (see Reference 5 for a more detailed discussion; for an article comparing antisense and ribozyme activities see Reference 6) is beyond the scope of this article. However, it is clear that numerous problems in the pharmacological development of antisense oligonucleotides and ribozymes must be overcome. Among these problems is the actual ability of antisense oligonucleotides to hybridize with their target DNA or RNA sequences *in vivo*. Some oligonucleotides are better inhibitors of gene expression than others for reasons presumably related to sequence-dependent susceptibility of the oligonucleotides to degradation, the accessibility of the target gene or message, or to the ability of ribosomes to displace oligonucleotides binding to coding regions. As information is lacking concerning details of the DNA-chromatin microenvironment, or the secondary and tertiary structure of mRNA as it exists in cells and the effect of these on hybridization, the selection of target sequences on the gene or mRNA is largely made by trial and error.

Other problems relate to the stability and intracellular transport of oligonucleotides. Since complementarity of base pairs between oligonucleotides and target RNA or DNA is the basis of information transfer in nature, and appears to occur with a minimum of pathologic consequences, native DNA and RNA would appear to be optimal agents to confer that complementarity for pharmacological purposes. The difference between physiologic and pharmacological polynucleotides is that the former are synthesized in the cell, while the latter have to be delivered there. Biologic fluids contain nucleases and phosphodiesterases which are capable of degrading exogenously administered oligonucleotides.

C. STRUCTURAL MODIFICATIONS

The structure of nucleotides may be modified for the purpose of augmenting their stability. The "first generation" modifications include replacement of oxygen molecules in the phosphate groups with sulfur (phosphorothioates and phosphorodithioates) or methyl groups (methyl phosphonates) (for review, see Reference 7); other, more extensive modifications are under investigation. Unfortunately, while increasing oligonucleotide stability these changes may also modify complementarity, as demonstrated by non-sequence-specific inhibitory effects of phosphorothioate oligonucleotides on the replication of the human immunodeficiency virus *in vitro*.[8] It is to be hoped that this fortuitous antiviral effect may be useful in the clinic. However, the inhibition of a viral sequence by a modified oligonucleotide lacking sequence complementarity raises the question of the effects of acute or chronic administration of these agents on uninfected cells. It may be naive to expect that inhibitory effects will be limited to viral replication with no toxic effects due to interaction with host sequences, especially when the majority of experimental systems studied to date consist primarily of *in vitro* models of limited complexity. The potential for short- or long-term toxicity must thus be carefully evaluated in clinical trials. Neither should the possibility of secondary effects of phosphorothioate or other non-natural nucleotide analogues blind us to the possibility of similar problems associated with native phosphodiester structures. It is premature to say that a proposed antisense sequence is specific for a pathogen because no corresponding human sequence exists in the databases when only a small fraction of human genome has been sequenced; nor are potential pharmacological or toxic effects necessarily limited to hybridization with totally or partially complementary sequences. In fact, one of the first studies using liposome-encapsulated oligonucleotides demonstrated the antiviral and antiproliferative effects of certain naturally occurring oligonucleotides in the unusual 2'-5' linkage (discussed in a subsequent section). The activity of these oligonucleotides does not depend on hybridization with complementary DNA or RNA.[9]

Improving the resistance of oligonucleotides to degradation solves only one of the problems limiting their successful clinical application. The most important problem in the application of antisense technology relates to the ability of the oligonucleotides to arrive at their intracellular sites of action in order to exert their pharmacological activity. This may be achieved by further modifying sequences to augment penetration into cells, but increasing the extent to which these molecules differ with respect to native molecules is likely to increase the problems of toxicity cited above.

III. LIPOSOME-BASED STRATEGIES

A. OVERVIEW

The delivery of antisense agents constitutes an attractive application of liposome technology (reviewed in References 10 and 11). Liposome encapsulation of oligonucleotides, as for conventional drugs, is interesting for several reasons: the potential for protecting encapsulated contents against degradation, since liposomes are impermeable to nucleases; improvement of pharmacokinetics of the encapsulated molecules and restriction of their access to sites of potential toxicity; enhanced delivery to macrophages; augmentation of uptake *in vivo* by accessible cells other than macrophages, as recently demonstrated for antibody-bearing liposomes containing drugs;[12] and increased retention of the oligonucleotides in cells.[10,13] Since the liposomes protect and transport the oligonucleotides, their use may permit native or minimally modified nucleotide structures to be used as antisense agents, limiting potential toxicity. An inconvenient feature of liposomes is that they are particles of considerable (virus) size, which will reduce access of the complex to some target cells, though improvement in pharmacokinetics may compensate for this problem. Other strategies for protecting or targeting antisense agents (nanoparticles, peptide carriers) exist, and have been recently reviewed.[14]

B. CATIONIC LIPOSOMES

Liposomes used for the purpose of transport of nucleotides may encapsulate these agents or bind them on their surfaces. Liposomes composed with a part of their phospholipids as cationic lipophiles bind to negatively charged oligo- and polynucleotides by electrostatic interaction, depending to some extent on the length of the nucleotide sequence. Residual positive charge on these liposomes permits binding to cells. In fact, some methods using such "cationic liposomes" employ compounds that may be considered to be liposomes, while others are distinct entities (discussed in this book by Remy et al. and Debs and also recently reviewed by Felgner[15]).

Cationic liposomes have the capacity to mediate the entry of associated nucleic acids into the cytoplasm by a poorly understood process involving perturbation of the plasma membrane and/or the membrane of endocytic vesicles. These agents have been extensively used for the delivery of genes and mRNA into cells; their use for the delivery of oligonucleotides has been reported in only a limited number of papers.[16-21] The activity of a cationic liposome-associated phosphorothioate antisense oligonucleotide that hybridizes to the initiation codon of the human intracellular adhesion molecule-1 (ICAM-1) expressed by cultured endothelial cells was increased by three orders of magnitude relative to the nonliposome-associated oligonucleotide. This was presumably achieved by modifying intracellular distribution, since there was only a 6- to 8-fold increase in the amount of oligonucleotide which became cell associated. In the presence of the cationic liposomes fluorescent oligonucleotides were seen in the cytoplasm and nucleus; the oligonucleotides were confined to endocytic vesicles in their absence.[16,17] The activity of the cationic liposomes depended on their composition, on the cell type, and on the extent of confluence of the cells. Possible differences in the quantity of ICAM-1 message or its stability in these cells were not evaluated. The results also depended on the chemical nature of the antisense agent, since phosphodiester antisense oligonucleotides to ICAM-1 were inactive in this model, and the same group had previously reported only modest effects with 2-O-methyl-modified antisense oligonucleotides, which do not activate RNase H.[18] Cationic liposomes were used in a similar experimental system to transport an antisense oligonucleotide to basic fibroblast growth factor into cultured bovine aortic endothelial cells, resulting in inhibition of their proliferation.[19] Ribozymes composed entirely of RNA or of DNA for the antisense sequences and RNA for the enzymatic core were introduced into human T lymphocytes via cationic liposomes. The stability of the DNA-RNA hybrid was superior, though no biological effects were reported.[21] Ribozymes specific for tumor necrosis factor RNA were introduced into human promyelocytic leukemia cells and mediated a 85–90% reduction in levels of both TNF message and protein. A modified ribozyme with a bacteriophage T7 transcription terminator at its 3' end was more stable than one lacking this sequence. These early studies are likely to be rapidly extended to diseases including pathologic inflammation, vascular endothelial proliferation, HIV infection, and cancer.

Nevertheless, the extent to which cationic liposomes may be useful for oligonucleotide delivery *in vivo* is unclear at present. These liposomes may be destabilized by serum and attach to the first biological membrane they encounter, which may preclude dissemination *in vivo*. If their level of toxicity is acceptable, these agents may still be useful for a large number of clinically important diseases, as they may be administered topically or via catheters. This question is being evaluated in animal studies. The recent successful demonstrations of intratracheal delivery into mice of cationic liposome-bound plasmids containing the gene encoding the cystic fibrosis transmembrane conductance regulator (CFTR) are quite encouraging,[22,23] but it is too early to say if the correction is sufficiently stable to be clinically useful for treatment of genetic diseases. Some cationic lipids have considerable toxicity for cells and because the oligo- or polynucleotides are bound to the outside of the liposomes these are less effective at protecting associated nucleic acids against degradation than conventional liposomes, in which the contents are protected by a phospholipid bilayer.

C. "CONVENTIONAL" LIPOSOMES: GENERALITIES

Liposomes which encapsulate oligonucleotides may be used as such or may be coupled to antibodies or other ligands for the purpose of promoting their binding to selected cell-surface determinants: these are often called immunoliposomes. In addition, liposomes may be made using lipids (usually phosphatidyl-ethanolamine) that undergo phase transitions at acid pH, which permit them to rapidly release their contents intracellularly in association with membrane perturbation after endocytosis (reviewed in Chapter 13 of this book and also recently by Litzinger and Huang[24]). The additional effort required for encapsulation of nucleic acids and the relative inefficiency of this process is compensated for by protection of the nucleic acids against degradation and the possibility of their dissemination in the body. The encapsulation strategy requires that oligonucleotides do not rapidly diffuse out of liposomes. This has been demonstrated for both native or modified charged and uncharged (methylphosphonate) liposomes.[25]

1. Non-targeted liposomes

In addition to widely disseminated techniques for liposome formation, several "non-classical" methods have been used for forming liposomes in order to increase the efficiency of oligonucleotide-liposome association, which rarely exceeds 10% of available oligonucleotide by most methods. These techniques include increasing encapsulation of oligonucleotides, or associating the oligonucleotide with a lipophilic molecule which becomes part of the liposome structure.

One method for increased encapsulation uses liposomes composed of the acidic phospholipid phosphatidylserine, which will aggregate in the presence of 10 mM calcium. These have been used to encapsulate phosphorothioate antisense oligonucleotides specific for the c-*myc* oncogene,[26] or for the human or mouse IL-1 receptors.[27] In both cases antisense effects were seen. The entry of the liposome contents required fusion of the liposomes and cells mediated by polyethylene glycol in serum-free medium. This application is thus limited to certain *in vitro* applications and can be used only for cells resistant to high calcium concentrations and to the toxic properties of phosphatidylserine.

Another technique, termed "minimal volume entrapment" (MVE) uses liposomes containing cardiolipin which, when hydrated in the presence of oligonucleotides, encapsulates them efficiently. This technique was reported to be useful for only small amounts (micrograms) of oligonucleotides, apparently precluding large scale or *in vivo* applications.[10] Liposomes prepared via this technique permitted uptake of oligo-nucleotides by lung carcinoma and T leukemia cells *in vitro*, while unencapsulated oligonucleotides were not taken up (because of rapid degradation) under the same conditions. The uptake was energy dependent and could be competed by an excess of empty liposomes of the same composition, so was presumably mediated by a cell surface molecule having affinity for cardiolipin, since liposomes lacking cardiolipin are not taken up by lymphoid cells under similar incubation conditions.[28] Biological effects have been demonstrated for MVE liposomes encapsulating 15-mer phosphodiester oligonucleotides having two phosphorothioate residues at each end in complementary (antisense) and sense configurations to the p-glycoprotein encoded by the multidrug resistance gene (*mdr-1*). Expression of the p-glycoprotein was reduced and sensitivity to doxorubicin increased for two multidrug-resistant cell lines for antisense (and to a lesser degree, sense) oligonucleotides in liposomes, but not empty liposomes or the same oligonucleotides free in solution.[29]

The intracellular fate of free and liposome-encapsulated oligonucleotides was studied by the same group using radiolabeled and fluorescein-modified liposomes in conjunction with confocal microscopy.[13] Liposome encapsulation substantially improved the uptake by Molt-3 cells of phosphodiester and phosphorothioate oligonucleotides, and also reduced their rate of efflux from the cells relative to free oligonucleotides, implying a certain level of intracellular stability of the liposomes. Liposome encapsulation also resulted in increased delivery to the cytoplasm and the nucleus. Phosphorothioate oligonucleotides delivered in liposomes were more stable intracellularly than were phosphodiester oligonucleotides. Though no biological effects of these antisense agents were reported, these results correlate with biological results seen for phosphorothioate and phosphodiester oligonucleotides in an *in vitro* model of infection of human T cells with the human immunodeficiency virus (HIV) (discussed in Section III.C.3).

Liposomes have been made with cholesterol linked covalently to 2'-*O*-methyl oligoribonucleotides via a disulfide bond by reacting pyridylthio-modified oligonucleotide with thiocholesterol. The modification of the oligonucleotide resulted in a large increase of the incorporation in liposomes relative to the encapsulation of the same oligonucleotide by standard techniques, and a much higher association of the oligonucleotide with cells when liposome-associated than when free in solution. The purpose of the disulfide linkage is to permit the oligonucleotide to be released from its anchoring moiety once taken up by cells. Biological effects of these cholesterol-associated oligonucleotides in liposomes have not yet been reported.[30] Cholesterol-bound oligonucleotides may also be taken up by cells without association in liposomes.[31]

Oligonucleotides complementary to the *env* gene of the Friend retrovirus, which were without effect on viral replication, became active when encapsulated in pH-sensitive or, to a lesser extent, in non-pH-sensitive liposomes.[32]

2. Targeted Liposomes

While targeted liposomes are usually considered to be synonymous with immunoliposomes (those coupled to an antibody or other protein ligand), some liposomes may have increased affinity to certain cells via lipid-lipid interactions. One example is the cardiolipin-containing liposomes discussed above. Two other examples have been reported for liposomes containing oligonucleotides. The first used liposomes made with an ethylmaleimide derivative of distearoylphosphatidyl-ethanolamine. These liposomes bound to and were endocytosed by both lymphocytes and monocytes *in vitro*. This formulation was used to encapsulate oligonucleotides specific for the 5' *tat* splice acceptor site of HIV in sense and antisense configurations. The sense, but not the antisense configuration mediated an inhibition of viral p24 production. The mechanism of this inhibition is unknown.[33] A second liposome preparation was made with 50% of constituent phospholipids composed of phosphatidylglycerol, and contained a ribozyme designed to cleave the *gag* message of HIV-1 in the region of the initiation codon. These liposomes were taken up by H9 cells, as shown by co-labeling liposomal lipid with ^{3}H cholesteryl hexadecyl ether and ribozymes with ^{32}P. Autoradiographic evidence for nuclear transport was obtained. Further, when ribozymes and an oligonucleotide containing the *gag* target sequence were encapsulated in separate liposomes and incubated with cells, some RNA cleavage was observed, though only at high ribozyme concentrations. These experiments demonstrated that at least a fraction of the ribozymes may be delivered to an intracellular environment suitable for cleavage activity.[5] Other examples of *in vitro* uptake by different cells of liposomes containing negatively charged phospholipids have been reported.[34] This uptake depends on the cell type and the liposome charge density.

3. Immunoliposomes

Liposomes have been coupled by a number of techniques to antibodies or other proteins which may mediate cell-specific binding (see Chapter by Schuber). Antibody-targeted liposomes were used to deliver the interferon-induced molecule (2'–5')(A)$_n$, which was discussed in Section II.C, into cells. (2'–5')(A)$_n$ is an activator of RNase L, which subsequently cleaves mRNA at single-stranded UA, UU, or UG sequences. This molecule is thought to mediate at least part of the antiviral activity of interferon, which up-regulates its synthesis. The hypothesis that $(2'-5')(A)_n$ may have antiviral activity in the absence of interferon could not be tested by incubating cells in the presence of $(2'-5')(A)_n$ because it is not taken up by cells in its free form as a consequence of its polarity. The ability of liposomes containing $(2'-5')(A)_n$ to deliver active product into mouse lymphoid L1210 cells was tested by challenging the cells at various times with VSV. Incubation of L1210 cells with $(2'-5')(A)_n$-containing liposomes targeted by antibodies specific for surface H2K molecules caused a small but significant reduction of VSV titer. A phosphodi-esterase-resistant variant of $(2'-5')(A)_n$ exerted a more pronounced antiviral activity, in agreement with its enhanced metabolic stability. In the same experiments, neither targeted liposomes containing a fluorescent marker, nor non-encapsulated $(2'-5')(A)_n$ nor its derivative exert any antiviral activity. The kinetics of development of an antiviral effect revealed a 1- to 2-hour lag, which probably corresponds to the time necessary for endocytosis of encapsulated oligonucleotides and their intracellular release. This was one of the first experiments which indicated that phosphorylated molecules entered into the cytoplasm after being taken up into endocytic vesicles in liposomes, and that their activity could be affected by their intracellular stability.

Antibody-targeted liposome-encapsulated antisense oligonucleotides were studied in a system using mouse L cells and 15-mer phosphodiester oligonucleotides complementary to the 5'-end region of the mRNA encoding the N protein of VSV. When incubated with immunoliposomes the cells became less permissive for multiplication of that virus and a reduction of more than 95% in viral multiplication was achieved. No reduction was seen for liposomes containing a random oligonucleotide sequence, or liposomes containing a sequence complementary to the 5'-end of c-*myc* protooncogene mRNA targeted by the same antibody. Antibody-targeted antisense oligonucleotides thus demonstrated a double specific-ity: a particular cell selected by the targeting antibody on the liposomes and a particular RNA in the cells selected by sequence complementarity with the liposome-encapsulated oligonucleotides. Oligonucle-otides encapsulated in liposomes resisted DNase and were active in amounts 1-2 orders of magnitude lower than for those reported for unencapsulated oligonucleotide sequences.[35]

A study on the inhibition of HIV used antisense RNA to the viral *env* region (synthesized *in vitro* using T7 and SP6 RNA polymerase) delivered in immunoliposomes targeted by antibodies specific for the T cell receptor molecule CD3. The antisense constructions were poly- rather than oligo-nucleotides: the viral *env* segment covered a part of exon II of HIV-1 *tat* gene and was greater than 1 kilobase in length. These liposomes almost completely inhibited HIV-1 production in the HTLV-IIIB/H9 system *in vitro*. No anti-HIV activity could be detected with similarly targeted liposome-encapsulated sense *env* RNA or with *pol* RNA synthesized in the sense or antisense orientations, or with *env* region antisense RNA free in solution or encapsulated in liposomes in the absence of the targeting antibody. A semiquantitative evaluation revealed that 4000–7000 RNA molecules became cell-bound in targeted liposomes. The half-life of the intracellularly hybridizable antisense *env* RNA was approximately 12 h. Western blots showed that antisense *env* RNA suppressed *tat* gene expression by approximately 90% and gp160 production by 100%. These data were confirmed by immunoprecipitation studies. Northern blots (using an *env* probe) demonstrated the existence of all major HIV RNA species (9.3-, 4.3-, and 2.0-kb mRNA) in HIV-infected cells treated with antisense *env* RNA although at reduced levels. The authors concluded that the antisense *env* RNA inhibited viral protein production at the translational level.[36]

The effects of short (15–28 base pair) HIV-1 *rev* and *tat* gene-specific DNA antisense phosphodiester or phosphorothioate oligonucleotides were compared, either free in solution or encapsulated in immunoliposomes, on acutely or chronically HIV-1 infected cells. Phosphodiester antisense oligonucle-otides were inactive in their free form on acutely and chronically infected cells (up to a concentration of 50 µM), because of rapid degradation in the medium. When encapsulated in immunoliposomes directed to HLA-class I molecules expressed by targeted cells, they inhibited viral replication (at a concentration of 0.5 µM) in a sequence-specific manner on acutely infected cells. The same phosphodiester antisense oligonucleotides in liposomes had no antiviral activity in chronically infected cells. In acutely infected cells, phosphorothioate oligonucleotides free in solution inhibited the replication of HIV without se-quence specificity and had slightly greater activity, also non-specific, when encapsulated in liposomes. Phosphorothioate antisense (anti-*rev*) oligonucleotides specifically blocked HIV-replication in chroni-cally infected cells. Since HIV replication is viral reverse transcriptase independent in chronically infected cells (because the viral genome is integrated), this enzyme is unavailable for the sequence non-specific inhibitory mechanism of phosphorothiate oligonucleotides seen in acutely infected cells. When encapsulated in targeted liposomes the efficiency of inhibition for chronically infected cells was increased by at least 60-fold relative to the same oligonucleotide free in solution. This study indicates that phosphodiester (or other minimally modifed oligonucleotides) may have some utility *in vivo*, but that the sensitivity to degradation remains limiting, even if protected in liposomes.[37]

IV. CONCLUSIONS, PERSPECTIVES

The logic of oligonucleotides is so powerful that it seems almost inconceivable that they will not be effective as drugs. The only question relates to the importance of the diseases they can be used to treat. The current state of the art suggests that the concept falls short of expectations for reasons principally related to the fragility of the oligonucleotides and of problems associated with their transport, resulting in modifications which progressively distort the specificity of the oligonucleotides and, as a consequence, the essence of the concept. The association with or encapsulation by liposomes can protect the oligonucleotides from nuclease degradation and effect delivery into cells. However, there remain the problems associated with the efficient transport of the liposomes to their sites of action within the body: liposomes must avoid non-specific uptake by the reticuloendothelial system, traverse endothelia, and rest all the while invisible to the host immune system. It is perhaps most likely that the conceptual purity of both the antisense technique and of liposomes as a delivery system will need to be compromised to optimize the procedure. This will result in a hybrid strategy in which both the oligonucleotide and the liposome carrier will be designed and constructed not only in conjuction with each other, but also with a view to the precise disease being treated. Time and a considerable amount of effort will lead to a better understanding of the interplay between these different requirements, and to the practical realization of the potential offered by liposome-mediated delivery of antisense oligonucleotides as a new form of drug.

ACKNOWLEDGMENTS

I thank Ken Crook and Patrick Machy for criticism of the manuscript. Work in my laboratory is supported by grants from the Association pour la Recherche contre le Cancer, the Ligue National Française contre le Cancer, the Agence Nationale de Recherches sur le SIDA, and by institutional grants from the Centre National de la Recherche Scientifique and the Institut National de la Santé et de la Recherche Médicale.

REFERENCES

1. Weinstein, J. N., Liposomes in the diagnosis and treatment of cancer, in *Liposomes From Biophysics to Therapeutics,* Ostro, M. J., Ed., Marcel Dekker, New York, 1987, 277.
2. Toulmé, J. J. and Hélène, C., Antimessenger oligodeoxyribonucleotides: an alternative to antisense RNA for artificial regulation of gene expression. A review, *Gene,* 729, 51, 1988.
3. Bishop, J. M., Molecular themes in oncogenesis, *Cell,* 64, 235, 1991.
4. Saison-Behomoaras, T., Tocqué, B., Rey, I., Chassignol, M., Thoung, N. T., and Hélène, C., Short modified antisense oligonucleotides directed against Ha-*ras* point mutation induce selective cleavage of the mRNA and inhibit T24 cells proliferation, *EMBO J.,* 10, 1111, 1991.
5. Rossi, J. J., Elkins, D., Zaia, J. A., and Sullivan, S., Ribozymes as anti-HIV-1 therapeutic agents: principles, applications, and problems, *AIDS Res. Human Retroviruses,* 8, 183, 1992.
6. Cotten, M., Schaffner, G., and Birnstiel, M., Ribozyme, antisense RNA, and antisense DNA inhibition of U7 small nuclear ribonucleoprotein-mediated histone pre-mRNA processing in vitro, *Mol. Cell. Biol.,* 9, 4479, 1989.
7. Uhlmann, E., and Peyman, A., Antisense oligonucleotides: a new therapeutic principle, *Chem. Rev.,* 90, 543, 1990.
8. Agrawal, S., Goodchild, J., Civeira, M. P., Thornton, A. H., Sarin, P. S., and Zamecnik, P. C., Oligodeoxynucleoside phosphoramidates and phosphorothioates as inhibitors of human immunodeficiency virus, *Proc. Natl. Acad. Sci. U.S.A.,* 85, 7079, 1988.
9. Bayard, B., Leserman, L. D., Bisbal, C., and Lebleu, B., Antiviral activity in L1210 cells of antibody-targeted liposomes containing (2′–5′) oligo (adenylate) analogues, *Eur. J. Biochem.,* 151, 319, 1985.
10. Thierry, A. R., Rahman, A., and Dritschilo, A., Liposomal delivery as a new approach to transport antisense oligonucleotides, in *Gene Regulation: Biology of Antisense RNA and DNA,* Erickson, R. P., and Izant, J. G., Eds., Raven Press, New York, 1992, 147.
11. Juliano, R. L., and Akhtar, S., Liposomes as a delivery system for antisense oligonucleotides, *Antisense Res. Develop.,* 2, 165, 1992.

12. Ahmad, I., Longenecker, M., Samuel, J., and Allen, T. M., Antibody-targeted delivery of doxorubicin entrapped in sterically stabilized liposomes can eradicate lung cancer in mice, *Cancer Res.*, 53, 1484, 1993.

13. Thierry, A. R., and Dritschilo, A., Intracellular availability of unmodified, phosphorothioated and liposomally encapsulated oligodeoxynucleotides for antisense activity, *Nucl. Acids Res.*, 20, 5691, 1992.

14. Leonetti, J.-P., and Leserman, L., Targeted delivery of oligonucleotides, in *Antisense Research and Applications*, Crooke, S. T., and Lebleu, B., Eds., CRC Press, Boca Raton, FL, in press.,

15. Felgner, P. L., Cationic lipid/polynucleotide condensates for *in vitro* and *in vivo* polynucleotide delivery—the cytofectins, *J. Liposome Res.*, 3, 3, 1993.

16. Bennett, C. F., Chiang, M.-Y., Chan, H., Shoemaker, J. E., and Mirabelli, C. K., Cationic lipids enhance cellular uptake and activity of phosphorothioate antisense oligonucleotides, *Mol. Pharmacol.*, 41, 1023, 1992.

17. Bennett, C. F., Chiang, M.-Y., Chan, H., and Grimm, S., Use of cationic lipids to enhance the biological activity of antisense oligonucleotides, *J. Liposome Res.*, 3, 85, 1993.

18. Chiang, M.-Y., Chan, H., Zounes, M. A., Freir, S. M., Lima, W. F., and Bennett, C. F., Antisense oligonucleotides inhibit intercellular adhesion molecule 1 expression by two distinct mechanisms, *J. Biol. Chem.*, 266, 18162, 1991.

19. Itoh, H., Mukoyama, M., Pratt, R. E., and Dzau, V. J., Specific blockade of basic fibroblast growth factor gene expression in endothelial cells by antisense oligonucleotide, *Biochem. Biophys. Res. Commun.*, 188, 1205, 1992.

20. Sioud, M., Natvig, J. B., and Forre, O., Preformed ribozyme destroys tumor necrosis factor mRNA in human cells, *J. Mol. Biol.*, 20, 831, 1992.

21. Taylor, N. R., Kaplan, B. E., Swiderski, P., Li, H., and Rossi, J. J., Chimeric DNA-RNA hammerhead ribozymes have enhanced *in vitro* catalytic efficiency and increased stability *in vivo*, *Nucl. Acids Res.*, 20, 4556, 1992.

22. Hyde, S. C., Gill, D. R., Higgins, C. F., Trezise, A. E. O., MacVinish, L. J., Cuthbert, A. W., Ratcliff, R., Evans, M. J., and Colledge, W. H., Correction of the ion transport defect in cystic fibrosis transgenic mice by gene therapy, *Nature*, 362, 250, 1993.

23. Yoshimura, K., Rosenfeld, M. A., Nakamura, H., Scherer, E. M., Pavirani, A., Lecocq, J.-P., and Crystal, R. G., Expression of the human cystic fibrosis transmembrane conductance regulator gene in the mouse lung after *in vivo* intratracheal plasmid-mediated gene transfer, *Nucl. Acids Res.*, 20, 3233, 1992.

24. Litzinger, D. C., and Huang, L., Phosphatidylethanolamine liposomes: drug delivery, gene transfer, and immunodiagnostic applications, *Biochim. Biophys. Acta*, 1113, 201, 1992.

25. Akhtar, S., Basu, S. E. W., and Juliano, R. L., Interaction of antisense DNA oligonucleotide analogs with phospholipid membranes (liposomes), *Nucl. Acids Res.*, 19, 5551, 1991.

26. Loke, S. L., Stein, C., Zhang, X., Avigan, M., Cohen, J., and Neckers, L. M., Delivery of c-*myc* antisense phosphorothioate oligodeoxynucleotides to hematopoietic cells in culture by liposome fusion: specific reduction in c-*myc* protein expression correlates with inhibition of cell growth and DNA synthesis, *Curr. Top. Microbiol. Immunol.*, 141, 282, 1988.

27. Burch, R. M., and Mahan, L. C., Oligonucleotides antisense to the interleukin 1 receptor mRNA block the effects of interleukin 1 in cultured murine and human fibroblasts and in mice, *J. Clin. Invest.*, 88, 1190, 1991.

28. Machy, P., and Leserman, L. D., Small liposomes are better than large liposomes for specific drug delivery *in vitro*, *Biochim. Biophys. Acta*, 730, 313, 1983.

29. Thierry, A. R., Rahmann, A., and Dritschilo, A., Overcoming multidrug resistance in human tumor cells using free and liposomally encapsulated oligonucleotides, *Biochem. Biophys. Res. Commun.*, 190, 952, 1993.

30. Oberhauser, B., and Wagner, E., Effective incorporation of 2′-*O*-methyl-oligoribonucleotides into liposomes and enhanced cell association through modification with thiocholesterol, *Nucl. Acids Res.*, 20, 533, 1992.

31. Krieg, A. M., Tonkinson, J., Matson, J., Zhao, Q., Saxon, M., Zhang, L. M., Bhanja, U., Yabukov, L., and Stein, C. A., Modification of antisense phosphodiester oligodeoxynucleotides by a 5′ cholesteryl moiety increases cellular association and improves efficacy, *Proc. Natl. Acad. Sci. U.S.A.*, 90, 1048, 1993.

32. Ropert, C., Lavignon, M., Dubernet, C., Couvreur, P., and Malvy, C., Oligonucleotides encapsulated in pH sensitive liposomes are efficient towards Friend retrovirus, *Biochem. Biophys. Res. Commun.*, 183, 49, 1992.

33. Sullivan, S. M., Gieseler, R. K. H., Lenzner, S., Ruppert, J., Gabrysiak, T. G., Peters, J. H., Cox, G., Richer, L., Martin, W. J., and Scolaro, M. J., Inhibition of human immunodeficiency virus-1 proliferation by liposome-encapsulated sense DNA to the 5' *TAT* splice acceptor site, *Antisense Res. Develop.*, 2, 187, 1992.

34. Lee, K. D., Hong, K., and Papahadjopoulos, D., Recognition of liposomes by cells: *in vitro* binding and endocytosis mediated by specific lipid headgroups and surface charge density, *Biochim. Biophys. Acta*, 1103, 185, 1992.

35. Leonetti, J.-P., Machy, P., Degols, G., Lebleu, B., and Leserman, L., Antibody-targeted liposomes containing oligodeoxyribonucleotides complementary to viral RNA selectively inhibit viral replication, *Proc. Natl. Acad. Sci. U.S.A.*, 87, 2448, 1990.

36. Renneisen, K., Leserman, L., Matthes, E., Schröder, H. C., and Müller, W. E. G., Inhibition of expression of human immunodeficiency virus-1 *in vitro* by antibody-targeted liposomes containing antisense RNA to the *env* region, *J. Biol. Chem.*, 265, 16337, 1990.

37. Zelphati, O., Zon, G., and Leserman, L., Inhibition of HIV replication with antisense oligonucleotides encapsulated in immunoliposomes, *Antisense Res. Develop.*, 3, 323, 1993.

Chapter 15

Imaging Tools: Liposomal Agents for Nuclear Medicine, Computed Tomography, Magnetic Resonance, and Ultrasound

Colin Tilcock

CONTENTS

I. INTRODUCTION

A. THE CLINICAL CONTEXT

What is the sensitivity of nuclear medicine (NM), computed tomography (CT), magnetic resonance (MR), or ultrasound (US) for the detection of a 5 mm lesion in the liver or spleen, with or without contrast enhancement? There are no simple answers to this question, but to pick numbers that few would argue with: zero for NM and greater than 50% for CT, MR and US depending upon the type of lesion, location, orientation, blood supply, regional perfusion, type and dose of contrast, hardware and software sophistication, post-processing, and not least, the skill of the technologist as well as the expertise of the clinician. Increase the size of the lesion and the sensitivity increases for each modality, however, the point is that with no modality can 100% detection be guaranteed.

Over the past decade CT has become widely used for the detection of hepatosplenic metastases which, because of enlarged interstitial volumes and increased water content, typically exhibit lower attenuation than surrounding parenchyma.[1] However, this is a generalization and certain tumors exhibit similar attenuation to parenchyma and so cannot be detected without use of contrast media,[1,2] which typically is a water-soluble small molecular weight radiopaque material. There is no question that these agents certainly do improve the detection of a wide range of tumors,[3] however, because these small molecular weight materials rapidly equilibrate with the extravascular space following bolus injection, tumors with similar blood flow to surrounding tissue can actually be obscured;[4,5] similar considerations apply to small molecular weight paramagnetic contrast media such as gadolinium-diethylene triamine pentaacetic acid (Gd-DTPA) for MR.[6-8] One attempt to alleviate this problem has been to design agents that target not the tumor but rather the reticuloendothelial system (RES) itself, and it is here that liposome-based agents may find one area of application.

B. PARTICULATES AS RES IMAGING AGENTS

In order to increase the contrast between metastasis and healthy tissue one of two general approaches may be used. Either contrast is directed specifically to tumor or else is directed specifically to normal, nontumorous tissue. In principle, in the first approach it is possible to borrow from the techniques of

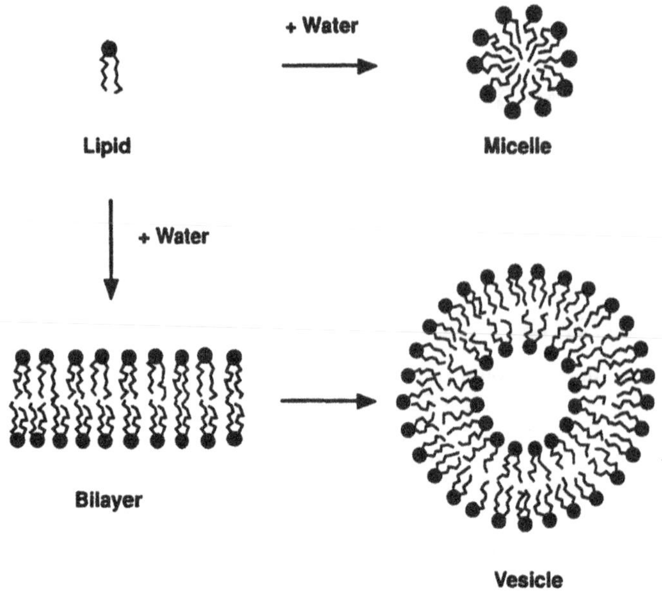

Figure 1 Examples of phase structures formed by lipids upon dispersal in water. At concentrations above the critical micelle concentration (of the order 10^{-9}–10^{-10} M) lipid monomers self-assemble to form both micelles and bilayers. In order to minimize exposure of hydrocarbon to water the lipid bilayers spontaneously seal to form closed vesicular structures.

immunoscintigraphy and tag a suitable monoclonal directed to a specific antigen expressed on the surface of the tumor cell.[9] Neglecting the technical difficulties of monoclonal production and labeling, there exist several physical barriers that complicate this approach including heterogeneous blood supply to tumors, nonspecific binding, large solute diffusion distances within the interstitium, amongst others.[10] Although these structural barrier issues are of lesser importance in well perfused or very small tumors, it then becomes debatable whether it is possible to incorporate sufficient label on the antibody. For nuclear medicine imaging, target concentrations in the subnanomolar range suffice for detection. For MR, a submillimolar concentration of paramagnetic label is required to induce changes in signal intensity. For CT, an iodine concentration of approximately 10 mM is required in the parenchyma. Thus, while for immunoscintigraphy direct labeling of antibodies is practicable, for CT direct-labeling is not a viable approach. As to MR, the current literature suggests that it is not possible to label a monoclonal with a sufficient number of paramagnetic centers to appreciably affect the target to nontarget signal ratio by MR,[11] however, other approaches such as coupling paramagnetic cations to a polymeric ligand which is then attached to a monoclonal are worth investigating.

From several perspectives it is easier to target not the tumor but the surrounding healthy tissue. Tumors tend to have either no or reduced phagocytic function, thus it is possible to exploit the ability of the reticuloendothelial system (RES) to ingest particulates tagged with a suitable contrast agent. In this way it is the surrounding parenchyma and not the tumor that will change in intensity. This is the basis for the use of radioactive sulfur colloid in scintigraphy,[12] radiopaque lipid-vesicles in CT,[13] and paramagnetically labeled vesicles in MR.[14] By having an agent that goes to and remains within the parenchyma for an extended period, but is eventually metabolized, obviates the need to perform rapid scanning during the initial bolus phase after administration and allows a detailed examination of the whole liver and spleen.

C. LIPOSOME STRUCTURE AND NOMENCLATURE

Under conditions approximating the physiological regime, the vast majority of lipids, either in isolation or in mixtures that mimic the composition found within cell membranes, will adopt a bilayer configuration when dispersed in water.[15] In order to minimize the energy of the system, the hydrophobic hydrocarbon tails of the lipid orient together away from the water while the hydrophilic phospholipid headgroups orient outward toward the aqueous milieu to form either micelles or bilayer structures (Figure 1). In order to avoid nonfavorable water-hydrocarbon interactions these bilayers will in turn spontaneously seal to form closed vesicular systems that define an interior aqueous space. Water-soluble materials may be entrapped within the aqueous core of the vesicle or else fat-soluble materials may be dissolved within the hydrocarbon matrix of the lipid. Lipids may be made from spin-labeled, radioactive, or radiopaque constituents or else may be synthesized with a suitable chelator attached to the lipid headgroup. Liposomes are therefore versatile tools for the solubilization and transport of various kinds of drug molecules, including imaging contrast agents.

Systems with a single lipid bilayer surrounding an aqueous core are termed unilamellar vesicles, whereas liposomes that contain multiple closely opposed bilayers are termed multilamellar vesicles (MLVs). The term small unilamellar vesicle (SUV) generally refers to structures less than approximately 0.1 μm (100 nm) diameter, large unilamellar vesicle (LUV) encompasses the range 0.1-1 μm micron and giant unilamellar vesicle (GUV) refers to micron-sized structures or larger; the boundary between designations is vague. Liposomes may be designated in terms of composition, e.g., nonionic surfactant vesicles (NSV), or by the method of manufacture, e.g., reverse-phase evaporation (REV), dehydration/rehydration vesicles (DRV), or interdigitation/fusion (IF) vesicles. See Gregoriadis[16] for a recent detailed discussion of lipid self-assembly and liposome formulation methodologies.

D. RES UPTAKE AND AVOIDANCE

When injected into the circulation, lipid vesicles, by virtue of their size, are relatively confined to the vasculature, exiting the circulation at sites of fenestrated or discontinuous epithelia and being taken up by the free and fixed macrophages that constitute the RES. Injected vesicles are thus delivered principally to the Kupffer cells and (depending on vesicle size) to hepatocytes of liver and macrophages of spleen as well as, to a much lesser extent, macrophages in lung and bone marrow.[17,18] Circulating peripheral blood monocytes may also endocytose liposomes and then infiltrate tissues to become fixed macrophages and histiocytes, providing a method to passively target liposome to sites of tumor, infection, or inflammation.[19] Low lipid doses, the use of large vesicles, and the presence of negative charge or other specific recognition molecules such as lectins on the vesicle surface, are all factors that favor RES uptake.[18] If the target organ is the liver or spleen, this is eminently desirable. If not, nonspecific RES uptake is a major operational consideration and until recently has been a limitation of parenteral vesicle-based imaging agents. Lipid vesicles, because of their size, do not rapidly equilibrate with the extravascular space like small molecular weight CT or MR contrast media. Vesicles are therefore potentially excellent markers of the vasculature so long as nonspecific RES uptake can either be avoided or markedly decreased. RES uptake can be suppressed by the use of high lipid doses that effectively saturate the ability of the RES to ingest further particulates, or else by predosing with, e.g., methyl cellulose or empty vesicles[18] (Figure 2). Both of these approaches are workable but neither can be considered optimal, in particular because it is not desirable to further challenge a patient who may already have compromised phagocytic function (e.g., as in diffuse hepatocellular carcinoma).

The aggregation of particles in suspension can be prevented by coating them with long, flexible surfactant or polymer molecules. When two particles approach, chains in each of the stabilizing layers must necessarily be bent aside and their motion restricted; aggregation is therefore entropically disfavored.[20] In a recent incarnation of this idea, various studies have shown that by covalently coupling the neutral polymer polyethylene glycol (PEG) to the surface of vesicles, the lifetime of the vesicles in the circulation can be greatly extended (up to several hours) and uptake by the RES suppressed.[21-27] PEG is not unique in this regard as it has been shown that attachment of ganglioside GM1 to the surface of lipid vesicles is similarly effective,[28] however PEG is favored for several reasons. PEG has been previously shown to reduce the immunogenicity and antigenicity of several proteins,[29,30] PEG is inexpensive and readily available, the chemistry of PEG is well understood and several methods of derivatization exist[31] and, in addition, PEG itself has extremely low toxicity.[32] Although it is not clear exactly how PEG attached to the membrane surface causes changes in the biodistribution behavior of lipid vesicles, it is reasonable to conjecture that the polymer acts in part as a steric barrier to the close approach and intercalation of circulating proteins into the vesicles surface, perhaps interfering in the opsonization process.[33] Presumably, other neutral polymers such as polypropylene glycol or polyvinylpyrrolidone would be equally as effective as PEG. A detailed discussion of the effect of PEG on RES uptake is outside the scope of this article, but the interested reader is directed to a recent excellent review by Woodle and Lasic.[34]

II. APPLICATIONS OF LIPOSOMES IN IMAGING

A. NUCLEAR MEDICINE

The traditional liver imaging agent in nuclear medicine for many years has been technetium-99m (^{99m}Tc)-labeled sulfur colloid particulate, however detection of metastases with this agent has been largely superseded by CT and MR because of their superior resolution. Sulfur colloid is still useful for the detection of hemangioma in liver (T2-weighted MR scans may be ambiguous, is this hemangioma or cyst?), as well as the detection of shunts and lower GI bleeds. Can radiolabeled lipid vesicles provide

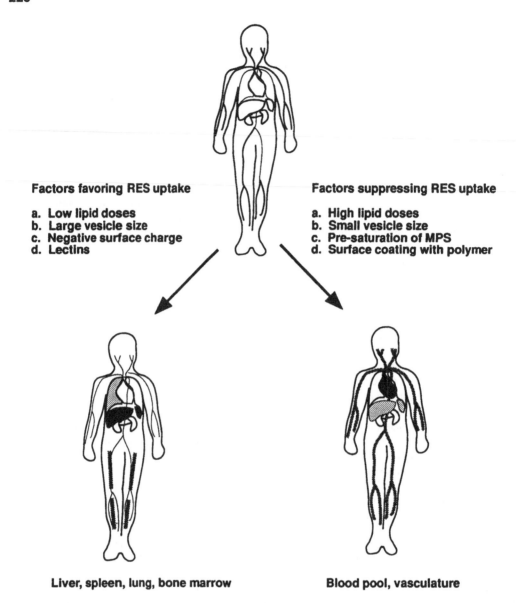

Factors favoring RES uptake

a. Low lipid doses
b. Large vesicle size
c. Negative surface charge
d. Lectins

Factors suppressing RES uptake

a. High lipid doses
b. Small vesicle size
c. Pre-saturation of MPS
d. Surface coating with polymer

Liver, spleen, lung, bone marrow

Blood pool, vasculature

Figure 2 Factors affecting uptake of lipid vesicles injected intravenously into the reticuloendothelial system (RES).

more or alternate information than radiolabeled sulfur colloid? Certainly the biodistribution behavior will be similar thus there is little merit to using radiolabeled vesicles as a liver imaging agent *per se*. The only potential benefit is operational, in that lipid vesicles can be prepared of well-defined size and size-distribution, stored for extended periods in a freeze-dried form, and subsequently rehydrated with little or no change in vesicle size.[35] This would remove any potential artifacts with altered biodistributions caused not by the disease state but by variation in size or aggregation of the sulfur colloid.

There is good evidence in the literature from several groups that radiolabeled vesicles are useful for imaging deep-seated tumors and sites of inflammation,[36-46] not via direct targeting but through macrophage loading.[19] It has been demonstrated that radiolabeled lipid vesicles can be used to image Lewis lung carcinoma, melanoma, and mammary adenocarcinoma in mice.[36,37] These studies have been extended to humans where it has been reported that ^{111}In-labeled vesicles can detect with 85% sensitivity breast, kidney, pancreatic, and ovarian tumors not usually visualized with conventional scanning agents.[38-40] It has also been shown that lipid vesicles injected into the circulation accumulate in the phagocyte-rich synovium of inflamed arthritic joints[41,42] and hence may prove useful for both imaging and therapeutic applications for delivery of anti-inflammatory drugs.[45,46] Polymer-coated lipid vesicles with extended circulation half-life can extravasate through the relatively leaky endothelium to locate within the interstitial space around tumor cells and thereby increase the therapeutic efficacy of liposome-associated drug;[47] the same approach may be valid for imaging agents. A problem common to all these agents is their lack of specificity for tumor vs. sites of infection or inflammation. It has been suggested that specific targeting may be achievable by coupling monoclonals or other recognition molecules to the surface of the liposome.[48] These complex conjugate molecules will be subject to the same kinds of interactions and physical barriers that presently are a problem in radioimmunotherapy; problems of physical access may be exacerbated with liposomes because of their size. To date, specific targeting of liposomes to tumor in vivo in humans has yet to be demonstrated.

An application that has received scant attention is the possibility of using radiolabeled vesicles as a marker of the blood pool and as a tool in the measurement of, e.g., cardiac ejection fraction, detection of left-right shunts, vascular malformations, or GI bleeds. Currently the tool of choice is the patients' own red blood cells (RBCs) labeled either in vivo or in vitro with ^{99m}Tc. While cardiac-gated, rapid-scan or echo-planar MR could conceivably complement or supplant nuclear medicine for certain of these procedures, the continued use of RBCs would seem assured because of their perfect patient compatibility, unique pharmacokinetic properties in terms of half-life, and the inherent sensitivity of nuclear medicine procedures for screening purposes. Labeling of the patients RBCs may be accomplished in vitro or in vivo and is a relatively straightforward procedure although not always trouble-free — many drugs such as heparin, digoxin, methyldopa, and others can interfere in the labeling procedure.[49] In several ways it would be beneficial to take what is currently a procedure and replace it with an off-the-shelf pharmaceutical. ^{99m}Tc-labeled lipid vesicles with an extended half-life in the circulation could usefully substitute for RBCs in many of these procedures.

Although RBCs when undamaged have a half-life in the circulation of approximately 60 days, it is the decay of the ^{99m}Tc label rather than the clearance of the RBCs that is the limiting factor in their use. By 6 half-lives (approximately 36 hr for ^{99m}Tc) only 1% of the original counts remain, thus even for RBCs with their extended half-life there is little point in imaging the patients later than a day postadministration. If it were possible to generate a radiolabeled lipid vesicle with an extended circulation half-life of 12 hr or greater this would be adequate to the task. The best work in this area published to date comes from Phillips and co-workers[50] who recently showed that highly saturated lipid vesicles can be labeled with pertechnetate using hexamethylpropylene amine oxime (HMPAO = exametazime) to transport technetium across the lipid bilayer. The labeling method is efficient, with >90% of added pertechnetate transported across the vesicle membrane. These vesicles retain >95% of their contents after 12 hr incubation in serum in vitro. In vivo studies also demonstrate an extended circulation half-life, in part due to the high lipid concentration employed leading to RES saturation. Recent studies have shown that radiolabeled lipid vesicles coated with PEG can also function as a radiolabeled blood pool agent,[27] only in this approach technetium is coupled to a lipophilic chelator intercalated into the membrane itself. The position of the label (internal or external) makes no difference because it is a gamma emitter, however, each approach has unique advantages and disadvantages.

By attaching the radionuclide to the membrane surface preparation is simplified in that binding to the membrane surface is immediate and quantitative — there is no nonentrapped chelating agent to be removed prior to radiolabeling and, additionally, there is no material to be lost if the vesicles should rupture in the circulation.[51] However the biggest drawback of this approach is that the radiolabel is exposed to circulating proteins and can be exchanged away from the vesicle surface. If the material to which the label is transferred, e.g., circulating serum lipoproteins, also has an extended circulation half-life, this effect may actually be advantageous. By entrapping the radionuclide inside the vesicle, interaction with circulating serum proteins is eliminated so long as the vesicles remains intact; this is not trivial. Preparation is also more complex than for surface chelation. Ionophoretic loading techniques developed for the transport of radioactive cations, such as ^{111}In or ^{67}Ga across a lipid bilayer,[52-54] do not work with technetium because of nonspecific binding of the reduce pertechnetate to the membrane surface and subsequent rapid removal in serum.

B. COMPUTED TOMOGRAPHY

Conceptually, the use of liposomes to deliver radiopaque material to the liver and spleen is straightforward: they deliver as much contrast (iodine) as possible to the parenchyma (to maximize changes in attenuation) for the least amount of lipid (to minimize both lipid-associated toxicity and cost) leaving the

Table 1 **Preparation of vesicles with entrapped radiopaque contrast material**

Method	% Entrapped	I/L ratio (g/g)	Advantages	Disadvantages
MLV	0.5	1:20	Simple	Variable sizes and size distribution. Low trapping efficiency
MEL	0.5–2	1:4	Automatable Reproducible Large batches	Low trap efficiency
REV	10–30	1:1	Efficient entrapment	Difficult to make procedure reproducible, hard to scale-up
DRV	5–20	1:1	Efficient entrapment	Leaky
IF	10–50	4:1	Highest entrapment yet described	Difficult to define procedure and scale-up

MLV = Multilamellar vesicle, MEL = Microemulsified liposomes, REV = Reverse-phase evaporation vesicles, DRV = Dehydration/Rehydration vesicles, IF = Interdigitation/Fusion vesicles, I/L = iodine/lipid ratio

tumor as a region of low attenuation. This means using large or giant unilamellar vesicles composed of lipids which form a highly stable membrane and the use of nonionic contrast media. Large vesicles give the highest trap volume and similarly unilamellar systems minimize the amount of lipid required, a stable lipid composition minimizes the potential for disruption of the vesicles in the circulation and premature release of entrapped contrast, while nonionic media will minimize the osmotic stresses to which the vesicle will be subject once introduced into the circulation. Despite early proof of the principle,[13] the production of large batches of sterile, pyrogen-free lipid vesicles with high and reproducible trapping efficiency has been a major technical hurdle in the development of this work.[55-73] Table 1 illustrates some essential features of the varying liposomal systems that have been employed over the years and some of the associated problems with those approaches. A unifying trend of all preparation methods used to date is that a final step such as centrifugation or chromatography is used to separate liposomes with internal contrast from nonentrapped material. This seems an unnecessary complication. So long as the contrast material is well tolerated and is rapidly excreted it would only be necessary to wait 1–2 hr for the nonentrapped material to be renally excreted before examining the patient within the extended time-window provided by the liposomal agent. Indeed it may be argued that benefits would accrue by imaging within both the initial vascular phase as well as at extended times after administration.

The best system to date appears to be the interdigitated-fusion (IF) vesicles developed by Janoff and colleagues.[70,71,73] As illustrated in Figure 1, lipids in each half of a lipid bilayer may move independently because the hydrocarbon tails of the constituent lipids touch but do not intercalate. In the presence of ethanol certain saturated lipids such as distearoyl phosphatidylcholine or hydrogenated soya lipids can form a new rigid phase structure in which the acyl chains of the lipids become interdigitated. Small unilamellar vesicles dispersed in the presence of contrast media and high ethanol concentrations form an extended matrix of gel-like sheets. Evaporation of ethanol at high temperature results in the formation of large vesicles of a diameter of 1–5 μm which have a very high trapping efficiency for contrast material. Vesicles prepared by this method exhibit a reported iodine/lipid ratio of 4:1 or greater, superior to reverse-phase evaporation methods. An example of the liver and spleen opacification achievable on CT images after injection of contrast-carrying IF vesicles is presented in Figure 3. In order to achieve a 50 HU enhancement of parenchyma a total iodine dose of approximately 0.1 g/kg would be required within the parenchyma, corresponding to a 1.5–2 g lipid dose for a typical patient. Issues of toxicity and cost effectiveness have yet to be determined.

C. MAGNETIC RESONANCE

In magnetic resonance imaging an ensemble of nuclear spins within a defined region of space is excited by a brief pulse of radiofrequency (r.f.) radiation equivalent to a continuum of oscillating magnetic fields

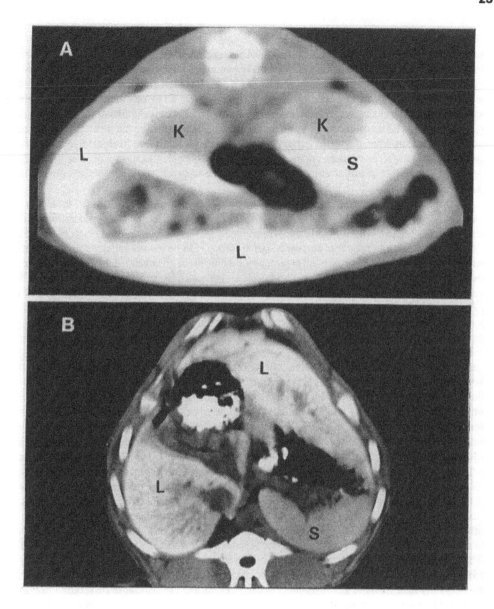

Figure 3 CT images; **(A)** CT scan through the upper abdomen of a prone rat, 60 minutes after administration of 1000 mg I/kg of iotrolan-IF vesicles. Liver (L) and spleen (S) demonstrate striking opacity, while the kidney (K) contains little free iotrolan. **(B)** CT scan through the upper abdomen of a supine dog, 60 min after administration of 250 mg I/kg of iotrolan-IF vesicles. Liver (L) and spleen (S) are markedly opaque. (Reprinted from Seltzer, S. E., Janoff, A. S., Blau, M., Adams, D. F., Minchey, S. R., and Boni, L. T., *Inv. Radiol.,* 26, S169, 1991, with permission.)

of differing frequency. When the frequency of the oscillating magnetic field matches the natural precession frequency of the nucleus (Larmor frequency), energy is absorbed and resonance achieved. After the r.f. pulse is turned off, the excited nuclei give off energy (relax) by a variety of processes including spin-lattice relaxation, in which thermal energy is given off to surrounding molecules as well as spin-spin relaxation in which nuclei exchange energy. In the same way that the nucleus must experience oscillating magnetic fields of the appropriate frequency in order to achieve resonance, it is necessary that the nucleus also experience oscillating magnetic fields of the same frequency (or harmonic thereof) in order to relax. Generally, these oscillating magnetic fields are provided by random motions of the surrounding molecules which in turn generate fluctuating magnetic fields of various frequency. However, relaxation may be enhanced by the interaction of nuclei with unpaired electrons in paramagnetic substances such as

Gd(III), Mn(II), and Fe(III) ions. These materials cause more efficient relaxation because their large magnetic moment of an unpaired electron generates larger oscillating magnetic fields than diamagnetic materials. All magnetic resonance (MR) contrast materials, including liposomal agents, function by causing both spin-lattice (T1) and spin-spin (T2 or T2*) relaxation enhancement for the imaged nuclei (usually water protons). At low concentrations of contrast agent T1 effects are dominant, resulting in an increase in signal intensity, whereas at high concentrations T2 effects dominate, resulting in a decrease in signal intensity. Relaxation caused by diffusion (of water) through high magnetic field gradients at the surface of small superparamagnetic ferrite particles also cause a decrease in signal intensity and often a signal void at high ferrite concentrations. A detailed discussion of relaxation processes in MR and their relevance to the design of MR contrast media is outside the scope of this article, however many excellent books and reviews exist.[74-76]

As with radiopaque liposomal CT agents discussed in the previous section, small molecular weight paramagnetically tagged chelates (e.g., Gd-DTPA, Gd-DO3A, etc.) may be entrapped within the aqueous core of a lipid vesicle.[14,77-80] Alternatively, paramagnetic species such as nitroxide radicals (which would be of particular value in magnetic resonance spectroscopy) can be incorporated within the hydrocarbon matrix of the membrane[81,82] or else the paramagnetic nucleus may be chelated by a suitable ligand at the membrane surface.[83-85] In addition to paramagnetic materials, superparamagnetic materials such as ferrimagnetic ferrites may be entrapped[86] or grown inside[87] lipid vesicles. The rationale behind the use of liposomal MR agents for detection of hepatosplenic metastases is exactly the same as that for CT agents, normal parenchyma takes up the contrast whereas tumor tends to have reduced phagocytic function, thus improving the detectability of lesions under situations where small paramagnetic chelates are ineffective (Figure 4).

Entrapment of small molecular weight chelates inside lipid vesicles is the simplest approach to the design of these systems in that preapproved chelates such as gadolinium-diethylene triamine pentaacetic acid (Gd-DTPA) may be employed. Entrapment does not affect the stability constants of the chelates and methods for the large-scale production vesicles appropriate to this application already exist, e.g., microemulsification procedures, and the resulting vesicles can be terminally sterilized by filtration. Best of all, such vesicles have been clearly demonstrated to facilitate the visualization of hepatic metastases by MR in situations where small molecular weight chelates such as Gd-DTPA were either ineffective or positively detrimental.[9] However there are certain limitations to this approach. It is clear that if paramagnetic cations entrapped within the vesicle are to affect water outside the vesicle then water must cross the lipid bilayer. If transport of water across the membrane is the rate-determining step in the relaxation process, then it would be expected that the relaxivity (relaxation rate per unit concentration) of the entrapped paramagnetic species would also scale linearly with the surface area to volume ratio[79] consistent with experiment.[80] Thus, for entrapped chelates the smallest vesicles would be optimal because they have the highest surface area to volume ratio, exhibit the greatest water flux across the lipid bilayer, and have the highest relaxivity.[79,88] The water flux is maximal for vesicles whose constituent lipids are in the liquid-crystal rather than the gel state, unfortunately the former are also more susceptible to lysis when introduced into the circulation. Stability can be improved by incorporation of saturated lipids or sterols,[18] but only at the cost of decreased relaxivity.[87] Similarly, the vesicles should also be unilamellar rather than multilamellar because the additional water barriers imposed by multiple internal lamellae will lead to relaxation becoming exchange limited.[88,89] The main disadvantage of encapsulation is that although the smallest vesicles have the highest relaxivity they also have the lowest trapping efficiency and therefore have the highest ratio of lipid to entrapped chelate. This is expensive of lipid and wasteful of chelate which must be either discarded or reclaimed by other methods. It is necessary to trap the chelate at high concentrations in order to be able to deliver sufficient contrast to the target organ, however, this raises the possibility of hypertonic lysis once introduced into the bloodstream. Additionally, at high internal concentrations of paramagnetic chelate the observed relaxation rate is dominated by the life time of water within the vesicle.[89] Thus, as the internal chelate concentration is increased the relaxivity does not increase proportionally. From a practical viewpoint, because the relaxivity varies linearly with the vesicle size it is very important to carefully control not only the vesicle size but also size distribution; this is technically very demanding.

An alternate approach is to attach the paramagnetic species to the surface of the lipid vesicle via a ligand such as DTPA.[83-85] There are certain advantages to this approach including:

1. Optimization of water-paramagnetic nucleus interaction,
2. Increased relaxivity due to an increase in the rotational correlation time for the complex,

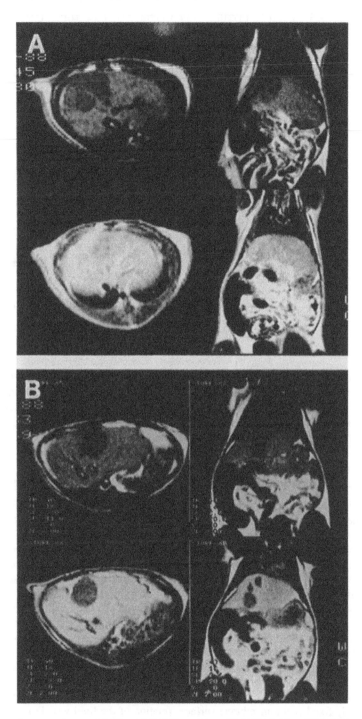

Figure 4 **(A)** Transverse and coronal T1-weighted images at 0.5 T of rats with intrahepatic tumors before and 15 min after administration of 0.1 mmol/kg of Gd-DTPA. Tumor is hypointense relative to parenchyma pre-contrast. Post-contrast both tumor and surrounding parenchyma are enhanced because of the Gd-DTPA, however tumor and parenchyma are isointense and cannot be differentiated. **(B)** Transverse and coronal T1-weighted images at 0.5 T of rats with intrahepatic tumors before and 30 min after administration of 0.1 mmol/kg of Gd-DTPA entrapped in 70 nm lipid vesicles. Tumor is hypointense relative to parenchyma pre-contrast and increases only slightly in intensity post-contrast. By comparison, the surrounding parenchyma is greatly enhanced and consequently the tumor is easier to detect.

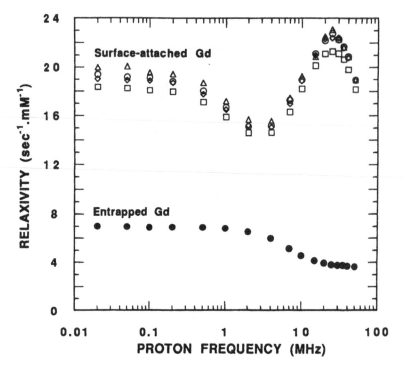

Figure 5 The magnetic-field dependence of the spin-lattice relaxivity at 35°C for Gd either (●) entrapped in 100 nm diameter lipid vesicles or attached to the surface of vesicles of (△) 50, (○) 100, (◇) 200, and (□) 400 nm average diameter. The relaxivity of Gd attached to the surface of the vesicle is approximately 4-6-fold greater than for encapsulated Gd at clinically used field strengths (0.03–2 T) and is invariant with vesicle size within experimental error.

   3. Relaxivity is invariant with vesicle size over a wide regime, and
   4. Vesicle lysis in the circulation is immaterial because the chelate is attached to the membrane.

The relaxation of small paramagnetic chelates such as Gd-DTPA is dominated by their rotational correlation time τ_r (typically about 6×10^{-11} s), which is too short for optimal relaxation at field strengths in clinical use (approximately 0.03–2 T). Because τ_r varies approximately as the cube of the particle radius, by attaching the paramagnetic nucleus to the surface of a larger particle (protein, polymer, lipid vesicle) which is rotating relatively more slowly, then the correlation time for the complex will increase, spin-lattice relaxation becomes more efficient, and the relaxivity will be increased. Figure 5 shows nuclear magnetic resonance dispersion data for lipid vesicles, both with Gd-DTPA entrapped within the vesicles and also chelated to the vesicle surface. For relaxation of vesicles with entrapped Gd-DTPA the relaxivity disperses from 10 τ_c to 3 τ_c in the regime $\omega_c^2\tau_c^2 > 1$, where τ_c is the overall correlation time for the complex, whereas when Gd is attached to the membrane surface there is a peak to high field caused by the field dependence of the electron spin correlation time at long τ_r.[76,90] The end result is that at clinically used field strengths, attaching Gd to the membrane surface increases the relaxivity four- to fivefold, depending on field strength, with corresponding decreases in the amount required to give the same changes in signal intensity. Additionally, in situations where the paramagnetic nucleus is not completely rigidly fixed at the membrane surface, rotation of the lipid vesicle makes only a relatively minor contribution to the overall relaxation process[90] such that relaxivity is invariant with vesicle size over a wide range. This has clear practical benefits in terms of the reproducibility of different batches and the ease of scale-up. As with nuclear medicine agents, attachment of polymer to the surface of the vesicle can greatly extend the circulation half-life of paramagnetically labeled vesicles significantly without affecting the relaxivity of either entrapped or surface-bound paramagnetics.[91]

D. ULTRASOUND

When ultrasound waves pass through tissue they are attenuated by scattering (reflection and refraction) as well as by energy absorption. Scattering from an interface is primarily dependent upon the size of the

Table 2 **Are vesicle based imaging agents valuable for imaging RES and blood pool?**

	RES	Vascular
Nuclear Medicine	No	Yes
Computed Tomography	Yes	For quantization of blood flow
Magnetic Resonance	Yes	For quantization of blood flow
Ultrasound	Unknown	Unknown

scatterer relative to the wavelength of the ultrasound beam, the angle between the beam and scatterer, as well as the difference in acoustic impedance between the scatterer and surrounding tissue. The greater the acoustic impedance mismatch the greater the extent to which the ultrasound beam is reflected at the interface. The acoustic impedance of most tissue is approximately 1.5–1.7 Rayls; only fat (1.4 Rayls), bone (7.5 Rayls), and air (0.0004 Rayls) differ markedly and, as may be expected, fat layers, bone, and gas bubbles within the GI tract give rise to bright specular echoes in the US image. The large acoustic impedance difference between soft tissue and air makes gas bubbles a natural choice as a parenteral US contrast agent. Early studies with high-rate hand injection of various solutes such as saline, water, etc., demonstrated cavitation with transient increased echogenicity,[92] however, a limitation of all such hand-injection techniques is that although the bubbles are capable of enhancing the right heart, they do not survive passage through the capillary bed of the lung.[93] Apart from a continued interest in perfluorocarbons,[94] most research has concentrated on the development of microbubble systems stabilized by various proteinaceous and lipid surfactants.[95,96] Lipid micelles have long been known to be able to solubilize a wide range of gases and low molecular weight hydrocarbons,[97,98] however, long-chain saturated monoglyceride/sterol mixtures, which are known to spontaneously form stable micron-sized gas-in-water emulsions upon dispersal,[96] are more echogenic and have clear clinical applications.[99] Preliminary indications also suggest that pressurization/depressurization is a valid approach to the formation of lipid vesicles with associated bubbles, which would be valuable for both vascular and RES imaging.[100,101]

III. SUMMARY AND FUTURE DIRECTIONS

Do liposomal agents have a role to play in these various imaging modalities? Certainly from a viewpoint of the scientific feasibility, the answer is unequivocally yes. Liposomal agents certainly improve detection of hepatic metastases by CT or MR. Outside the RES there is evidence that radiolabeled vesicles can image tumors or synovium. Lipid vesicles with entrapped gas are also of potential utility for vascular ultrasound applications. The ability to decrease RES uptake by masking the surface of the vesicle with polymer opens one path to the development of long-lived blood pool agents suitable for either nuclear medicine, CT, or MR. Many of the technical limitations that precluded the development of liposomal systems even as little as five years ago, have now largely disappeared. Table 2 synopsizes application areas for vesicle-based agents in each of the four principle imaging modalities discussed herein.

Apart from the fact that detection of liver lesions by scintigraphy has been almost completely supplanted by CT, radiolabeled lipid vesicles have biodistribution behavior similar to sulfur colloid[102] and there is therefore little benefit in developing radiolabeled vesicles for purely RES applications. However, radiolabeled polymer-coated vesicles with extended circulation half-lives are of potential utility as vascular markers and might usefully substitute for ^{99m}Tc-labeled RBCs in various nuclear medicine procedures. Radiolabeled vesicles have been touted as tumor imaging agents by various authors over the past decade, however, unless there are improvements in targeting to specific tissue, low specificity rather than sensitivity will limit the usefulness of such agents. For CT there is no question that the reproducible passive entrapment of iodinated contrast agents at high iodine/lipid ratios is now achievable and that such agents improve detection of liver lesions.

Is there a need for a vascular liposomal CT agent? Current helical-scan technology allows detection of both arterial and venous vascular phases over the entire volume of the liver within 20 s using conventional contrast media; it is not clear what benefit would accrue through the use of a liposomal agent for vascular phase studies, excepting where accurate measurement of blood flow is required. Similarly for MR, improved detection of hepatic lesions has been demonstrated using vesicles with entrapped chelates, although for a variety of technical reasons surface-chelated vesicles would be preferable. The same methodologies that have been used to prepare radiolabeled vesicles with extended half-life in the circulation can be applied to gadolinium or manganese-labeled liposomal MR agents. Such agents may prove useful for quantization of regional myocardial blood volume or regional myocardial perfusion. Lastly, ultrasound represents a largely unexplored area and it is therefore impossible to state with any certainty the roles for liposomal agents. Certainly, ultrasound is by far the most common and least expensive radiological exam, and a truly effective vascular agent would have greatly utility —witness the interest of Mallinckrodt and Schering in this field. There is no reason in principle why a stabilized microbubble could not be used as an RES agent for US,[101] although excessive scattering from parenchyma may defeat the very purpose of the agent.

ACKNOWLEDGMENTS

The financial support of the Medical Research Council of Canada (MRC) and the Sterling-Winthrop Imaging Research Institute are gratefully acknowledged. Thanks to Dr. Steven Seltzer for a critical appraisal of the manuscript and for providing CT images. The author is an MRC Scholar.

REFERENCES

1. Moss, A. A., Schrumpf, J., Schnyder, P., Korobkin, M., and Shimshak, R. R., Computed tomography of focal hepatic lesions: a blind clinical evaluation of the effect of contrast enhancement, *Radiology*, 131, 427, 1979.
2. Berland, L. L., Lawson, T. L., Foley, W. D., Melrose, B. L., Chintapalli, K. N., and Taylor, A. J., Comparison of pre- and post-contrast CT in hepatic masses, *AJR*, 138, 853, 1982.
3. Young, S. W., Turner, R. J., and Castellino, R. A., A strategy for the contrast enhancement of malignant tumors using dynamic computed tomography and intravascular pharmacokinetics, *Radiology*, 137, 147, 1980.
4. Zornoza, J. and Ginaldi, S., Computed tomography in hepatic lymphoma, *Radiology*, 138, 405, 1981.
5. Neumann, C. H. and Castellino, R. A., CT assessment of splenic involvement by Hodgkin's disease and non-Hodgkin's lymphoma, *Tumor Diagn. Ther.*, 3, 113, 1984.
6. Carr, D. H., Graif, M., Niendorf, H. P., Brown, J., Steiner, R. E., Blumgart, L. H., and Young, I. R., Gadolinium-DTPA in the assessment of liver tumors by magnetic resonance imaging, *Clin. Radiol.*, 37, 346, 1986.
7. Ferrucci, J. T., MR imaging of the liver, *AJR*, 147, 1113, 1986.
8. Ohtomo, K., Itai, Y., Yoshikawa, K., Kokubo, I., Yashioro, N., Ito, M., and Furukawa, K., Hepatic tumors: dynamic MR imaging, *Radiology*, 163, 255, 1987.
9. Unger, E. C., Winokur, T., MacDougall, P., Rosenblum, J., Clair, M., Gatenby, R., and Tilcock, C., Hepatic metastases: liposomal Gd-DTPA-enhanced MR imaging, *Radiology*, 171, 81, 1989.
10. Jain, R. K., Physiological barriers to delivery of monoclonal antibodies and other macromolecules in tumors, *Cancer. Res.*, Suppl. 50, 814s, 1990.
11. Unger, E. C., Totty, W. G., Neufeld, D. M., Otsuka, F. L., Murphy, W. A., Welch, M. S., Connett, J. M., and Philpott, G. W., Magnetic resonance imaging using gadolinium labeled monoclonal antibody, *Invest. Radiol.*, 20, 693, 1985.
12. Radt, P., Eine neue Methode zur rontgenologischen Kontrastdarstellung vom Milz und Leber, *Klin. Wochen.*, 8, 2128, 1929.
13. Havron, A., Seltzer, S. E., Davis, M. A., and Shulin, P., Radiopaque liposomes a promising new contrast material for computed tomography of the spleen, *Radiology*, 140, 507, 1981.
14. Navon, G., Pangiel, R., and Valensin, G., Liposomes containing paramagnetic molecules as MRI contrast agents, *Magn. Reson. Med.*, 3, 876, 1986.
15. Tilcock, C., Lipid polymorphism, *Chem. Phys. Lipids*, 40, 109, 1986.
16. Gregoriadis, G., *Liposome Technology*, Vol. 1, Gregoriadis, G., Ed., CRC Press, Boca Raton, FL, 1992.
17. Poste, G., Kirsh, R., and Koestler, T., The challenge of liposome targeting in vivo, in *Liposome Technology*, Vol. 3, Gregoriadis, G., Ed., CRC Press, Boca Raton, FL, 1984, 1.
18. Hwang, K. J., Liposome pharmacokinetics, in *Liposomes: From Biophysics to Therapeutics*, Ostro, M., Ed., Marcel Dekker, New York, 1987, 109.
19. Fidler, I. J., Incorporation of immunomodulators on liposomes for systemic activation of macrophages and therapy of cancer metastasis, in *Liposome Technology*, 2nd ed., Vol. 2, Gregoriadis, G., Ed., CRC Press, Boca Raton, FL, 1992, 45.

20. Berkovsky, B., *Thermomechanics of Magnetic Fluids. Theory and Applications,* Hemisphere, New York, 1978.
21. Blume, G. and Cevc, G., Liposomes for sustained drug release in vivo, *Biochim. Biophys. Acta,* 1029, 91, 1990.
22. Terumo, K., Agents for inhibiting adsorption of proteins on the liposome surface. EPA 0 354 855 A2 (1990).
23. Klibanov, A. L., Maruyama, K., Torchilin, V. P., and Huang, L., Amphipathic polyethylene glycols effectively prolong the circulation time of liposomes, *FEBS Lett.,* 268, 235, 1990.
24. Allen, T. M., Austin, G. A., Chonn, A., Lin, L., and Lee, K. C., Uptake of liposomes by cultured mouse bone marrow macrophages: influence of liposome composition and size, *Biochim. Biophys. Acta,* 1061, 56, 1991.
23. Senior, J., Delgado, C., Fisher, D., Tilcock, C., and Gregoriadis, G., Influence of surface hydrophilicity of liposomes on their interaction with plasma proteins and clearance from the circulation: studies with polyethylene glycol-coated vesicles, *Biochim. Biophys Acta,* 1066, 29, 1991.
24. Klibanov, A. L., Maruyama, K., Beckerleg, A. M., Torchilin, V. P., and Huang, L., Activity of amphipathic poly(ethylene glycol) 5000 to prolong the circulation time of liposomes depends on the liposome size and is unfavorable for immunoliposome binding to target, *Biochim. Biophys. Acta,* 1062, 142, 1991.
25. Allen, T. M., Hansen, C., Martin, F., Redemann, C., and Yau-Young, A., Liposomes containing synthetic lipid derivatives of poly(ethylene glycol) show prolonged circulation half-lives in vivo, *Biochim. Biophys. Acta,* 1066, 29, 1991.
26. Litzinger, D. C. and Huang, L., Amphipathic poly(ethylene glycol) 5000-stabilized dioleoylphosphatidylethanolamine liposomes accumulate in spleen, *Biochim. Biophys. Acta,* 1127, 249, 1992.
27. Tilcock, C., Ahkong, Q. F., and Fisher, D., Polymer-derivatized technetium-99m-labeled liposomal blood pool agents for nuclear medicine applications, *Biochim. Biophys. Acta,* 1993 (in press).
28. Allen, T. M. and Chonn, A., Large unilamellar liposomes with low uptake into the reticuloendothelial system, *FEBS Lett.,* 223, 42, 1987.
29. Abuchowski, A., van Es, T., Palczuk, N. C., and Davis, F. F., Alteration of immunological properties of bovine serum albumin by covalent attachment of polyethylene glycol, *J. Biol. Chem.,* 252, 3578, 1977.
30. Wieder, K. J., Palczuk, N. C., van Es, T., and Davis, F. F., Some properties of polyethylene glycol:phenylalanine ammonia-lyase adducts, *J. Biol. Chem.,* 25, 12579, 1979.
31. Harris, J. M., Struck, E. C., Case, M. G., Paley, M. S., Yalpani, M., van Alstine, J. M., and Brooks, D. E., Synthesis and characterization of poly(ethylene glycol) derivatives, *J. Polym. Sci.,* 22, 341, 1984.
32. Smyth, H. F., Jr., Carpenter, C. P., and Shaffer, C. B., The toxicology of high molecular weight polyethylene glycols; chronic oral and parental administration, *J. Am. Pharm. Assoc.,* 64, 157, 1947.
33. Patel, H. M. and Moghimi, S. M., Techniques to study the opsonic effect of serum on uptake of liposomes by phagocytic cells from various organs of the RES, in *Liposome Technology,* 2nd ed., Vol. 3, CRC Press, Boca Raton, FL, 1992, 43.
34. Woodle, M. C. and Lasic, D. D., Sterically stabilized liposomes, *Biochim. Biophys. Acta,* 1113, 171, 1992.
35. Crowe, J. H. and Crowe, L. M., Preservation of liposomes by freeze-drying, in *Liposome Technology,* 2nd ed., Vol. 1, Gregoriadis, G., Ed., CRC Press, Boca Raton, FL, 1992, 229.
36. Patel, K. R., Tin, G. N., Williams, L. E., and Baldeschweiler, J. D., Biodistribution of phospholipid vesicles in mice bearing Lewis lung carcinoma and granuloma, *J. Nucl. Med.,* 26, 1048, 1985.
37. Williams, L. E., Proffitt, R. T., and Lovaisatti, L., Possible applications of phospholipid vesicles (liposomes) in diagnostic radiology, *J. Nucl. Med.,* 28, 38, 1984.
38. Turner, A. F., Presant, C. A., Proffitt, R. T., Williams, L. E., Winsor, D. W., and Werner, J. L., In-111 labelled liposomes: dosimetry and tumor depiction, *Radiology,* 166, 761, 1988.
39. Presant, C. A., Proffitt, R. T., Turner, F., et al., Successful imaging of human cancer with indium-111 labeled phospholipid vesicles, *Cancer,* 46, 951, 1989.
40. Briele, B., Graefen, M., Brockish, A., et al., Indium-111 labeled liposomes: first clinical results, *Eur. J. Nucl. Med.,* 16, 411, 1990.
41. Williams, B. D., O'Sullivan, M. M., Saggu, G. S., Williams, K. E., Williams, L. A., and Morgan, J. R., Synovial accumulation of technetium labelled liposomes in rheumatoid arthritis, *Ann. Rheum. Dis.,* 46, 324, 1987.

42. Love, W. G., Amos, N., Kellaway, I. W., and Williams, B. D., Specific accumulation of cholesterol rich liposomes in the inflammatory tissue of rats with adjuvant arthritis, *Ann. Rheum. Dis.*, 49, 611, 1990.

43. Ogihara, I., Kojima, S., and Jay, M., Differential uptake of gallium-67-labelled liposomes between tumors and inflammatory lesions in rats, *J. Nucl. Med.*, 27, 1300, 1986.

44. Caride, V. J., Technical and biological considerations on the use of radiolabelled liposomes for diagnostic imaging, *Nucl. Med. Biol.*, 17, 35, 1990.

45. Cleland, L. G., Shandling, M., Percy, J. S., and Poznansky, M., Liposomes, a new approach to gold therapy, *J. Rheumatol.*, 6, 154, 1979.

46. Zalutsky, M. R., Noska, M. A., Gallagher, P. W., Shortkroff, S., and Sledge, C. B., Uses of liposomes for radiation synovectomy, *Nucl. Med. Biol.*, 15, 151, 1988.

47. Papahadjopoulos, D., Allen, T. M., Gabison, A., et al., Sterically stabilized liposomes: improvements in pharmacokinetics and antitumor therapeutic efficacy, *Proc. Natl. Acad. Sci. U.S.A.*, 88, 11460, 1991.

48. Leserman, L., Suzuki, H., and Machy, P., Comments on the application of liposome technology to specific cell targeting, in *Liposome Technology*, 2nd ed., Vol. 3, Gregoriadis, G., Ed., CRC Press, Boca Raton, FL, 1992, 139.

49. Chilton, H. M., Callahan, R. J., and Thrall, J. H., Radiopharmaceuticals for cardiac imaging: myocardial infarction, perfusion, metabolism and ventricular function (blood pool), in *Pharmaceuticals in Medical Imaging*, Swanson, D. P., Chilton, H. M., and Thrall, J. H., Eds., Macmillan, New York, 1990, 419.

50. Phillips, W. T., Rudolph, A. S., Goins, B., Timmons, J. H., Klipper, R., and Blumhardt, R., A simple method for producing a technetium-99m-labeled liposome which is stable in vivo, *Nucl. Med. Biol.*, 19, 539, 1992.

51. Hnatowich, J., Friedman, B., Clancy, B., and Novak, M., Labelling preformed liposomes with 67Ga and 99mTc by chelation, *J. Nucl. Med.*, 22, 810, 1981.

52. Hwang, K. J., Modes of interaction of (In^{3+})-8-hydroxyquinoline with membrane bilayer, *J. Nucl Med.*, 19, 1162, 1978.

53. Mauk, M. R. and Gamble, R. C., Preparation of lipid vesicles containing high levels of entrapped radioactive cations, *Anal. Biochem.*, 94, 302, 1979.

54. Baumier, P. L. and Hwang, K. J., An efficient method for loading indium-111 into liposomes using acetylacetone, *J. Nucl. Med.*, 23, 810, 1982.

55. Caride, V. J., Sostman, H. D., Twickler, J., Zacharis, H., Urphanoukis, S. C., and Jaffe, C. C., Brominated radiopaque liposomes: contrast agent for computed tomography of liver and spleen — a preliminary report, *Inv. Radiol.*, 17, 381, 1982.

56. Rozenberg, O. A. and Hanson, K. P., Radiopaque liposomes for imaging of the spleen and liver, *Radiology*, 149, 877, 1983.

57. Ryan, P., Davis, M. A., and Melchior, D. L., The preparation and characterization of liposomes containing X-ray contrast agents, *Biochim. Biophys. Acta*, 756, 106, 1983.

58. Zherbin, E. A., Ein neues Verfahren zur Kontrastdarstellung von Leber und Milz unter Anwendung wasserloslicher Rontgenkontrastmittel in Liposomen, *Radiol. Diagn.*, 24, 507, 1983.

59. Ryan, P., Davis, M. A., DeGeata, L. R., Woda, B., and Melchior, D. M., Liposomes loaded with contrast material for image enhancement in computed tomography, *Radiology*, 152, 759, 1984.

60. Seltzer, S. E., Davis, M. A., Adams, D. F., Shulkin, P. M., Landis, W. J., and Havron, A., Liposomes carrying diatrizoate: characterization of biophysical properties and imaging applications, *Inv. Radiol.*, 19, 142, 1984.

61. Benita, S., Radiopaque liposomes: effect of formulation conditions on encapsulation efficiency, *J. Pharm. Sci.*, 73, 1751, 1984.

62. Rozenberg, O. A., Diagnostische Moglichkeiten des Kontrastmittels Verografin in Liposomen, *Radiol. Diagn.*, 26, 285, 1985.

63. Jendrasiak, G. L., Frey, G. C., and Hein, R. C., Liposomes as carriers of iodiolipid radiocontrast agents for CT scanning of the liver, *Inv. Radiol.*, 9, 995, 1985.

64. Payne, N. I. and Whitehouse, G. H., Delineation of the spleen by a combination of proliposomes with water-soluble contrast media: an experimental study using computed tomography, *Br. J. Radiol.*, 60, 535, 1987.

65. Zalutsky, M. R., Noska, M. A., and Seltzer, S. E., Characterization of liposomes containing iodine-125-labeled radiographic contrast agents, *Inv. Radiol.*, 22, 141, 1987.

66. Henze, A., Friese, J., Magerstedt, P., and Majewski, A., Radiopaque liposomes for the improved visualization of focal liver disease by computerized tomography, *Comput. Med. Imag. Graph.*, 13, 455, 1989.

67. Revel, D., Corot, C., Carrillon, Y., Dandis, G., Eloy, R., and Amiel, M., Ioxaglate-carrying liposomes. Computed tomographic study as hepatosplenic contrast agent in rabbits, *Inv. Radiol.*, 25, S95, 1990.
68. Adzamli, I. K., Seltzer, S. E., Slifkin, M., Blau, M., and Adams, D. F., Production and characterization of improved liposomes containing radiographic contrast media, *Inv. Radiol.*, 25, 1217, 1990.
69. Passariello, R., Pavone, P., Patrizio, G., Di Renzi, P., Mastantuono, M., and Giuliani, S., Liposomes loaded with nonionic contrast media; hepatosplenic computed tomographic enhancement, *Inv. Radiol.*, 25, S92, 1990.
70. Janoff, A. S., Minchey, S. R., Perkins, W. R., Boni, L. T., Seltzer, S. E., Adams, D. F., and Blau, M., Interdigitation fusion vesicles: a new approach to selective opacification of the RES, *Inv. Radiol.*, 26, S167, 1991.
71. Seltzer, S. E., Janoff, A. S., Blau, M., Adams, D. F., Minchey, S. R., and Boni, L. T., Biodistribution and imaging characteristics of iotrolan-carrying interdigitation-fusion vesicles, *Inv. Radiol.*, 26, S169, 1991.
72. Krause, W., Sachse, A., Wagner, S., Kollenkirchen, U., and Rossling, G., Preclinical characterization of iopromide-carrying liposomes, *Inv. Radiol.*, 26, S172, 1991.
73. Boni, L. T., Minchey, S. R., Perkins, W. R., Ahl, P. L., Slater, J. L., Tate, M. W., Gruner, S. M., and Janoff, A. S., Curvature dependent induction of the interdigitated gel phase in DPPC vesicles, *Biochim. Biophys. Acta*, 1146, 247, 1993.
74. Matwiyoff, N. A., *Magnetic Resonance Workbook*, Raven Press, New York, 1990.
75. Jackels, S. C., Enhancement agents for magnetic resonance imaging: fundamentals, in *Pharmaceuticals in Medical Imaging*, Swanson, D. P., Chilton, H. M., and Thrall, J. H., Eds., Macmillan, New York, 1990, 645.
76. Koenig, S. H., From the relaxivity of Gd(DTPA)2- to everything else, *Magn. Reson. Med.*, 22, 183, 1991.
77. Parasassi, T., Bombieri, G., Conti, F., and Croatto, U., Paramagnetic ions trapped in phospholipid vesicles as contrast agents in NMR imaging. I. Mn-citrate in phosphatidylcholine and phosphatidylserine vesicles, *Inorg. Chim. Acta*, 106, 135, 1985.
78. Magin, R. L., Wright, S. M., Niesman, M. R., Chan, H. C., and Swartz, H. M., Liposome delivery of NMR contrast agents for improved tissue imaging, *Magn. Reson. Med.*, 3, 440, 1986.
79. Bacic, G., Niesman, M. R., Bennett, H. F., Magin, R. L., and Swartz, H. M., Modulation of water permeation rates by liposomes containing paramagnetic materials, *Magn. Reson. Med.*, 6, 445, 1988.
80. Tilcock, C., Unger, E., Cullis, P., and MacDougall, P., Liposomal Gd-DTPA: preparation and characterization of relaxivity, *Radiology*, 171, 77, 1989.
81. Grant, C. W., Barker, K. R., Florio, E., and Karlik, S., A phospholipid spinlabel used as a liposome associated MRI contrast agent, *Magn. Reson. Med.*, 5, 371, 1987.
82. Bennett, H. F., Swartz, H. M., Brown, R. D., and Koenig, S. H., Modification of relaxation of lipid protons by molecular oxygen and nitroxides, *Inv. Radiol.*, 22, 502, 1987.
83. Kabalka, G. W., Buonocore, E., Hubner, K., Davis, M., and Huang, L., Gadolinium-labeled liposomes containing paramagnetic amphipathic agents, *Radiology*, 163, 255, 1987.
84. Schwendener, R. A., Wuthrich, R., Duewell, S., Westera, G., and von Schultess, G. K., Small unilamellar liposomes as magnetic resonance contrast agents loaded with paramagnetic Mn-, Gd-, and Fe-DTPA-stearate complexes, *Int. J. Pharm.*, 49, 249, 1989.
85. Kabalka, G. M., Davis, M. A., Moss, T. H., Buonocore, E., Hubner, K., Holmberg, E., Maruyama, M., and Huang, L., Gadolinium-labeled liposomes containing various amphiphilic Gd-DTPA derivatives: targeted MRI contrast enhancement agents for the liver, *Magn. Reson. Med.*, 19, 406, 1991.
86. Margolis, L. B., Namiot, V. A., and Kljukin, L. M., Magnetoliposomes: another principle of cell sorting, *Biochim. Biophys. Acta*, 735, 193, 1983.
87. Mann, S. and Hannington, J. P., Formation of iron oxides in unilamellar vesicles, *J. Coll. Interf. Sci.*, 122, 326, 1988.
88. Tilcock, C., MacDougall, P., Unger, E., Cardenas, D., and Fajardo, L., The effect of lipid composition on the relaxivity of Gd-DTPA entrapped in lipid vesicles of defined size, *Biochim. Biophys. Acta*, 1022, 181, 1990.
89. Koenig, S. H., Ahkong, Q. F., Brown, R. D., III, Lafleur, M., Spiller, M., Unger, E., and Tilcock, C., Permeability of liposomal membranes to water: results from the magnetic field dependence of 1/T1 of solvent protons in suspensions of vesicles with entrapped paramagnetic ions, *Magn. Reson. Med.*, 23, 275, 1992.
90. Tilcock, C., Ahkong, Q. F., Koenig, S. H., Brown, R. D., III, Davis, M., and Kabalka, G., The design of liposomal paramagnetic MR agents: effect of vesicle size upon the relaxivity of surface-incorporated lipophilic chelates, *Magn. Reson. Med.*, 27, 44, 1992.

91. Tilcock, C., Ahkong, Q. F., Koenig, S. H., Brown, R. D., Kabalka, G., and Fisher, D., Nuclear magnetic relaxation dispersion and 31P NMR studies of the effect of covalent modification of membrane surfaces with poly(ethylene glycol), *Biochim. Biophys. Acta,* 1110, 193, 1992.

92. Ziskin, M. C., Bonakdapour, A., Weinstein, D. P., and Lynch, P. R., Contrast agents for diagnostic ultrasound, *Inv. Radiol.,* 67, 500, 1972.

93. Goldberg, S. J., Valdez-Cruz, L. M., Feldman, M., Sahn, D. J., and Allen, H. D., Range-gated Doppler ultrasound detection of contrast echographic microbubbles for cardiac and great vessel flow patterns, *Am. Heart J.,* 101, 793, 1981.

94. Mattrey, R. F., Scheibe, F. W., Gosink, B. B., Leopold, G. R., Long, D. M., and Higgins, C. B., Perfluoroctylbromide: a liver/spleen specific and tumor-imaging ultrasound contrast material, *Radiology,* 145, 759, 1982.

95. Swanson, D. P., Enhancement fundamentals for ultrasound, in *Pharmaceuticals in Medical Imaging,* Swanson, D. P., Chilton, H. M., and Thrall, J. H., Eds., Macmillan, New York, 1990, 682.

96. D'Arrigo, J. S., Stable gas-in-liquid emulsions. Production in natural waters and artificial media, in *Studies in Physical and Theoretical Chemistry,* Elsevier, Amsterdam, 1986.

97. Matheson, I. B. C. and King, A. D., Solubility of gases in micellar solutions, *J. Coll. Interf. Sci.,* 66, 464, 1978.

98. Miller, K. W., Hammond, L., and Porter, E. G., The solubility of hydrocarbon gases in lipid bilayers, *Chem. Phys. Lipids,* 20, 229, 1977.

99. D'Arrigo, J. S., Ho, S.-Y., Wakefield, A. E., Quatrocelli, A. D., Simon, R. H., and Davis, T., Lipid-coated, uniform microbubbles for earlier ultrasonic detection of brain tumors, 199th ACE Meeting, Boston, Symposium on Drug Delivery in Colloidal Systems, Abstr. 82, 1990.

100. Lund, P., Fuller, L., Fritz, T., Kulik, B., Herres, B., and Tilcock, C., "Aerosomes" liposomes as ultrasound contrast agents, *J. Ultrasound Med.,* 10, S44, 1991.

101. Richardson, V. J., Ryman, B. E., and Jewkes, R. F., 99mTc-labelled liposomes; preparation of radiopharmaceutical and its distribution in a hepatoma patient, *Int. J. Nucl. Med. Biol.,* 5, 118, 1978.

102. Matsuda, Y. and Yabuuchi, I., Hepatic tumors: US contrast enhancement with CO_2 microbubbles, *Radiology,* 161, 701, 1986.

Chapter 16

Non-Phospholipid Liposomes: Principles and Bulk Applications

Donald F. H. Wallach and Rajiv Mathur

CONTENTS

I. INTRODUCTION

The background of non-phospholipid liposomes made of membrane-mimetic amphiphiles is presented in Chapter 3. Here we describe the principles and practice of producing non-phospholipid liposomes on a scale of liters to thousands of liters. The manufacture of non-phospholipid liposomes, as here described, relies on the conversion of micellar solutions of membrane-mimetic amphiphiles to liposomal structures by manipulation of environmental variables (e.g., temperature, hydration) in an appropriate temporal sequence.

As the concentration of a membrane-mimetic amphiphile in water increases, a sudden transition in the properties of the system occurs, representing the conversion of a molecular solution to a micellar solution. The transition concentration is known as the critical micelle concentration. The critical micelle concentrations of the membrane-mimetic amphiphiles that we employ lie well below 10^{-6} M.

Micellar solutions consist of amphiphile molecules in small clusters (50-200 molecules) dispersed in water. The shapes of micellar clusters depends on the amphiphile type (e.g., head group, apolar chain) and solution conditions (e.g., concentration, electrolytes, temperature). Micelle shape is determined by the surface area of the amphiphile molecule, the nature of the hydrophobic/hydrophilic interface, and the curvature of that interface. Micelles form in various phases. Hexagonal phases consist of long, normal or reversed, rod-shaped structures with hexagonally packed arrays, and spherical micelles have been classified into two types of cubic packing: body- or face-centered.

The concentration of free amphiphile [A] is related to the proportion of amphiphile in micellar aggregates, $[A_n]$ by a constant, K, which equals:

$$[A_n]/[A]^n = K \qquad (1)$$

When the concentration of free amphiphile [A] becomes equal to $K^{-1/n}$, equivalent to the critical micelle concentration, formation of micelles begins and, as the value of [A] continues to increase above $K^{-1/n}$, the concentration, $[A_n]$, of micellar aggregates rises rapidly. An increase in total amphiphile concentration above the critical micelle concentration thus implies an increase in the size and/or number of micelles, not the concentration of free amphiphile. (Highly concentrated micellar solutions, for example of soaps/detergents in some shampoos, does not signify a high concentration of free amphiphile, but a large reservoir of micellar amphiphile.)

There is some exchange of amphiphile between micelles. Random thermal motion may shake off individual amphiphile molecules, but the hydrophobic effect,[1] minimizing water exposure of apolar residues, favors the association of amphiphile molecules in micelles.

The methods to be discussed for the manufacture of non-phospholipid liposomes involves the formation of micellar aggregates of membrane-mimetic amphiphiles and the controlled coalescence of these aggregates into membrane-bounded vesicles. The membranes of such vesicles are stabilized by a number of interactions, including the hydrophobic effect[1] driving apolar moieties out of water, van der Waals attractions between ordered amphiphile residues, and hydrogen-bonding of amphiphile head groups to the water at the hydrophilic surfaces of each bilayer.

II. LARGE-SCALE MANUFACTURE OF NON-PHOSPHOLIPID LIPOSOMES

To our knowledge, no methods have been devised for the large-scale production of non-phospholipid liposomes, except for the manufacture of Niosomes by L'Oreal, France and of the non-phospholipid liposomes made by Micro Vesicular Systems, Inc., U.S. Since the manufacture of Niosomes is proprietary to L'Oreal, we will focus this manuscript on the production of non-phospholipid liposomes using the techniques published by Micro Vesicular Systems, Inc.[2-21] These were originally developed for the bulk production of poultry vaccines, but have also allowed the application of liposome techniques to areas such as cosmetics, personal care products, paints and coatings, agrichemicals, and other bulk products. The same technology may be applicable to the non-phospholipid liposomes hitherto produced on a small scale.[22-27] As shall be seen, our methods allow the production of non-phospholipid liposomes in amounts ranging from small, pharmaceutical bedside doses to large industrial volumes, at costs equivalent to making simple emulsions.

A. GENERAL PRINCIPLES

The generation of non-phospholipid liposomes by Micro Vesicular Systems, Inc. is based on the known phase behavior of single-chain lipid amphiphiles[28-31] but also extends to some phospholipids and glycolipids. In the Micro Vesicular Systems approach, anhydrous, membrane-forming lipid (amphiphile, plus auxiliary molecules such as sterols, and ionogenic amphiphiles) are heated (if necessary) to yield a homogeneous liquid. In continuous flow applications, this liquid is then injected into a 5- to 15-fold excess of heated aqueous phase, through tangentially placed nozzles into a small, cylindrical mixing chamber.[2,3,6] One or two nozzles are used for the lipid phase and one or two alternating ones for the aqueous phase[2,3,6] (Figure 1). The two liquids are driven by flow-controlled, positive displacement pumps. Depending on the hydraulic and viscous properties of the lipid phase the injection velocity can range from 10 to 50 m/s through a 0.1–1.5 mm diameter nozzle. Initiation of hydration occurs at the nozzles within less than one millisecond. Photon correlation counting indicates an initial particle (micelle) size of less that 0.1 micron diameter. Cooling of the mixture from a typical initial value of 50°C to about 25°C under conditions of high turbulence causes the micelles to fuse into paucilamellar liposomal vesicles.[2,3,6,19] The process can also be conducted on a small (milliliters) scale,[2,3,6] using hand or motor-driven syringes. Vesicles formed by this process are heat stable under nearly all storage conditions, and survive mixing procedures other than those involving very high shear (e.g., sonication).

We have approximated the aggregation rate of the primary particle created by injection of the lipid using the Smoluchowski equation,

$$-dn_1/dt = 8\pi DRn_1^2 \tag{2}$$

in which n_1 is the number of primary particles, in the system, R is the collision radius, approximately twice the radius, a, of the colliding particle, and D is the diffusion coefficient given by the Einstein formula

$$D = kT6\pi\eta \; a \tag{3}$$

where k is the Boltzman constant and μ is the viscosity of the medium. Equation 3 can be converted to:

$$t_{1/2} = 3\eta/4kTn_0 \tag{4}$$

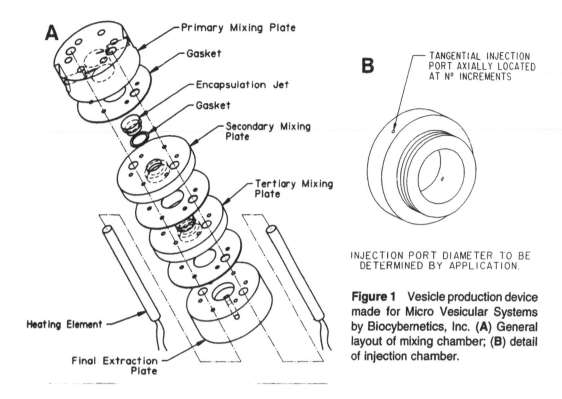

A
- Primary Mixing Plate
- Gasket
- Encapsulation Jet
- Gasket
- Secondary Mixing Plate
- Tertiary Mixing Plate
- Heating Element
- Final Extraction Plate

B
TANGENTIAL INJECTION PORT AXIALLY LOCATED AT N° INCREMENTS

INJECTION PORT DIAMETER TO BE DETERMINED BY APPLICATION.

Figure 1 Vesicle production device made for Micro Vesicular Systems by Biocybernetics, Inc. (**A**) General layout of mixing chamber; (**B**) detail of injection chamber.

For water $\mu = 0.01$ poise, so that at 25–$50°$, $t_{1/2}$ approximates

$$t_{1/2} = 2\text{–}4 \times 10^{11}/n_0 \tag{5}$$

If 1 ml of wall lipid is injected into 4 ml water, it will yield approximately 10^{15} 0.1 micron primary particles. According to Equation 5, their aggregation would be approximately $10^{11}/10^{15}$sec, or 0.1 millisec.

It is probable that the aggregation of primary particles is slower because of an energy barrier **W** (e.g., hydration, electrical charge) which make many of the collisions inelastic. To a first approximation one can describe such a stabilized system by:

$$t_{1/2} = (3\eta 4kTn_0)e^{W/RT} \tag{6}$$

However, even with substantial stabilization values (10–15 RT), the aggregation of primary particles micelles into the stable liposome structures that are ultimately observed must occur in a fraction of a second. In other words, vesicle formation is a rapid process and, conversely, many vesicles can be made in a very short time.

B. VESICLES WITH A PREDOMINANTLY AQUEOUS CORE

Theoretical captured water volumes (calculated for spheres with 1 to 5 bilayer shells 6 nm thick and with interlamellar water computed as a bilayer component) are shown in Figure 2. Experimental data obtained with high- and low-capture volume vesicle populations with diameters recorded by photon correlation counting, indicate that most of the vesicles in these populations have 2-3 bilayers. Freeze-fracture electron microscopy supports this. Capture of water-soluble solute parallels the entrapment of water into the vesicle core. By repeated washing of vesicles over dextran, Ficoll, or albumin density gradients, we arrive at a capture efficiency of 35–45% for water-soluble macromolecules.[21]

1. "Humectant" Non-Phospholipid Liposomes

Suspensions of paucilamellar non-phospholipid lipsomes have three aqueous compartments. The first, consisting of the extravesicular aqueous phase is immediately available after deposition on a surface such

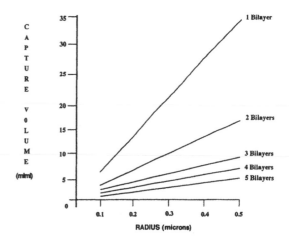

Figure 2 Calculated variation of water capture with vesicle radius for vesicles with 2–5 bilayer shells.

as the skin. The second, consisting of entrapped water, is released more slowly, depending on the osmotic activity of any solutes in the vesicle core. The third, interlamellar water, is released very slowly. These facts make paucilamellar non-phospholipid liposomes very effective skin humectants.

An example of humectant liposomes made on a bulk scale is given below (Table 1):

Table 1 Composition of lipid and aqueous phases for non-phospholipid humectant liposomes.

Lipid Phase	w/w (%)	Aqueous Phase	w/w (%)
Glyceryl monostearate	20.8	Glycerine	11.2
Cholesterol	12.7	Diglycerine	11.2
Stearyl alcohol	8.6	Sodium lauryl sulfate	1.2
Polysorbate 60	8.6	Deionized water	75.1
Mineral oil	49.1	Preservatives	As required

The ingredients of the lipid phase are combined in a steam-jacketed, stainless steel kettle to give a homogeneous solution at 70–73°C. The ingredients of the aqueous phase are similarly combined at 65–68°C. The two phases are then combined at a rate of 5–7 liters per minute using a Micro Vesicular Systems' high-capacity, continuous flow machine. Upon cooling, we obtain a lotion with the following properties (Table 2):

Table 2 Properties of the humectant liposome lotion

	w/w (%)		
Glyceryl monostearate	3.6	Vesicle diameter	0.270–0.570 microns
Cholesterol	2.2	Viscosity	50–350 cps
Stearyl alcohol	1.5	(Brookfield, RVT)	
Polysorbate 60	1.5	Specific gravity	0.970–1.100
Mineral oil	8.5	pH	6.5–7.5
Glycerine	9.8		
Diglycerine	9.8		
Sodium lauryl sulfate	1.0		
Deionized water	62.1		
Preservatives	As required		

C. VESICLES WITH A WATER-IMMISCIBLE CORE

The bilayers of phospholipid vesicles have a low capacity for most apolar molecules because partition of such substances into bilayers is between a structured and an isotropic phase rather than between two isotropic phases.[32] For example, the partition coefficients of noble gases are 2- to 15-fold lower into a bilayer than into a bulk organic phase. Moreover, solute uptake declines with increasing chain density and factors that increase chain density of the bilayer (e.g., lowering temperature through the T_c; incorporation of cholesterol above the T_c). Finally, the uptake of excessive amounts of apolar cargo can disrupt the bilayer. For the above reasons there has been little success in achieving stable incorporation of lipophilic substances (other than certain steroids) into phospholipid liposomes at levels greater than 0.5% dry weight.

Our non-phospholipid system, in contrast, allows transport of apolar materials within the central core of the liposomes. Two processes allow loading of the appropriate non-phospholipid liposomes core with a wide variety of water-immiscible materials (including a broad range of oils, waxes, resins, drugs, nutrients, biocides, and perfluorocarbon liquids) and solutions or suspensions of substances in water-immiscible carriers.

Mixtures of core material can be employed to adapt the melting temperature of the lipid core to specific needs. Selection of appropriate proportions of wall-forming amphiphile (and indifferent surfactant) provides control over vesicle wall thickness and vesicle diameter. With less than maximal loading of water-immiscible material, an important capacity for carrying hydrophilic substances can be retained, allowing the vesicles to act as multifunctional carriers.

1. One-Step Process

In this process[17,19] the water-immiscible, apolar substance is incorporated into the liquid lipid used to make vesicle membranes. During injection of the liquid lipid into the aqueous phase containing a small amount of "indifferent surfactant" — surfactant that does not form bilayers — the water-immiscible substance forms microdroplets (<0.1 μm diameter) stabilized at the surface by the indifferent surfactant around which the bilayers of membrane-forming surfactants coalesce. The captured apolar volume and encapsulation efficiency of this process are potentially very high and can range from traces to near theoretical (over 90% in the encapsulation of alkyd resins for water-borne paints[20]).

Vitamin E-Loaded Non-Phospholipid Liposomes

The production of vitamin E-containing non-phospholipid liposomes is an example of the one-step process for encapsulation of water-insoluble materials for cosmetic use. The same process has been employed in cosmetic applications with cosmetic oils such as triglycerides and silicon fluids, as well as for insecticides and fungicides, as we shall see. The components are as follows (Table 3):

Table 3 Composition of lipid and aqueous phases for production of vitamin E-loaded non-phospholipid liposomes.

Lipid Phase	w/w (%)	Aqueous Phase	w/w (%)
Glyceryl distearate	11.1	POE 20 sorbitan monooleate	1.5
Stearyl alcohol	9.4	Deionized water	~98.5
POE 10 stearyl alcohol	14.3	Preservatives	As required
Cholesterol	8.0		
Vitamin E	57.1		

The ingredients of the lipid phase are mixed in a steam-jacketed stainless steel kettle at 78°C to give a homogeneous solution. The components of the aqueous phase are similarly combined at 70°C. The two phases are reacted at high velocity as before using Micro Vesicular Systems' high-capacity rapid flow machine. Cooling to room temperature yields a cream with the following characteristics (Table 4).

Table 4 Composition and specifications of vitamin E-loaded non-phospholipid liposomes.

	w/w (%)	Vesicle Diameter	0.900–1.3 microns
Glyceryl distearate	3.9	pH (10% suspension)	3.0–4.0
Stearyl alcohol	3.3	Specific gravity	0.96–1.09
POE 10 stearyl alcohol	5.0		
Cholesterol	2.8		
Vitamin E	20.0		
POE 20 sorbitan monooleate	1.0		
Deionized water	~64.0		
Preservatives	As required		

Fungicide-Loaded Non-Phospholipid Vesicles

In this section we describe the encapsulation and delivery of 2-(thiocyanomethyl-thio)benzothiazole (TCMTB), a common agricultural fungicide for the pregermination treatment of cotton seed against fungal pathogens,[33] avoiding the use of organic solvents. The encapsulating agent, DMATO, is composed of N,N-dimethylamides of tall oil fatty acids, with the general formula

$$R - \underset{\underset{O}{\|}}{C} - N(CH_3)_2$$

where R represents the alkyd chains derived from C_{12}–C_{18} straight chain carboxylic acids such as oleic and linoleic acids. DMATO has been used in the formulation of products that act as penetrants, dispersants, plasticizers, and solvents. DMATO[33-36] has also been used as a special-purpose solvent for agricultural applications, including the delivery of TCMTB[34] and carbamate pesticides.[37] DMATO has heretofore been used in organic solution or emulsified with surfactants, but we have found that DMATO can form paucilamellar vesicles which can serve as delivery vehicles for water-insoluble fungicides such as TCMTB. The head group of DMATO and other N,N-dimethlymides (e.g., the soy bean oil equivalent of DMATO) consist of two methyl residues on the amide nitrogen, with the oxygen on the neighboring carbon having the potential for hydrogen bonding. Molecular modelling suggests that the oxygen of the dimethylamide groups of adjacent DMATO molecules may be joined through hydrogen-bonding to a water molecule. This type of interaction between adjacent DMATO molecules would favor interdigitation of the hydrocarbon chains, rather than the typical end-to-end configuration found in lamellar structures composed of amphipathic molecules.

Technical grade TCMTB (80% active) and DMATO were obtained from Buckman Laboratories International, Inc. (Memphis, TN), Busan® 30 E.C., containing 30% TCMTB, and Nu-Flow ND, a combination product containing 23.5% 1,4,-dichloro-2,5-dimethoxybenzene (Chloroneb), another fungicide, plus 9.0% TCMTB were supplied by Wilbur-Ellis (Fresno, CA). Paucilamellar vesicles were formed by combining a lipophilic phase consisting of DMATO or DMATO/TCMTB with an aqueous phase containing 20% glycerine or propylene glycol in 1.5% sodium lauryl sulfate. The lipid and aqueous phase were combined at 55°C as described in the previous example, yielding a fluid vesicle suspension.

One pound batches of Chembred DES 119 or Acala SJ-2 cotton seed were treated using treatment solutions prepared in 50 ml batches. Following treatment, the seeds were dried before bagging and storage. Greenhouse testing was conducted by the Plant Pathology Department, University of California, Davis. Flats containing autoclaved Yolo loam:sand (1:1) were infested with the fungal pathogens *Rhizoctonia solani* or *Pythium ultimum*. Each test involved six randomized seed treatments with the test formulations in randomized flats. Twenty seeds per flat were planted for each treatment. After one week, emergence and post emergence damping-off were recorded. Percent survival and percent damping-off were determined after two weeks. The significance of results was determined by an analysis of variance and Duncan's multiple range test.

Optical microscopy showed empty DMATO liposomes to consist of spherical vesicles with a clearly defined aqueous center. DMATO vesicles containing TCMTB encapsulated within the central cavity were more than twice the diameter of empty DMATO vesicles. Particle sizes determined using a Coulter N4SD submicron sizing instrument, are given in Table 5.

Table 5   **Characteristics of DMATO/TCMTB vesicles**

Vesicle Composition	w/w %	Mean Diameter (nm)
1. DMATO	9.25	209
2. DMATO + TCMTB	9.25 + 44.0	522

After centrifugation of DMATO/TCMTB vesicles (3000 rpm, 30 min.), no separation of free TCMTB was observed, indicating complete entrapment of this fungicide. No free TCMTB was observed when concentrated DMATO/TCMTB vesicles (30%) were either stored at elevated temperature (50°C) or diluted 1:10 with water. Stable mixtures were also obtained when DMATO/TCMTB vesicles were combined with powdered Chloroneb, a second fungicide, to give final active ingredient concentrations of 9.9% and 23.5%, respectively. Table 6 shows the results of greenhouse testing of cotton seeds treated with four different formulation directed against *Pythium ultimum* and *Rhizontonia solani.* Formula I is Nu-Flow ND, containing emulsified TCMTB with Chloroneb added as a wettable powder. Identical results were obtained with aqueous suspensions of DMATO/TCMTB vesicles in the presence of Chloroneb (Formula II). The formulations containing Chloroneb gave better protection against *Rhizoctona* since Chloroneb acts primarily against this fungus. Formula III is Busan 30 EC, which contains TCMTB and DMATO formulated with aromatic solvents (Tenneco 500-100) and ethylene glycol butyl ether. Formula IV, containing TCMTB/DMATO vesicles but no organic solvents, gave similar results against fungal pathogens as Nusan 30 EC, showing that the efficacy of TCMTB is not modified by its encapsulation in DMATO vesicles.

Table 6 **Treatment of cotton seed with DMATO/TCMTB vesicles**

Treatment	Rate[1] fl oz/cwt	*Pythium Ultimum*		*Rhizoctonia*	
		Percent Emergence	Percent[2] Survival	Percent Emergence	Percent[2] Survival
Formula I	14.5	99.0	80.0[2]	83.9	31.9
Formula II	14.5	94.4	78.8	81.9	32.5
Formula III	4.35	85.6	59.9	0.6	0
Formula IV	3.04	83.8	32.5	3.8	0.6
Water	24.0	0	0	0	0

[1]Adjusted to give equivalent concentrations of TCMTB (and Chloroneb) per cwt seed.
[2]Duncan's multiple range values at 0.05 level of significance.

Greenhouse testing of cotton seeds which had been stored for six months following treatment with TCMTB/DMATO vesicles gave essentially the same results as the nonvesicular commercial formulations, confirming the stability of the TCMTB/DMATO vesicle formulations.

Delivery of TCMTB has previously required a combination of organic solvents and/or surfactants. Stability, ease of dilution, and impact on the environment are among the factors that determine the final composition of the formulation. The fact that TCMTB can be delivered in stable DMATO vesicles that are freely dispersible in water and biodegradable after application, avoids the use of organic solvents. The successful treatment of cotton seed with aqueous dispersions of TCMTB-laden liposomes highlight the commercial utility of vesicular delivery systems.

Water-Borne Alkyd Resin Paint

To create water-borne alkyd paints containing no or little organic solvent, we developed membrane-mimetic amphiphiles of the alkyd type.[20] For this, unsaturated fatty acids are conjugated with polyols such as glycerol or pentaerythritol and further with an anhydride such as phthalic or trimellitic anhydrides, yielding anionic, "monomeric" alkyds (Table 7). Oils at low pH, these molecules form vesicles at neutral pH and above, and can then be used to efficiently encapsulate conventional liquid alkyd resins in non-phospholipid liposomes suspended in aqueous media. Preparations with 50% solids can be readily achieved. When painted out and dried, the "wall alkyds" copolymerize with the cargo alkyd resin to form coherent alkyd films.

Table 7 **Examples of alkyd membrane-mimetic surfactants[20]**

C_{12}-C_{18} (0–4 unsaturations)	–O–CO	⬡–COO⁻
C_{12}-C_{18} (0–4 unsaturations)	–CO–O	–CH–C–O–CO–⬡(–COO⁻)(–COO⁻)

An example of a water-borne alkyd resin non-phospholipid liposome system which has desirable polymerization characteristics, a particle size less than 2 microns, and shows little degradation at 50°C for 1 month is as follows (Table 8):

Table 8 **Composition of lipid and aqueous phases used to form a liposomal alkyd paint.**

Lipid Phase

Cargo resin (num 86.5, visc. 3500 cps, acid value 7, OH value 40, containing 15% mineral spirits and built in "drier")	87.5% (v/v)
Wall resin (num 97, visc. 7600 cps, acid value 23–29, OH value 20)	12.5% (v/v)

Aqueous Phase

NH_4 lauryl sulfate	6.3% (w/v)
NH_4OH	2.0% (w/v)
Deaerated, deionized water	91.7% (v/v)

Seven parts of cargo resin are combined with 1 part of the wall resin at 70°C, are thoroughly mixed then 2 parts of the aqueous phase added at 65°C using a Micro Vesicular Systems, Inc. motor-driven syringe mixer. The product contains 70% v/v solids. After standing 16 hours, the product is diluted to 61% v/v resin using deaerated, deionized water.

Important considerations for the preparation are (1) the resins should be kept in a totally anhydrous state until the encapsulation process, (2) the resins and product should have minimal contact with air, (3) vesicles should be allowed to fully form before dilution with aerated, deionized water. The product has the following characteristics (Table 9):

Table 9 **Properties of alkyd resin laden-liposomal suspension.**

Cargo resin	52.1% (v/v)	Specific gravity	~0.92 g/ml
Wall resin	8.8% (v/v)	pH—6.5	
Mineral spirits	9.2% (v/v)	Solids—60.9% (w/w)	
NH_4 lauryl sulfate	1.1% (w/v)		
NH_4OH	~0.3% (w/v)		
Deionized, deaerated water	28.5% (v/v)		

The diluted product is a thick-flowing beige water-miscible material. Microscopic examination shows small spherical vesicles with an occasional larger sphere. No free resin is released upon centrifugation. The product produces a smooth, clear film on glass which sets to touch in 8 hours and becomes hard in 16 hours.

Anthralin-Loaded Non-phospholipid Liposomes

Anthralin is the most effective drug for the treatment of psoriasis, a chronic or recurrent skin disease. The efficacy of anthralin-containing non-phospholipid liposomes, in which the drug is carried in stable form as microcrystals embedded within the paraffin/petrolatum core of the liposomes, was evaluated and compared to a commercially available anthralin ointment (Anthraderm®).[38]

The anthralin liposomes were prepared from the following two phases (Table 10):

Table 10 **Composition of lipid and aqueous phases used to make anthralin-laden non-phospholipid liposomes.**

Lipid Phase	w/w (%)	Aqueous Phase	w/w (%)
White petrolatum	38.2	Sodium metabisulfite	0.09
Paraffin wax	40.4	Deionized, deaerated water	99.91
Salicylic acid	0.6		
Cetyl alcohol	15.1		
POE 10 cetyl alcohol	3.0		
Cholesterol	1.5		
Anthralin	1.2		

The paraffin and petrolatum were heated to 65°C and the salicylic acid dissolved in the mixture. The resulting solution was then combined at 70°C with a liquid mixture of the membrane-forming components (cetyl alcohol/POE 10 cetyl alcohol/cholesterol), the resulting mixture was deoxygenated with nitrogen and anthralin added until in complete solution. The lipid phase was then hydrated with water (containing sodium metabisulfite) at 65°C at a volume ratio of 1/1, and cooled under conditions of turbulent shear to give lipid vesicles with an average diameter near 0.8 microns and a solid lipid core. Microscopic examination shows that, at the anthralin concentrations used, the drug exists as microcrystals within the lipid vesicle core.

The product consists of a cream with the following composition given in Table 11:

Table 11 **Composition of liposomal anthralin cream.**

	w/w (%)
Anthralin	0.55
White petrolatum	17.3
Paraffin wax	18.3
Salicylic acid	0.2
Cetyl alcohol	6.8
POE 10 cetyl alcohol	1.4
Cholesterol	0.7
Sodium metabisulfite	0.05
Deionized, deaerated water	54.7

A preliminary clinical trial[38] showed the following results.

At the start of the study, the mean score for severity of psoriasis was $5.75 \pm$ (SEM) 0.25 for the lesions about to receive anthralin liposomes, and 5.75 ± 0.24 for the lesions to be treated with Anthraderm®. The mean score of the control (to be untreated) lesions was 4.92 ± 0.29. After 7 days, the liposome-treated lesions scored 2.85 ± 0.25, the Anthraderm®-treated lesions scored 3.43 ± 0.27, and the control lesions 5.03 ± 0.25. At the end of the treatment period, the liposome-treated lesions scored 2.07 ± 0.26, the Anthraderm®-treated lesions 2.65 ± 0.28, and the control 5.14 ± 0.30. The differences between the mean scores for the liposome and Antraderm®-treated lesions was highly significant ($p < 0.01$ by the Student "t" test, using a one-tail hypothesis). After a further one-week follow-up, the mean scores for anthralin liposomes and Anthraderm®-treated lesions were 2.64 ± 0.35 and 3.32 ± 0.35 and the controls 4.88 ± 0.37, the difference between the first two remaining statistically significant.

The data show that anthralin delivered by suitably constructed, non-phospholipid liposomes is more effective for the treatment of psoriasis than anthralin ointment. Such liposomes represent a vehicle that offers maximum stability for the drug and may enhance the penetration of anthralin into psoriatic skin.

2. Two-Step Process

This process is particularly useful for the loading of fragile materials (e.g., flavor or fragrance oils, reactive resins) and substances (such as diethyltoluamide and certain solvents) that interfere with micelle formation or fusion at higher temperatures.

In the two-step process,[8] appropriately compositioned vesicles are loaded at ambient or near-ambient temperatures. The vesicles are mixed under low shear conditions with the water-immiscible cargo in the presence of an indifferent surfactant. Microdroplets of the apolar material then enter the central vesicle space within seconds.

Cyclomethicone-Loaded Non-Phospholipid Liposomes

Because of its volatility, cyclomethicone is most effectively encapsulated by the two-step process, as shown in the following example (Table 12):

Table 12 **Composition of lipid and aqueous phase for the preparation of liposomes to be loaded with cyclomethicone**

First Lipid Phase	w/w (%)	First Aqueous Phase	w/w (%)
Glyceryl distearate	57.3	Deionized water	~100
POE 10 stearyl alcohol	28.4	Preservatives	As required
Cholesterol	14.2		

The components of the first lipid phase are combined in a steam-jacketed stainless steel kettle at 72°C to give a clear homogeneous solution. The components of the aqueous phase are similarly combined at 65°C, and the two phases are then combined using the Micro Vesicular Systems rapid flow device. After cooling, the resulting vesicle suspension has a viscosity of 270,000–380,000 cps with a particle diameter of 0.460–0.650 microns and a specific gravity of 0.973–0.993. After about 24 hours the lipsome suspension is transferred to a suitable stainless steel container with an appropriate stirring device, cyclomethicone is added (Table 13) and the mixture stirred for 1 hour at room temperature, causing all the cyclomethicone to become internalized within the liposomes. The resulting lotion has a viscosity of 36,000–52,000 cps, a specific gravity of 0.983–1.003, and a particle diameter of 0.240–0.36 microns.

Table 13 Loading of non-phospholipid liposomes with cyclomethicone.

Second Lipid Phase	w/w (%)	Second "Aqueous" Phase	w/w (%)
Cyclomethicone	37.5	Liposomes from first step	62.5

III. SUMMARY

This paper summarizes processes used for the bulk production of non-phospholipid liposomes and provides examples from the agrichemical, cosmetic, paint, and pharmaceutical fields.

ACKNOWLEDGMENTS

We want to express thanks to Ms. Ornellis for her work on the alkyd vesicles, Mr. S. Celeste for providing the figures of the liposome production machine made for Micro Vesicular Systems by Biocybernetics Laboratories, Inc., and to Ms. D. Welch and Ms. L. Bannister for preparing the manuscript.

REFERENCES

1. Tanford, C., *The Hydrophobic Effect*, Wiley, New York, 1980.
2. Wallach, D. F. H., Method of producing high-aqueous volume multilamellar vesicles. U.S. Patent 4,855,090, 1989.
3. Wallach, D. F. H. and Yiournas, C., Method and apparatus for producing lipid vesicles. U.S. Patent 4,895,452, 1990.
4. Wallach, D. F. H., Paucilamellar lipid vesicles. U.S. Patent 4,911,928, 1990.
5. Wallach, D. F. H., Lipid vesicles formed of surfactants and steroids. U.S. Patent 4,197,951, 1990.
6. Wallach, D. F. H. and Yiournas, C., Method and apparatus for producing lipid vesicles. U.S. Patent 5,013,497, 1990.
7. Wallach, D. F. H., Protein coupling to lipid vesicles. U.S. Patent 5,000,960, 1991.
8. Wallach, D. F. H., Removing oil from surfaces with liposomal cleaner. U.S. Patent 5,019,174, 1991.
9. Wallach, D. F. H., Paucilamellar lipid vesicles using charge localized single chain, nonphospholipid surfactants. U.S. Patent 5,032,457, 1991.
10. Tabibi, E. and Wallach, D. F. H, Theoretical consideration of lipid vesicle formation by Novamix. INTERPHEX-USA, PROC. OF 1991 Technical Program, 61, 1991.
11. Tabibi, E., Sakura, J. D., Mathur, R., Wallach, D. F. H., Schulteis, D. T., and Ostrom, J. K., The delivery of agricultural fungicides in paucilamellar amphiphile vesicles, in *Pesticide Formulations and Application Systems*, Vol. 12, ASTM STP 1146, Devisetty, B. N., Chasin, D. G., and Berger, P. D., Eds., American Society for Testing and Materials, Philadelphia, 155, 1991.
12. Wallach, D. F. H., Encapsulation of parasiticides. U.S. Patent 5,019,392, 1991.
13. Wallach, D. F. H., Encapsulation of ionophore growth factors. U.S. Patent 5,019,392, 1991.
14. Wallach, D. F. H., Paucilamellar lipid vesicles using charge-localized, single chain, nonphospholipid surfactant. U.S. Patent 5,032,457, 1992.
15. Wallach, D. F. H., Reinforced paucilamellar vesicles. U.S. Patent 5,104,736, 1992.
16. Wallach, D. F. H., Paucilamellar lipid vesicles. U.S. Patent 5,147,723, 1992.
17. Wallach, D. F. H., Method of making oil filled paucilamellar lipid vesicles. U.S. Patent 5,160,669, 1992.

18. Wallach, D. F. H., Mathur, R. M., Reziniak, G. J. M., and Tranchant, J. F., Some properties of N-acyl sarcosinate lipid vesicles, *J. Cosmet. Chem.*, 43, 113, 1992.
19. Wallach, D. F. H. and Philippot, J., New type of lipid vesicle: Novasome®, in *Liposome Technology,* 2nd Ed., Gregoriadis, G., Ed., CRC Press, Boca Raton, FL, 141, 1992.
20. Wallach, D. F. H., Mathur, R., Chang, A. C., and Tabibi, E., Lipid vesicles having an alkyd as wall-forming material. U.S. Patent 5,164,191, 1992.
21. Vandergriff, K., Wallach, D. F. H., and Winslow, R. K., Encapsulation of hemoglobin in non-phospholipid vesicles, *Biomat. Art. Cells Immob. Tech.*, in press, 1993.
22. Kano, K., Romero, A., Djermouni, B., Ache, H. J., and Fendler, J. H., Characterization of surfactant vesicles as membrane mimetic agents. 2. Temperature-dependent changes of the turbidity, viscosity, fluorescence polarization of 2-methylanthracene, and positron annihilation in sonicated dioctadecyldimethylammonium chloride, *J. Am. Chem. Soc.*, 101, 4030, 1979.
23. Murakami, Y., Nakano, A., and Fukuya, K., Stable single-compartment vesicles with zwitterionic amphiphile involving an amino acid residue, *J. Am. Chem. Soc.,* 102, 4253, 1980.
24. Murakami, Y., Nakano, A., and Ikeda, H., Preparation of stable single-compartment vesicles with cationic and zwitterionic amphiphiles involving amino acid residues, *J. Org. Chem.*, 47, 2137, 1982.
25. Ishigami, Y. and Machida, H., Vesicles from sucrose fatty acid esters, *J. Am. Oil Chem. Soc.*, 66, 599, 1989.
26. Schenk, P., Ausborn, M., Bendas, F., Nuhn, P., Arndt, D., and Meyer, H. W., The preparation and characterization of lipid vesicles containing esters of sucrose and fatty acids, *J. Microencapsulation,* 6, 95, 1989.
27. Kaler, E. W., Murthy, K., Rodriguez, B. E., and Zasadzinski, J. A., Spontaneous vesicle formation in aqueous mixtures of single-tailed surfactants, *Science,* 245, 1371, 1989.
28. Mitchell, J., Tiddy, G. J. T., Waring, L., Bostock, T., and McDonald, M. P., Phase behavior of polyoxyethylene surfactants with water, *J. Chem. Soc. Faraday Trans.*, 1, 79, 975, 1983.
29. Adam, C. D., Durrant, J. A., Lowry, M. R., and Tiddy, G. J., Gel and liquid-crystal phase structures of the trioxyethyleneglycol monohexadecyl ether/water system, *J. Chem. Soc. Faraday Trans.*, 1, 80, 789, 1984.
30. Randall, K. and Tiddy, G. J. T., Interaction of water and oxyethylene groups in lyotropic liquid-crystalline phases of polyoxyethylene n-dodecyl ether surfactants studied by ^{2}H nuclear magnetic resonance spectroscopy, *J. Chem. Soc. Fraraday Trans.*, 1, 80, 3339, 1984.
31. Carvell, M., Hall, D. Y., Lyle, I. G., and Tiddy, G. J. T., Surfactant water interactions in lamellar phases, *Faraday Discuss. Chem. Soc.*, 81, 223, 1986.
32. De Young, L. R. and Dill, K. A., Solute partitioning into lipid bilayer membranes, *Biochemistry,* 27, 5281, 1988.
33. Pulido, M. L., 2-(Thiocyanomethylthio)benzothiazole: An effective treatment for cottonseed, Proceedings of the 1971 World Agricultural Conference, October 11-15, 1-12, 1971.
34. Lutey, R. W., King, V. M., and Cleghorn, M. Z., Mechanisms of action of dimethylamides as a penetrant/dispersant in cooling water systems, 50th Annual Meeting of the International Water Conference, Pittsburgh, PA, October 23-25, 1989.
35. Buckman, S. J., Pera, J. D., and Purcell, W. P., Pitch control in pulp and papermaking. U.S. Patent 3,274,050, 1966.
36. Mod, R. R., Magne, F. C., and Skau, E. L., The plasticizing characteristics of some N,N-dimethylamides and ester amides of long-chain fatty acids, *J. Am. Oil Chem. Soc.*, 45, 385, 1968.
37. Kaufman, H. A. and Broderick, E. J., Solvents system for formulating carbamates. U.S. Patent 3,342,673, 1967.
38. Raychaudhuri, P., Katz, M., Wilkinson, D., Mathur, R., and Wallach, D. F. H., Increased efficacy in psoriasis of topical anthralin encapsulated in non-phosholipid liposomes, *Arch. Dermatol,* in press.

Chapter 17

Non-Phospholipid Vesicles as Experimental Immunological Adjuvants

Carole Varanelli, S. Kumar, and Donald F. H. Wallach

CONTENTS

I. INTRODUCTION

A. NATURE OF IMMUNOLOGICAL ADJUVANTS

Many adjuvants act as carriers but also serve as nonspecific immunological stimulants when they attract and activate macrophages.[1] The more potent nonspecific adjuvants include mycobacteria, bentonite particles, and mineral oil emulsions. Inserted into tissues, all of these substances initiate the primary function of macrophages — to remove a foreign object and to subject it to immunological surveillance. Macrophages are the principal "antigen presenting cells", APCs, in the tissues. But macrophages cannot efficiently ingest or digest oil emulsions, bentonites, etc. Instead they become converted into "angry macrophages" producing massive doses of superoxide anion and other oxidants leading to their own destruction and, with the release of lymphokines and other chemoattractants, to the attraction of additional macrophages in a "suicidal" cycle, ultimately leading to fibrosis and "granuloma". Such massive macrophage traffic nevertheless allows a sufficient number of antigen-bearing cells to reach the regional lymph nodes and initiate an effective antigenic response.

The severe tissue responses to oil emulsion adjuvants,[2,3] in particular Freund's complete adjuvant (containing mycobacterial peptidoglycan), have led to the search for alternate immunological adjuvants. These include polymeric surfactants composed of covalently linked blocks with polyoxyethylene (hydrophilic) or polyoxypropylene (hydrophobic) headgroups;[4-6] immune stimulatory complexes, iscoms, 40 nm cage-like structures composed of cholesterol, Quil-A (a saponin), and immunogen;[7-10] and liposomes.[11-16]

II. LIPOSOMES

A. PHOSPHOLIPID LIPOSOMES

A major reason for the use of liposomes in immunization is to "convert a poorly immunogenic or 'nonimmunogenic' protein into a highly immunogenic one."[12-14] Liposomes are vesicular lipid membrane structures enclosing a volume of water and can act as analogs of certain biological cell membrane systems. They were described — more than 30 years before the term "liposome" was coined — by X-ray diffraction studies[17-19] defining particles consisting of organized lipid bilayers enclosing a water space. The use of lipid vesicles as membrane analogs dates back to 1959, when Wallach et al.[20] showed that injection of organic solutions of purified phosphatidylethanolamine into aqueous media yielded pH-sensitive particles mimicking the thromboplastic function of blood platelets. Electron micrographs[21] of these particles revealed all categories of liposomes now recognized, including the multilamellar vesicles later studied by Bangham et al.[22] and given the name "liposome". Until lately, liposome technology has dealt mostly with vesicles composed of phospholipid (PL), and most publications continue to focus on PL.

There are doubts as to whether phospholipids are suitable structural materials for immunological adjuvants except on an experimental level. Phospholipids are labile and expensive to purify or synthesize; manufacture of phospholipid liposomes depends on organic solvents and is difficult and costly to scale up; PL exhibit poor loading, especially for lipophilic materials, and have short shelf lives unless refrigerated or lyophilized in the dark with antioxidants. In addition, PL placed in an *in vivo* environment are too rapidly degraded by phospholipases and altered by various phospholipid exchange proteins for most vaccine applications.

B. NON-PHOSPHOLIPID LIPOSOMES

For the reasons described above, there is increasing interest in non-phospholipid liposomes (NPL), made of non-phospholipid, "membrane mimetic" amphiphiles.[23] Indeed, liposomes have been made from oleic acid[24-26] and a number of long-chain soaps combined with nonionic surfactants,[27] single-tailed ether derivatives of polyglycerol,[28] two-headed ammonium amphiphiles,[29-37] double-tailed cationic surfactants,[38] cationic or zwitterionic two-chain amphiphiles involving amino acid residues,[39-40] two-tailed sucrose fatty acid esters,[41-42] and aqueous mixtures of single-tailed cationic and anionic surfactants.[43] NPL can be engineered for particular applications. This requires building up membrane prototypes from a series of membrane "modules", each module imparting a desired characteristic. The modular approach provides great breadth and flexibility in membrane design. It allows the combination of numerous single-tailed amphiphiles with each other, or with phospholipids, sphingolipids, or both, to yield membrane hybrids with novel properties.

Some of the major structural membrane amphiphiles currently used in NPL design include the ethoxylated fatty alcohols and fatty acids with 2–5 oxyethylene residues, polyoxyethylene glycerol fatty acid monoesters, fatty acid glycol esters, fatty acid glycerol monoesters, and fatty acid glycerol diesters.

The modular approach used in the design of NPL relies on the use of sterol molecules. Cholesterol can intercalate between amphiphile hydrocarbon chains in bilayers and thereby can allow the intermixing of different acyl chains without phase segregation, and can broaden temperature range of the crystal → liquid-crystal transition. Cholesterol interacts with the polar ends of the chains, leaving them less free to change conformation than the long apolar chain segments.

Ionogenic membrane amphiphiles can be used in conjunction with any of the nonionic modules ranging from 0.05 to 10 mol%. Of the anionic modules, the fatty acids must be used at bulk pH levels of less than 7. Diacylphosphates can be used over a broad pH range.

Spacer modules can be employed here primarily to modulate the accessibility of the external bilayer. Ethoxylated amphiphiles and propoxylated amphiphiles allow the placement, atop the outermost bilayer surface, of bulky residues acting as steric barriers. The use of such residues may be advantageous in modulating delivery of antigens to circulating or mucosal macrophages.

Provided there are no steric or other interferences between head groups, the most stable membranes are produced by matching the chain length of the membrane modules. However, the action of sterols allows assembly of membranes from modules of varying chain lengths.

1. Preparation of NPL

The formation of NPL vesicles has been described.[44] It is based on the known phase behavior of single-chain amphiphiles.[45-48] The membrane-forming amphiphile (plus auxiliary molecules such as sterols and ionogenic amphiphiles) are heated (if necessary) to give a liquid. This liquid is then injected at high velocity (5–50 m/sec) through small channels (0.1–1.5 mm radius) into excess, turbulent aqueous phase and immediately cooled. The injection step causes the lipid to be disrupted into minute droplets that are quickly hydrated into micelles. Cooling under conditions of high turbulence causes the micelles to fuse into paucilamellar vesicles within seconds. The vesicles formed have two to four bilayers, separated by hydrogen-bonded water, surrounding an amorphous core of aqueous phase. The typical size of such vesicles is 0.3–0.7 microns.

In practice NPL are prepared on a small scale using an automated mechanism, allowing for precise control of velocity and cooling. Sets of two sterile 5 ml disposable syringes are connected via a 12.5 mm long by 0.46 mm diameter orifice. One syringe contains the lipid phase, typically 1 gram, and is preheated in a dedicated oven to 65–85°C. The other syringe, containing the aqueous phase (generally 4 grams; phosphate buffered saline), is heated to 65–85°C and the syringes are connected. The lipid phase is injected into the aqueous phase over a period of 1 sec (linear velocity in connector is greater than 10m/sec) and the resulting lipid/aqueous mixture returned to the first syringe within 1 sec. Each cycle is repeated 50 times, with cooling initiated at stroke 15 (formulation temperature 45–50°C). The exit temperature of the NPL is at 33–36°C. NPL can also be made on an industrial scale by a continuous flow device.

2. Characterization of NPL

Observed by phase contrast or dark-field microscopy, NPL greater than 0.3 microns in diameter appear as discrete spherical objects. Moving vesicles tumble about their centers and exhibit flows typical of noninteracting spheres. Nevertheless, collisions between large vesicles may produce momentary surface deformations. Although the radial arrangement of amphiphile molecules in bilayers leads to optical birefringence, such is not seen with typical NPL because these contain only 2–4 bilayers.

High-resolution, negative-staining images of NPL show these to be bounded by two or more membranes, each with the "railroad track" or "unit membrane" image characteristic of lipid bilayers, and each about 6 nanometers thick, with an amorphous core typically 0.1–0.5 microns in diameter.[44]

Actual encapsulation efficiencies are estimated after removal of extravesicular antigen by centrifugation on a 5% dextran (mw 87K) gradient (2000 rpm for 15 minutes), which causes the NPL to pack at the air-water interface. The antigen content in the wash is measured by standard methods (e.g., protein analysis, gel electrophoresis).

III. LIPOSOME ADJUVANTS

A. PHOSPHOLIPID LIPOSOMES

In order to stimulate a specific immune response two components are required, namely the antigen or immunologically specific substance, and an adjuvant, a component augmenting the immune response to the antigen. Conventional extrinsic adjuvants can serve as vehicles or depots for the antigen, and also as nonspecific immunological stimulants. Liposomes can carry out these functions[11-16] and can, in addition, act as carriers for chemical adjuvants and as vehicles for unique presentation of antigens. Intercalation of fusogenic virus envelope glycoproteins in a PL lipid bilayer has been shown to enhance adjuvant/carrier properties when such a PL-virus fusion vesicle is used to deliver an encapsulated antigen.[49]

PL have been used experimentally as carriers of chemical adjuvants such as Gram-negative bacterial lipopolysaccharides (LPS), lipid A,[50] the lipid portion of LPS, muramyl dipeptide, MDP,[51,52] the minimal structure of mycobacterial peptidoglycan having adjuvant activity (accounting for much of the efficacy of Freund's complete adjuvant), and MDP derivatives such as 6-O-acyl muramyl dipeptide.[53] It is known, moreover,[54-57] that in vitro, liposome-associated antigens can stimulate T helper cells without the mediation of macrophages (or other APCs). The antigens, covalently coupled to liposomes containing solubilized class II major histocompatibility (MHC) molecules, cause MHC-restricted activation of antigen-specific T-cells.

B. NON-PHOSPHOLIPID LIPOSOMES

NPL should be more effective than PL as adjuvants, as carriers for chemical adjuvants, and vehicles for antigen presentation because of their stability, composition flexibility, and high capture volumes.

The NPL studied as adjuvants[44,58-65] and which will be treated here are paucilamellar liposomes with high cargo capture that can be formed from many nontoxic, biodegradable tailed amphiphiles. They have two to four amphiphile bilayer membranes surrounding a large, amorphous "cargo hold" for either water-soluble or water-immiscible substances.

NPL of the general type described herein have been used for more than three years as safe, sterile, endotoxin-free immunological adjuvants for two USDA-approved poultry vaccines prepared by Vineland Laboratories (Vineland, NJ). Because these vaccines are for use in food animals, they required clearance from a toxicological point of view by the U.S. Food Inspection Service. Similar NPL are being evaluated as adjuvants to be used in the production of polyclonal and monoclonal diagnostic antibodies and as adjuvant/antigen carriers for human vaccines.

NPL adjuvants can be used in several ways to stimulate an immune response, including use as (1) nonspecific immune stimulators, (2) adjuvant/carriers, (3) carriers of chemical adjuvants, and (4) unique antigen presenters.

Table 1 **Neutralizing antibody production in chickens by a subcutaneous immunization of NPL-adjuvanted avian reovirus**

Adjuvant	No. of Birds	Virus Neutralization Antibody Titers			
		Weeks After Vaccination			
		0	4	8	12
Unvaccinated	20	<8	<8	<8	<8
Type A NPL	20	<8	181	239	44
Oil emulsion adjuvant	20	<8	215	223	79

1. NPL Adjuvants as Nonspecific Immune Stimulators

Original adjuvanticity studies were carried out in chickens using anionic NPL (Type A) containing as the major structural membrane amphiphile an ethoxylated fatty alcohol of high purity, together with 25 mol% cholesterol. Such NPL function as nonspecific immune stimulators when mixed with antigens which are nonreactive with the vesicle lipid membrane and which are solubilized or suspended in the extravesicular phase.

Table 1 shows antibody production with an NPL-adjuvanted avian reovirus vaccine. Formalin-killed avian reovirus was mixed with previously prepared NPL; in this manner the antigen, in the form of a killed virus, remains in the extravesicular phase of the preparation. The antigen-NPL adjuvant mixture was administered in a single 0.5 ml subcutaneous injection in eight-week-old specific pathogen free (SPF) chickens, which had been previously primed at 3 and 5 weeks of age with Tenosynovitis vaccine. A second group of primed chickens was immunized with antigen emulsified in a standard water-in-oil emulsion adjuvant. Birds in the control group received no treatment. The NPL-adjuvanted neutralizing antibody response was equivalent to that of the oil emulsion adjuvant. NPL did not cause the granuloma formation typically observed with the oil emulsion.

The adjuvanticity of Type A NPL and Type B NPL (containing as the primary membrane amphiphile a fatty acid glycerol monoester, together with 25 mol% cholesterol), were compared to that of soybean lecithin (Emulpur N-P1, Lucas Meyer) PL. Each liposomal preparation was mixed with equal quantities of formalin-inactivated avian reovirus, such that the virus remained in the extravesicular phase. Then 0.5 ml of each mixture was injected subcutaneously into primed birds. Serum neutralization titers were measured at 4 weeks post administration. Birds which received NPL-adjuvanted avian reovirus generated strong neutralizing titers, while those immunized with PL-adjuvanted virus did not (Table 2).

Type A NPL have also been successfully used as nonspecific immune stimulators in the production of polyclonal antibodies in rabbits. New Zealand white rabbits were immunized with 30 µg of diphtheria toxoid, either mixed in the extravesicular phase of a Type A NPL, or in phosphate buffered saline. Animals were immunized intradermally in 10 sites, 0.1 ml per site, on days 0 and 21. Serum antibody endpoint titers were determined on day 35 (Table 3) in a solid phase enzyme immunoassay against diphtheria toxoid.

Table 2 **Serum neutralizing titers in chickens 4 weeks after administration of PL- and NPL-adjuvanted avian reovirus**

Adjuvant	Serum Neutralizing Titers 4 Weeks Post-Immunization
Type A NPL	234
Type B NPL	207
PL	15

2. NPL Adjuvant/Carriers

By encapsulating an antigen within the core of an NPL, the vesicle can not only act as a nonspecific immune stimulator, but can simultaneously act as a carrier for the antigen.

NPL of two basic types were used in studies designed to establish feasibility for the use of NPL as carrier for antigens. One of these studies was conducted with Hepatitis B Surface Antigen (HBsAg). HBsAg was encapsulated in Type A NPL and Type B NPL at a final antigen concentration of 10 µg/100 µl. Mice were immunized with 100 µl of inoculum subcutaneously in the hind limb on day 0. Serum antibody titers were measured over a period of 32 weeks after this single immunization (Table 4). Both types of NPL generated a strong antibody response by week 8, which persisted up to 32 weeks post immunization.

A comparison between NPL-adjuvanted and Freund's-adjuvanted *Haemophilus pleuropneumoniae* in mice, shown in Table 5, again demonstrates the advantages of the NPL adjuvant. In this study the *H. pleuropneumoniae* antigen was encapsulated within the NPL. Serum antibody titers were determined by ELISA on days 0 (preimmune), 21, and 49. NPL-bacterial extract inoculum generated antibody titers equivalent to those generated by Freund's incomplete adjuvant (IFA).

Table 3 Serum antibody production in rabbits after intradermal immunization of NPL-adjuvanted diphtheria toxoid

	Serum Antibody ELISA Titers[a] Mean Endpoint Titers

	Day 35
PBS	<500
Type A NPL	10200

[a]Animals immunized on day 0 and day 21.

Data summarized in Table 6 show that NPL can also act as an adjuvant/carrier for a conjugated peptide. Pigs were immunized with a porcine growth hormone peptide fragment coupled to Keyhole Limpet Hemocyanin (KLH) either encapsulated within NPL or emulsified with Freund's complete adjuvant (CFA). Similar antibody titers were obtained in both groups, but the severe tissue reaction of the Freund's adjuvant was absent in the NPL-adjuvanted animals.

3. NPL as Carriers of Chemical Adjuvants

Prior to the formation of NPL, hydrophilic chemical adjuvants such as muramyl dipeptide (MDP) can be incorporated into the aqueous phase, resulting in the encapsulation of the chemical adjuvant during liposome formation. This encapsulation serves to deliver the chemical adjuvant to antigen-presenting cells while minimizing the adjuvant's inherent toxicity. Hydrophobic adjuvants (e.g., lipid A) can be incorporated directly into the lipid phase or into an oil carrier prior to NPL production; the adjuvant is thus incorporated into the NPL lipid membrane or into the NPL core, respectively. Such NPL can be used to adjuvant unencapsulated antigens, encapsulated antigens, or antigens, presented on the liposomal surface.

Table 4 Antibody production in mice following administration of recombinant HBsAg adjuvanted and carried in NPL

Inoculum	Mean Serum ELISA Titers					
	3 wk	8 wk	12 wk	20 wk	28 wk	32 wk
HBsAg	2.7	18.6	17.2	20.6	6.1	4.3
Type A NPL/HBsAg	18.8	94.4	74.0	121.0	74.0	85.0
Type B NPL/HBsAg	30.6	150.0	136.0	146.0	103.0	147.0

Table 5 Comparison of NPL and Freund's adjuvant in the generation of antibody and toxicity in Balb/C mice following immunization with soluble extract of *H. pleuropneumoniae*[a]

Immunization	ELISA Titers[b]		
	Day 0	Day 21	Day 49
None	0.04	0.12	0.18
Bacterial extract	0.04	0.12	0.46
Bacterial extract in IFA	0.13	0.55	0.80
Bacterial extract in Type A NPL	0.12	0.78	0.97

[a]Dose = 20 ug/soluble *H. pleuropneumoniae* in 0.100 ul SC
[b]Titers (OD_{492}): preimmune (Day 0); 2nd inoculation (Day 21); 1 month post 2nd inoculation (Day 49)

Table 6 Comparison of NPL vesicles and Freund's complete adjuvant in the generation of serum antibodies in pigs against porcine growth hormone peptide fragments coupled to KLH[a,b]

Adjuvant	ELISA Titers			
	4 wks	11 wks	13 wks	16 wks
Buffer	0.159	0.287	0.228	0.241
CFA	1.660	1.069	1.158	1.286
NPL[c]	0.992	1.222	0.963	1.117

[a]Initial dose at 7 Wks age, 0.5 ml intramuscular, 2nd and 3rd doses at 11 and 13 weeks

[b]Severe abscess and granuloma formation from Freund's adjuvanted samples

[c]NPL encapsulated antigen

A series of concentrations of encapsulated MDP were prepared in Type A NPL in order to define the minimum MDP dose required to maximally enhance the immune response over that generated by NPL alone. The MDP-containing NPL were prepared by standard methods, mixed with formalin-inactivated Newcastle Disease Virus (NDV), and 100 µl of each of the resultant inocula were administered intramuscularly to Swiss albino mice. Serum hemagglutination inhibition titers on sera of immunized mice show that a dose of 0.02 µg encapsulated MDP was maximally effective in enhancing the immune response over NPL-adjuvanted controls (Table 7). This represents a more than 1000-fold enhancement over the mouse-effective dose of free MDP.

4. NPL in Specialized Antigen Presentation

Of the many biological processes that involve the fusion of biomembranes, one of the most efficient occurs when an enveloped virus introduces its genetic material into a target cell by the merging of its membrane envelope with a membrane of the target cell in a process mediated by a viral fusion protein.[66-68] The fusion protein of the influenza virus is the best characterized.[66] Its action has been clarified by the use of phospholipid liposomes[67-73] and has much in common with the fusion proteins of paramyxoviruses.[74] Collision of paramyxoviruses such as Sendai virus[75] and Newcastle disease virus[76] with phospholipid liposomes leads to fusion of the viral and liposomal membranes.

The ability of the membrane lipids of a Type A NPL to interchange or fuse with the membrane lipids of a PL has also been demonstrated, and is used as a model system for predicting NPL-virus fusion. NBD (2-(12-(7-nitrobenz-2-oxa-1,3-diazole-4-yl) amino)dodecanoyl-1-hexadecanoyl-ethanol) phospholipid fluorescent probe (NBD-PC) was incorporated in PL at a density high enough for self-quenching of its fluorescence. As these NBC-PC containing PL are mixed with NPL, dequenching, as evidenced by an increase in fluorescence, occurs with time (Table 8) indicating that the density of the NBD-PC probe decreases as the PL membrane lipids intermingle with the NPL membrane lipids. Membrane lipid intermingling was also demonstrated between NBD-PC micelles and NPL.

That Type A NPL membranes fuse rapidly with NDV with transfer of the viral RNA into the interior of the resulting fusion vesicle and redistribution of virus surface glycoproteins has been documented by electron microscopy. Suspensions of NDV ($10^{9-9.5}$ particles/ml) were diluted with an equal volume of NPL diluted to $10^{10-10.5}$ particles/ml at room temperature. After 10 minutes, 40 minutes, and 24–48 hours, small amounts of the mixtures were placed on Formvar and carbon-coated 300 mesh grids, negatively stained with 2% phosphotungstic acid, pH 7.5, dried after removing excess fluid, and examined immediately in a Phillips CM 10 electron microscope at 80 kV.

Table 7 Hemagglutination Inhibition (HI) titers in sera of mice immunized with NDV adjuvanted with NPL encapsulated MDP

3 Weeks Post-Immunization	
MDP/dose	HI Titer
0.000	89.8
0.001	67.3
0.002	113.1
0.010	269.0
0.020	380.5
0.040	380.5
0.200	28.3

Table 8 **Dequenched NBD-PC fluorescence of NBD-PC PL 24 hours after mixing with NPL**

	Fluorescence Units	Total Dequenching Units Fluorescence	% Quenching
NBD-PC/PL (83 nmol/ml)	18	190	91
NBD-PC/PL + NPL (37 μmol/ml)	41	140	68.5
NBD-PC/PL + NPL (10.9 μmol/ml)	34	110	69.1

Free NDV, like all paramyxoviruses,[74] have an outer 20 millimicron envelope consisting of membrane-anchored glycoproteins. The internal layer is thicker than expected from a lipid double layer and does not reveal a bilayer structure. The nucleocapsid structure of the virus core is rarely distinguishable. Liposomes alone exhibit two or three peripheral membrane bilayers.[44]

Within 10 minutes after mixing, continuity between the NPL and virus cores become evident. At this stage a virus fusing with a NPL gives a vesicle with a less spherical shape than usual, with one or more fusion sites (Figure 1). The fusion sites are usually convex and more electron-dense than the rest of the vesicle perimeter. The glycoprotein knobs remain externally disposed but are more widely separated than on the unfused virus. The nucleocapsid is usually seen uncoiled within the core of the NPL.[77]

Viral glycoprotein, presumably including the HN protein detected by hemagglutination inhibition assays, retains the orientation seen in the virus. The extent of fusion, and the effect of this NPL-viral fusion on the hemagglutinating activity of the fusion vesicle has been measured. One volume of inactivated virus diluted 1:10 was mixed at room temperature with one volume of NPL diluted 1:100 (10^{10-11} vesicles/ml), vortexed, and incubated for one hour at room temperature; the mixture was combined vigorously with an equal volume of 20% dextran (avg. mw 162,000) and centrifuged at 3,500

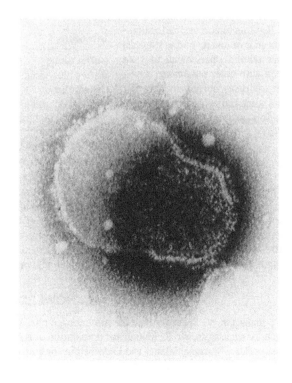

Figure 1 NPL-(Type A) NDV fusion vesicle with one fusion site. NDV glycoprotein knobs are externally disposed over the entire fusion vesicle and the nucleocapsid is partially uncoiled within its core. Magnification × 219,000.

rpm for 30 minutes in a countertop centrifuge, causing the liposomes to accumulate quantitatively as a firm layer at the surface, whereas free virus remains in the infranatant. The infranatant was sampled by piercing the side of the centrifuge tube and its hemagglutination titer determined. The neat virus-NPL mixture had a hemagglutination titer greater than 1:5120, whereas the infranatant titer was only 1:64. This indicates association of more than 95% of

Table 9 **Hemagglutination inhibition assays using NDV and NPL-NDV fusion vesicles as antigen**

Sample	NDV	NPL-NDV Fusion Vesicle
1	640	640
2	640	640
3	640	640
4	320	320
5	320	320
6	80	160

the virus with the NPL. Added washing caused no further elution of virus. Hemagglutination inhibition titers, obtained using sera from NDV-immune chickens, for virus and liposome-associated virus were not significantly different (Table 9), indicating that the HN protein was exposed on the NPL/virus fusion vesicle at least as efficiently as on free virus, while nucleocapsid antigens are sequestered in the vesicle interior.

In this way the NPL can successfully act as an adjuvant and unique antigen presenter, and effectively stimulate a protective immune response. A NPL-virus fusion vesicle preparation has successfully been developed as an effective poultry vaccine against Newcastle disease virus. Table 10 shows protection by a single intramuscular immunization of NPL-NDV fusion vesicles or oil-adjuvanted NDV.

Table 10 **Protection of chickens by a NPL-NDV fusion vesicle**

	% Protection
Unvaccinated	0
NPL-NDV	100
Oil emulsion	100

The capacity of the NPL-NDV fusion vesicle to stimulate an early immune response was tested using a mouse spleen focus-forming assay after intraperitoneal injection (Table 11). In the undiluted NPL samples there were 10^6 vesicles per virus particle. The NPL-NDV fusion vesicle stimulated the production of antibody-forming cells equivalent in number to that of the oil emulsion adjuvant.[78]

The capacity of NPL to fuse with NDV, and the ability of this NPL-virus fusion vesicle to stimulate an immune response, suggests that related phenomena may be seen with other paramyxoviruses, myxoviruses, coronaviruses, HIV, and some retroviruses which have fusion protein homologies to NDV.

Table 11 **Antibody-forming cells in spleens of Balb/C mice four days after vaccination with inactivated, unadjuvanted NDV, NDV adjuvanted with oil emulsion and NPL-NDV**[a]

Inoculum	Number of Antibody-Forming Cells per 10^6 Splenocytes[b]
Buffer alone	<10
Buffer plus virus	<10
Oil emulsion	50–170
NPL-NDV undiluted	210
NPL (diluted 1:100)-NDV	85

[a]Vaccine dose = 0.2 ml intraperitoneal

[b]Assayed after 4 days using sandwich antibody with gold/silver stain

IV. SUMMARY

It has been demonstrated that the major practical obstacles to the development of liposome adjuvants can be overcome by the substitution of non-phospholipid amphiphiles for conventional phospholipid amphiphiles, and that the resultant NPL can be effectively used as immunological adjuvants. The versatility of NPL composition, vesicle stability, fusogenic properties of the vesicles, and the inherent adjuvanticity of some of the primary amphiphiles, allow NPL to be successfully used as nonspecific immunogical stimulants, antigen and chemical adjuvant carriers, and antigen presenters.

ACKNOWLEDGMENTS

We thank Dr. V. Speth for the electron micrographs, Mr. W. Boclair and Dr. R. Liaw for the MDP and Splenocyte assays, Mr. R. Mathur for preparation of NPL and PL, Dr. A. Novikov for the NPL-PL fusion assays, Drs. Marcela Salazar and David Bing for assistance in the Hepatitis B animal trials, and Ms. D. Welch and Ms. L. Bannister for assistance in preparing the manuscript.

REFERENCES

1. Edelman, R., Vaccine adjuvants, *Rev. Infect. Dis.*, 2, 370., 1980.
2. Murray, R., Cohen, P., and Hardegree, M. C., Mineral oil adjuvants: biological and chemical studies, *Ann. Allergy*, 30 (3), 146, 1972.
3. Hilleman, M. R., Woodhour, A. F., Friedman, A., and Phelps, A. H., Studies for safety of adjuvant 65, *Ann. Allergy*, 30, 477, 1972.
4. Hunter, R. L. and Bennett, B., The adjuvant activity of nonionic block polymer surfactants. II. Antibody formation and inflammation related to the structure of triblock and octablock copolymers, *J. Immunol.*, 133, 3167, 1984.
5. Allison, A. C. and Byars, N., U. S. Patent 4,606,918, Polyoxypropylene-polyoxyethylene block polymer-based adjuvants; immunostimulating glycopeptide; emulsification, 1986.
6. Bennet, B., Check, I. J., Olsen, M. R., and Hunter, R. L., A comparison of commercially available adjuvants for use in research, *J. Immunol. Meth.*, 153, 31, 1992.
7. Lovgren, K. and Morein, B., The ISCOM: an antigen delivery system with built-in adjuvant, *Mol. Immunol.*, 28, 285, 1991.
8. Morein, B., The ISCOM: an immunostimulating system, *Immunol. Lett.*, 25, 281, 1990.
9. Morein, B., Iscoms, *Vet. Microbiol.*, 23, 79, 1990.
10. Ozel, M., Hoglund, S., Gelderblom, H. R., and Morein, B., Quaternary structure of the immunostimulating complex (iscom), *J. Ultrastruct. Mol. Struct. Res.*, 102, 240, 1989.
11. Allison, A. C., Mode of action of immunological adjuvants, *J. Reticuloendothelial Soc.*, 26, 619, 1979.
12. Alving, C. R., Liposomes as carriers for vaccines, *Liposomes from Biophysics to Therapeutics*, M. J. Ostro, Ed., Marcel Dekker, New York, 1987, 195.
13. Lowell, G., Proteosomes, hydrophobic anchors, iscoms and liposomes for improved presentation of peptide and protein vaccines, *New Generation Vaccines*, Woodrow, C. and Levine, M. M., Eds., Marcel Dekker, New York, 1990, 141.
14. Alving, C. R., Liposomes as carriers of antigens and adjuvants, *J. Immunol. Meth.*, 140, 1, 1991.
15. Tan, L., a study of the efficacy of liposomes in comparison to new and established adjuvants in potentiating the antibody response against hepatitis B virus surface antigen, *Asian-Pac. J. Allergy Immunol.*, 9, 83, 1991.
16. Fries, L. F., Gordon, D. M., Richards, R. L., Egan, J. E., Hollingdale, M. R., Gross, M., Silverman, C., and Alving, C. R., Liposomal malaria vaccine in humans: a safe and potent adjuvant strategy, *Proc. Natl. Acad. Sci. U.S.A.*, 89, 358, 1992.
17. Schmitt, F.O., Bear, R. S., and Clark, G. L., X-ray diffraction studies on nerve, *Radiology*, 25, 131, 1935.
18. Schmitt, F. O. and Bear, R. S., The ultrastructure of the nerve axon sheath, *Biol. Rev.*, 14, 27, 1939.
19. Bear, R. S., Palmer, K. J., and Schmitt, F. O., X-ray diffraction studies of nerve lipids, *J. Cell Comp. Physiol.*, 17, 355, 1941.
20. Wallach, D. F. H., Maurice, P. A., Steele, B. A., and Surgenor, D. M., Studies on the relationship between the colloidal state and the clot-promoting activity of pure phosphatidylethanolamines, *J. Biol. Chem.*, 234, 2829, 1959.
21. Surgenor, D. M. and Wallach, D. F. H., Biophysical aspects of platelet reaction mechanisms in clotting, *Henry Ford Symposium Blood Platelets*, Little, Brown, Boston, 1961.
22. Bangham, A. D., Standish, M. M., and Watkins, J. C., Diffusion of univalent ions across the lamellae of swollen phospholipids, *J. Mol. Biol.*, 13, 238, 1965.
23. Fendler, J., *Membrane Mimetic Chemistry*, John Wiley & Sons, New York, 1982, pp. 522.
24. Gebicki, J. M. and Hicks, M., Ufasomes are stable particles surrounded by unsaturated fatty acid membranes, *Nature*, 243, 232, 1973.
25. Gebicki, J. M. and Hicks, M., Preparation and properties of vesicles enclosed by fatty acid membranes, *Chem. Phys. Lipids*, 16, 142, 1976.
26. Hicks, M. and Gebicki, J. M., A quantitative relationship between permeability and the degree of peroxidation in Ufasome membranes, *Biochem. Biophys. Res. Commun.*, 80, 704, 1978.
27. Hargreaves, W. R. and Deamer, D. W., Liposomes from ionic, single chain amphiphiles, *Biochemistry*, 17, 3759, 1978.
28. Handjani-Vila, R. M., Ribier, A., and Vanlenberghe, G., Les niosomes, *Les Liposomes*, Puisieux, F. and Delattre, J., Eds., Techniques et Documentation Lavoisier, Paris, 1985, chap. 11.

29. Kunitake, T. and Okahata, Y., A totally synthetic bilayer membrane, *J. Am. Chem. Soc.*, 99, 3860, 1977.

30. Kunitake, T., Okahata, Y., Tamaki, K., Kumamura, F., and Takayanagi, M., Formation of the bilayer membrane from a series of quaternary ammonium salts, *Chem. Lett.*, 387, 1977.

31. Kunitake, T., Nakashima, N., Hayashida, S., and Yonemori, K., Chiral, synthetic bilayer membranes, *Chem. Lett.*, 1413, 1979.

32. Okahata, Y. and Kunitake T., Formulation of a stable monolayer membrane and related structures in dilute aqueous solutions from two-headed ammonium amphiphiles, *J. Am. Chem. Soc.*, 101, 5231, 1979.

33. Okahata, Y., Ihara, H., Shimomura, M., Tawaki, S., and Kunitake, T., Formation of disk-like aggregates from single chain phosphocholine amphiphiles in water, *Chem. Lett.*, 1169, 1980.

34. Kunitake, T. and Okahata, Y., Formation of stable bilayer assemblies in dilute aqueous solution from ammonium amphiphiles with the diphenylazomethine segment, *J. Am. Chem. Soc.*, 102, 549, 1980.

35. Kunitake, T., Nakashima, N., Takarabe, K., Nagai, M., Tsuge, A., and Yanagi, H., Vesicles of polymeric bilayer and monolayer membranes, *J. Am. Chem. Soc.*, 103, 5945, 1981.

36. Kunitake, T., Okahata, Y., Shimomura, M., Yasunami, S., and Takarabe, K., Formation of stable bilayer assemblies in water from single-chain amphiphiles. Relationship between the amphiphile structure and the aggregate morphology, *J. Am. Chem. Soc.*, 103, 5401, 1981.

37. Kunitake, T., Organization and functions of synthetic bilayers, *Ann. N.Y. Acad. Sci.*, 471, 70, 1986.

38. Kano, K., Romero, A., Djermouni, B., Ache, H. J., and Fendler, J. H., Characterization of surfactant vesicles as membrane mimetic agents. 2. Temperature-dependent changes of the turbidity, viscosity, fluorescence polarization of 2-methylanthracene, and positron annihilation in sonicated dioctadecyldimethylammonium chloride, *J. Am. Chem. Soc.*, 101, 4030, 1979.

39. Murakami, Y., Nakano, A., and Fukuya, K., Stable Single-compartment vesicles with zwitterionic amphiphile involving amino acid residue, *J. Am. Chem. Soc.*, 102, 4253, 1980.

40. Murakami, Y., Nakano, A., and Ikeda, H., Preparation of stable single-compartment vesicles with cationic and zwitterionic amphiphiles involving amino acid residues, *J. Org. Chem.*, 47, 2137, 1982.

41. Ishigami, Y. and Machida, H., Vesicles from sucrose fatty acid esters, *J. Am. Oil Chem. Soc.*, 66, 599, 1989.

42. Schenk, P., Ausborn, M., Bendas, F., Nuhn, P., Arndt, D., and Meyer, H. W., The preparation and characterization of lipid vesicles containing esters of sucrose and fatty acids, *J. Microencapsulation*, 6, 95, 1989.

43. Kaler, E. W., Murthy, K., Rodriguez, B. E., and Zasadzinski, J. A., Spontaneous vesicle formation in aqueous mixtures of single-tailed surfactants, *Science*, 245, 1371, 1989.

44. Wallach, D. F. H. and Philippot, J., New type of lipid vesicle: Novasome,® *Liposome Technology*, 2nd ed., Gregoriadis, G., Ed., CRC Press, Boca Raton, FL, 1992.

45. Mitchell, J., Tiddy, G. J. T., Waring, L., Bostock, T., and McDonald, M. P., Phase behaviour of polyoxyethylene surfactants with water, *J. Chem. Soc. Faraday Trans.*, 1, 79, 975, 1983.

46. Adam, C. D., Durrant, J. A., Lowry, M.R., and Tiddy, G. J. T., Gel and liquid-crystal phase structures of the trioxyethyleneglycol monohexadecyl ether/water system, *J. Chem. Soc. Faraday Trans.*, 1, 80, 789, 1984.

47. Randall, K. and Tiddy, G. J. T., Interaction of water and oxyethylene group in lyotropic liquid-crystalline phases of polyoxyethylene N-dodecyl ether surfactants studied by ^{2}H nuclear magnetic resonance spectroscopy, *J. Chem. Soc. Faraday Trans.*, 1, 80, 3339, 1984.

48. Carvell, M., Hall, D. Y., Lyle, I. G., and Tiddy, G. J. T., Surfactant water interactions in lamellar phases, *Faraday Discuss. Chem. Soc.*, 81, 223, 1986.

49. Gluck, R., Miechler, R., Brantschen, S., Just, M., Alhaus, B., and Cryz, S. J., Immunopotentiating reconstituted influenza virus virosome vaccine delivery system for immunization against Hepatitis A, *J. Clin. Invest.*, 90, 2491, 1992.

50. Alving, C. R., Liposomes containing lipid A: a potent nontoxic adjuvant for a human malaria sporozoite vaccine, *Immunol. Lett.*, 25, 275, 1990.

51. Schroit, A. and Fidler, I. J., Effects of liposome structure and lipid composition on the activation and tumoricidal properties by liposomes containing muramyl dipeptide, *Cancer Res.*, 42, 161, 1982.

52. Fidler, I. J., Brown, N. O., and Hart, I., Effects of lipsome structure and lipid composition on the activation of the tumoricidal properties of macrophages by liposomes containing muramyl dipeptide, *J. Biol. Response Modifiers*, 4, 298, 1985.

53. Tsujimoto, M., Kotani, S., Shiba, T., and Kusumoto, S., Adjuvant activity of 6-*O*-acyl-muramyl dipeptides to enhance primary cellular and humoral immune responses in guinea pigs: adaptability to various vehicles and pyrogenicity, *Infect. Immun.*, 53, 517, 1986.

54. Walden, P., Antigen presentation by liposomes as a model for T-B cell interactions, *Eur. J. Immunol.*, 18, 1951, 1988.

55. Walden, P., Nagy, Z. A., and Klein, J., Induction of regulatory T lymphocyte responses by liposomes carrying major histocompatibility complex molecules and foreign antigens, *Nature*, 315, 227, 1985.

56. Walden, P., Nagy, Z. A., and Klein, J., Major histocompatibility complex-restricted and unrestricted activation of helper T cell lines by liposome-bound antigen, *J. Mol. Cell. Immunol.*, 2, 191, 1986.

57. Walden, P., Nagy, Z. A., and Klein, J., Antigen presentation by liposomes: inhibition by antibodies, *Eur. J. Immunol.*, 16, 717, 1986.

58. Wallach, D. F. H., U.S. Patent 4,855,090, Method of producing high-aqueous volume multilamellar vesicles, 1989.

59. Wallach, D. F. H. and Yiournas, C., U.S. Patent 4,895,452, Method and apparatus for producing lipid vesicles, 1990.

60. Wallach, D. F. H., U.S. Patent 4,911,928, Paucilamellar lipid vesicles, 1990.

61. Wallach, D. F. H., U.S. Patent 4,197,951, Lipid vesicles formed of surfactants and steroids, 1990.

62. Wallach, D. F. H. and Yiournas, C., U.S. Patent 5,013,497, Method and apparatus for producing lipid vesicles, 1990.

63. Wallach, D. F. H., U.S. Patent 5,000,960, Protein coupling to lipid vesicles,1991.

64. Wallach, D. F. H., U.S. Patent 5,109,174, Removing oil from surfaces with liposomal cleaner, 1991.

65. Wallach, D. F. H., U.S. Patent 5,032,457, Paucilamellar lipid vesicles using charged localized single chain, nonphospholipid surfactants, 1991.

66. White, J. M., Membrane fusion, *Science*, 258, 917, 1992.

67. Wiley, D. C. and Skehel, J. J., Viral membranes, *Fundamental Virology*, Fields, B.N. and Knipe, D. M., Eds., Raven Press, New York, 1991, 63.

68. Fields, B. N. and Knipe, D. M., Eds., *Fundamental Virology*, Raven Press, New York, 1991, chap. 18–23, 27–28, 33–36.

69. Harter, C., Bachi, T., Semenza, G., and Brunner, J., Hydrophobic photolabeling identifies BHA2 as the subunit mediating the interaction of bromelain-solubilized influenza hemagglutinin with liposomes at low pH, *Biochemistry*, 27, 1856, 1987.

70. Samson, A. C. R. and Fox, C. F., Precursor protein for Newcastle disease virus, *J. Virology*, 12, 579, 1973.

71. Harter, C., James, P., Bachi, T., Semenza, G., and Brunner, J., Hydrophobic binding of the ectodomain of influenza hemagglutinin to membranes occurs through the "fusion peptide", *J. Biol. Chem.*, 264, 6459, 1989.

72. Brunner, J., Testing topological models for the membrane penetration of the fusion peptide of influenza virus hemagglutinin, *FEBS Lett.*, 257, 369, 1989.

73. Brunner, J., Zugliani, C., and Mischler, R., Fusion activity of influenza virus PR8/34 correlates with a temperature-induced conformational change within the hemagglutinin ectodomain detected by photochemical labeling, *Biochemistry*, 30, 2432, 1991.

74. Kingsbury, D. W., Paramyxoviridae and their Replication, *Fundamental Virology*, Fields, B. N. and Knipe, D. M., Eds., Raven Press, New York, 1991, 507.

75. Haywood, A. M., Characteristics of Sendai virus receptors in a model membrane, *J. Mol. Biol.*, 83, 427, 1974.

76. Lorge, P., Cabiaux, V., Long, L., and Ruysschert, J. M., Fusion of Newcastle disease virus with liposomes: role of the lipid composition of liposomes, *Biochim. Biophys. Acta*, 858, 312, 1980.

77. Wallach, D. F. H., Kumar, S., Varanelli, C., and Speth, V., Submitted for publication.

78. Wallach, D. F. H., Kumar, S., Varanelli, C., and Boclair W., To be published.

Index

INDEX

Pharmacodynamics, 184

Pharmacokinetics, 184

Pharmacological properties, 177–184

Phase separation, 9, 16

PHC, see Palmitoyl homocysteine

Phenylmethanesulfonyl fluoride (PMSF), 149

Phosphatidic acid (PA), 14, 108, 116, 117, 118, 121, 122, 143, 146

Phosphatidyl choline (PC), 4, 5, 6, 13, 15, 17, 47
  egg, 178, 181
  membrane fusion and, 110–113, 116–117, 120–121, 123, 142–144, 147, 149
  pH-sensitive liposomes and, 204, 205, 206
  polymerizable, 48
  "stealth" liposomes and, 181

Phosphatidyl ethanolamine (PE), 13, 14, 23, 26, 32, 46
  gene transfer and, 163
  membrane fusion and, 110–111, 115, 117–119, 121–123, 142–143, 147, 150
  pH-sensitive liposomes and, 201, 202, 204, 205, 206, 209, 211
  "stealth" liposomes and, 178, 179, 183
  unsaturated, 17

Phosphatidyl ethanolamine (PE)-muramyl tripeptide, 31

Phosphatidyl glycerol (PG), 13, 27, 108, 116, 117, 120, 147, 206

Phosphatidylinositol (PI), 27, 47, 108–109, 110, 122, 143, 146
  "stealth" liposomes and, 178, 181

Phosphatidyl serine (PS), 13–14, 106, 107, 108, 109
  membrane fusion and, 112, 115, 118–123, 143–144, 146–148, 150
  pH-sensitive liposomes and, 205

Phosphoglycerides, 12, see also specific types

Phospholipase, 94

Phospholipase D, 13

Phospholipids, 4, 5, 12–15, 41–51, 60, see also specific types
  acidic, 108–109, 110–111, 201
  anionic, 104, 108
  bis-diazo, 7
  brominated, 7
  fluorinated, 7
  glycero-, 13
  as immunological adjuvants, 253–254, 255–260
  ligand-coupling and, 31
  lyso-, 14, 64
  membrane fusion and, 108–115
  MHC restricted antigen presentation and, 94
  mixture of, 62
  molecules modifying membranes of, 46–51
  neutral, 110–111
  structure of, 5, 12, 13

Phosphorothioate, 216

Photoaffinity probes, 125

pH-Sensitive liposomes, 201–212
  in antigen processing, 208–209
  bacterial toxins and, 206–207
  cytoplasmic delivery with, 204–210
  defined, 201
  destabilization of, 202–204
  endocytic pathway and, 207–208
  in vivo, 210–211
  stabilization of, 202–204
  uptake of, 204

Phytosterol, 44

PI, see Phosphatidylinositol

Pinocytosis, 10

Plasma proteins, 210, see also specific types

PMSF, see Phenylmethanesulfonyl fluoride

Point defects, 112–115

Polyamines, 115–118, 160, 164, 166, 167, see also specific types

Polyanions, 51

Poly(aspartic acid), 116

Polycations, 48, 51

Polyethylene glycol (PEG), 5, 10, 17, 20, 92, 150
  imaging tools and, 227, 229
  membrane fusion and, 146
  in nuclear medicine, 229
  pH-sensitive liposomes and, 211
  "stealth" liposomes and, 178, 179, 181, 182, 183

Poly-ionic peptides, 116

Polylysines, 51, 116, 120, 162, see also specific types

Polymers, 159, 227, 229, see also specific types

Polyornithine, 51

Polyoxyethylene groups, 48

Polypeptides, 138, see also specific types

Polysaccharides, 5, 41–48, 64, 192, see also specific types

Polystyrene, 83

Polyvinylpyrrolidone, 227

POPC, see Palmitoyloleoylphosphatidyl choline

POPG, see Palmitoyloleoylphosphatidyl glycerol

Pore formation, 117, 120

Preparation, 8, 19–20, 31–33, see also specific methods

Preservative agents, 64, see also specific types

Probe diution assays, 105, 107

Probe mixing, 107

Probe mixing assays, 106

Processing, 61–64

Promega, 160

Properties of liposomes, 4–9, see also specific types

Propoxylated amphiphiles, 46

Proteases, 138, see also specific types

Protein A, 29–30

Proteins, 9–10, see also specific types
  acylation of, 31–33
  addition of, 79
  adsorption of, 9

Printed and bound by CPI Group (UK) Ltd, Croydon, CR0 4YY

22/10/2024

01777638-0011